SMALL AND MINI HYDROPOWER SYSTEMS

SMALL AND MINI HYDROPOWER SYSTEMS

Resource Assessment and Project Feasibility

Jack J. Fritz

McGRAW-HILL BOOK COMPANY

New York St. Louis San Francisco Auckland Bogotá Hamburg
Johannesburg London Madrid Mexico Montreal New Delhi
Panama Paris São Paulo Singapore Sydney Tokyo Toronto

Library of Congress Cataloging in Publication Data

Main entry under title:

Small and mini hydropower systems.

Includes bibliographical references and index.

1.Hydroelectric power plants—Design and construction. I.Fritz, Jack J.

TH4581.S53 1984 621.31′2134 83-13595

ISBN 0-07-022470-6

1234567890 KGP/KGP 89876543

ISBN 0-07-022470-6

The editors for this book were Patricia Allen-Browne and Peggy Lamb, the designer was Mark E. Safran, and the production supervisor was Reiko Okamura. It was set in Times Roman by University Graphics, Inc.

Printed and bound by The Kingsport Press

Contents

Contributors

Arndt, Roger E. A.
St. Anthony Falls Hydraulic Laboratory
Minneapolis, Minnesota

Brown, Peter W.
Franklin Pierce Law Center
Concord, New Hampshire

Cassidy, John J.
Bechtel Civil and Minerals, Inc.
San Francisco, California

Crawford, Norman
Hydrocomp
Mountain View, California

Farell, Cesar
St. Anthony Falls Hydraulic Laboratory
Minneapolis, Minnesota

Fritz, Jack J.
National Research Council
Washington, D.C.

Gladwell, John S.
Division of Water Sciences
UNESCO
Paris, France

Heitz, Leroy F.
Idaho Water Resources Research Institute
University of Idaho
Moscow, Idaho

Henry, Jean François
National Conservation Corp.
Baltimore, Maryland

Hughes, William L.
Engineering Energy Laboratory
Oklahoma State University
Stillwater, Oklahoma

Ichord, Robert F., Jr.
Agency for International Development
Washington, D.C.

Jacobsen, Willis E.
The Mitre Corp./Metrek Div.
McLean, Virginia

Jones, Herschel
CH_2M Hill
Sacramento, California

Knapp, Jerry W.
CH_2M Hill
Sacramento, California

Lingelbach, Dan
Engineering Energy Laboratory
Oklahoma State University
Stillwater, Oklahoma

Ramakumar, R. G.
Engineering Energy Laboratory
Oklahoma State University
Stillwater, Oklahoma

Warnick, Calvin C.
Idaho Water Resources Research Institute
University of Idaho
Moscow, Idaho

Wetzel, Joseph M.
St. Anthony Falls Hydraulic Laboratory
Minneapolis, Minnesota

Preface

The continually escalating energy prices experienced during the late 1970s necessitated a comprehensive reexamination of small and mini hydropower. In evaluating renewable and traditional hydrocarbon energy resources for their economics and technical feasibility, small hydropower emerged early as a clear winner. Since there was a dearth of in-depth technical and economic information available, United States federal agencies, such as the Department of Energy, U.S. Army Corps of Engineers, and the Department of Interior developed feasibility study methodologies and funded a number of their applications in the field. The results of these efforts are currently being evaluated but with project funding in doubt.

Internationally, small hydropower is coming to the attention of multilateral donors such as the World Bank, the Interamerican Development Bank, and the United Nations, as a viable alternative to diesel generators for rural areas of developing countries. Nevertheless, despite wide interest, a comprehensive volume on small and mini hydropower technology is not available. Based on his experience as manager of the Small Decentralized Hydropower Program of the Agency for International Development, the editor has assembled a collection of technical papers on this topic.

The contributors to this book were selected for their in-depth knowledge of such specialities as hydrology, turbine design, and government regulation. Each chapter addresses a critical element which requires serious consideration during project development. The approach is to guide the reader from the initial hydrologic resource assessment, through physical site design, turbine-generator selection, economics, and environmental impact, and culminating in case studies. This book therefore serves both as a technical primer and as a guidebook for energy specialists seeking to assess hydropower resource opportunities. It can also serve as a university level text in civil engineering. Economic and engineering design examples are given to aid the student reader.

The editor wishes to acknowledge the contributions made by the listed specialists as well as those by David Willer of Tudor Engineering Company, Judith Hughes and James Williamson of R. W. Beck and Assoc., Douglas Barr of Douglas Barr Consulting Hydraulic Engineers, and Dr. Edmund Leo of the U.N.'s Department of Technical Cooperation for Development in making case study materials available. Very special thanks go to Ms. Susan McCutchen and Mrs. Joan Polson for their patience during final typing, and to Ms. Adriane Wodey and Jean F. Henry for their editorial comments.

JACK J. FRITZ

Notes on Notation

The following list of symbols is a general guide to the notation used throughout this book. It is not absolute in that the same notation may be used for different concepts; however, there should be no confusion. Each time the symbol is used, it is defined.

Hydraulic/Hydrologic Symbols

A,a	A cross-sectional flow area
B,b	Width of a flow area or depth
C	Coefficient, drag, lift, discharge, velocity
c	Velocity of sound
D,d	Diameter, grain size
E	Energy
e	Efficiency
F	Force
f	Coefficient of friction
G	Submergence distance
g	Acceleration of gravity
H,h	Energy head, head loss, depth, height
J	Moment of inertia
K,k	Empirical coefficient

L,l	Wavelength, length
M	Moment
N	Specific speed, stability number, precipitation
n	Manning roughness coefficient
P	Power, perimeter, precipitation
p	Pressure
Q,q	Flow rate
R	Hydraulic radius, length
Re	Reynold's number
r	Radius, coefficient of correlation, reservoir reach
S	Slope, standard deviation, standard error, suction specific speed, percent turbine speed rise
s	Specific gravity, streamline coordinate
T	Wave period, dam width, torque
t	Time, thickness
u	Velocity
V,v	Velocity
W	Weight
w	Width
x	Horizontal distance, variable
Y	Depth, variable
z	Elevation, variable

Electrical Symbols

B	Magnetic flux density
C	Capacitance
$E,e\ (t)$	Voltage
f	Frequency
$I,i\ (t)$	Current
j	Imaginary number
k	Loss coefficient
L	Inductance
N,n	Number of coil turns, generator speed
$p\ (t)$	Power
p	Number of poles
$Q,q\ (t)$	Electrical charge
R	Resistance
s	Slip
$V,v\ (t)$	Voltage with inductance
X	Inductive reactance
Z	Impedance

Greek Symbols

α (alpha), β (beta)	Velocity angles in turbines, angles
Γ (gamma)	Circulation
γ (gamma)	Specific weight of water
η (eta)	Efficiency
θ (theta)	Angle
ν (nu)	Kinematic viscosity
ρ (rho)	Density of water
σ (sigma)	Cavitation index
ϕ (phi)	Velocity potential
Ω, ω (omega)	Angular velocity

About the Editor in Chief

JACK J. FRITZ, a Professional Associate with the National Research Council's Board on Science and Technology for International Development, has extensive experience in the renewable energy systems field. Prior to his joining the National Academy of Sciences, he was program manager for the Agency for International Development, managing a small hydropower program designed for application in less developed countries. In addition, he has more than ten years' experience in renewable energy systems design in both the private and the public sectors. A registered professional engineer in New York, he has a B.S. degree in mechanical engineering and a Ph.D. degree in civil engineering from the State University of New York at Buffalo and has published some two dozen papers on renewable energy technology, economics, and environmental engineering.

1

Introduction

Jack J. Fritz

Jean François Henry

The world is currently in the throes of an energy crisis that has reordered, and will continue to reorder, our perceptions of how to power the infrastructure on which we have come to depend. Increasing prices for petroleum products, projections that petroleum resources will be exhausted in a relatively short period of time, and use of fossil fuel resources for political purposes are adversely affecting worldwide economic and social development. The impact of the energy crisis is particularly felt in less developed countries (LDCs) where an ever-increasing percentage of national budgets earmarked for development must be diverted to the purchase of petroleum products. To reduce this dependence on imported fuels with high price volatility, most countries have initiated programs to develop alternative sources of energy based on domestic renewable resources. Among these resources are solar energy, wind, geothermal energy, biomass, and not least, hydropower.

In the United States, only a small percentage of electrical energy is produced from waterpower resources, primarily by the larger systems such as the Tennessee Valley Authority and the Bonneville Power Administration. However, most large sites in the United States have already been developed, making it highly unlikely that hydropower will offer significant new contributions to the national power grid in the future. This has given rise to a new trend: Small hydropower has reemerged as an energy source which is easily developed, cost competitive, and minimally disruptive to the environment.

Bringing hydroelectric power to a given site to serve a projected market is a complex proposition involving planning or feasibility studies, detailed analyses, construction, testing, and shakedown activities. Each step requires inputs from a variety of individuals or groups: planners; engineers knowledgeable in hydrology, site development, dams, turbines, and electrical equipment; environmental scientists; lawyers specializing in water rights, licensing, and regulatory matters; financial experts; and economists and market analysts.

Implementing a hydro project is a multidisciplinary effort. Few individuals have the in-depth knowledge of each aspect of hydropower to perform all the tasks involved in carrying out a project. It is, however, essential that the planners and decision makers ultimately responsible for overall implementation be aware in sufficient detail of the significance, constraints, and requirements attached to the various elements of the project. It is also essential that specialists involved in analysis be aware of the requirements and constraints imposed on their area of expertise by other components of the project. The most sophisticated dam design would be worthless if it did not include features to satisfy environmental regulations or if it ignored constraints resulting from downstream patterns of water usage.

The purpose of this book is to provide the reader with enough background to create that awareness. It is intended for engineers, economists, and planners as an introduction to a mature but neglected technology. Each chapter deals with essential system components, such as turbines and hydraulic structures, in enough depth to promote an understanding of their functions and principal design features as well as to serve as a planning and decision-making tool. The book will also supply engineers with the information they need to act as planners and managers of complete projects. It is further intended to familiarize generalists with the various aspects of hydro development as they relate to particular fields of expertise. Finally, the editor hopes that the book will influence students and newcomers to explore the potential of small hydropower as a cost-effective means of generating electric power.

1.1 THE PROMISE OF SMALL HYDROPOWER

Hydropower is among the most promising of the renewable energy resources. It is a clean source of power produced when water turns a hydraulic turbine. It provides the electricity essential to economic and social development. Although in the latter part of the nineteenth century many small (a few megawatts) hydroelectric plants were in operation, the trend in the twentieth century has been to develop large-scale systems (Fig. 1.1).

Typical large-scale installations include Boulder Dam in the United States, the Aswan Dam in Egypt, the Tarbela Dam in Pakistan, and numerous proj-

FIG. 1.1 Green Lake Dam on the Vodopad River near Sitka, Alaska, with a power capacity of 16.5 MW. Although considered small in capacity, this site is typical of medium- and large-scale projects. *(Courtesy of R. W. Beck & Assoc.)*

ects in the Alps of Western Europe. Such large undertakings benefit from economies of scale, low and essentially constant operating costs, and nondependence on fossil fuels. At the time these projects were constructed the attributes listed were the key factor in making them competitive with large-scale petroleum-fired power plants using low-cost and readily available fuels. In recent years, however, high construction costs, environmental concerns, and long delays in planning, construction, and licensing have reduced the attractiveness of such large-scale hydro projects.

By contrast, recent technical advances in water turbine design, construction, and efficiency, combined with increasing petroleum fuel costs have enhanced the competitive position of smaller hydroelectric projects in relation to diesel-powered generators of similar size. [*Small hydro* refers to projects with a capacity of less than 15 megawatts (MW).] When compared to large-scale hydro projects, small hydro can be planned and built in less time and are less likely to create extensive environmental problems. They are reliable and, within the limits of the water resources available, can be tailored to the needs of the end-use market. Although the unit cost per installed kilowatt of generating capacity is higher for small-scale projects, financing is often easier to obtain. These characteristics make small hydro particularly attractive for LDCs where near-term installation of dispersed energy systems is essential for economic and social development.

The potential impact of small and mini hydropower systems is being rediscovered by U.S. government agencies and international technical assistance institutions alike. Ellis Armstrong, former head of the U.S. Bureau of Reclamation, enumerated the advantages of small hydro power as follows:[1]

Small hydro:

1. is a non-consumptive generator of electrical energy, utilizing a renewable resource which is made continually available through the hydrologic cycle by the energy of the sun.
2. is essentally non-polluting and releases no heat. Adverse environmental impacts are negligible and, for small installations, may be totally eliminated.
3. plants can be designed and built within one or two years' time. Licensing requirements are minimal, equipment is readily available, and construction procedures are well-known.
4. plants require some type of water control, up to and including full regulation of watershed discharge. They are thus an important element in the multipurpose utilization of water resources and can reduce potential flood damage. Where storage facilities are involved, floodwaters are retained and can be better directed to agricultural production, river regulation, improved navigation, fish and wildlife protection, recreation, municipal use and better control of wastewater.
5. is a reliable resource within the hydrologic limitations of the site. The relative simplicity of hydraulic machinery makes energy instantly available as needed. Since no heat is involved, equipment has a long life and malfunctions are rare.
6. in remote areas using relatively simple technology can be a catalyst in mobilizing productive resources and creating enhanced economic opportunities for local residents.
7. is characterized by reliability and flexibility of operation, including fast start-up and shut-down in response to rapid changes in demand. It thus becomes a valuable part of any large electrical system, increasing overall economy, efficiency, and reliability.
8. has an excellent peak-power capability. While approximately four units of energy input are required for three units of output, the input is low-cost hydraulic energy and the output is high value electrical energy. In a large electrical system, the alternative for handling peak loads may be utilization or costly expansion of old and relatively inefficient thermal units.

9. technology is well-developed and proven, with turbine efficiencies running as high as 90 percent. Small units ranging from a few kilowatts to several megawatts have been in operation since the turn of the century. While the equipment must be adapted to the specific site for greatest efficiency, its performance will generally live up to manufacturer's claims.
10. facilities have a long life. As a rule, dams and control works will perform for a century or more with little maintenance.
11. requires few operating personnel. Some small-scale installations are operated entirely by remote control. Freedom from fuel dependence together with the long life of equipment make hydroelectric power installations resistant to inflation.
12. development can make maximum use of local materials and labor. When compared to thermal facilities, small hydro usually provides more local employment in the construction of civil works.
13. resources remain untapped, especially in the developing countries where less than seven percent of potential has been developed. In some countries the figure is less than two percent.
14. economic feasibility is improving when compared to other energy sources that use finite fuels. With more realistic methodologies for economic evaluation, including full recognition of the value of non-consumptive water use, freedom from fuel dependence and minimum environmental impact, small hydropower has become increasingly desirable.
15. potential in the industrialized countries can be developed to augment hydropower capacity at existing powerhouses and dams. The possibility of retrofits and additional turbines and generators makes the upgrading of present installations attractive.

1.2 DEFINITION OF SMALL, MINI, AND MICRO HYDRO PROJECTS

The definition of small hydro is somewhat arbitrary from a technical standpoint. In the United States *small hydro projects* were defined as systems of 15 MW or less capacity. This capacity limit has been adopted in Public Utilities Regulatory Policies Act, or PURPA (P.L. 95-617, November 1978) for special handling in licensing, loans, incentives, and other supporting federal, state, and local programs.[2] Provisions of the law specifically relating to small hydro projects address only additions to existing facilities. Recent legislation increased the definition to projects of 30 MW capacity or less. Mini and micro hydro projects can be defined in much the same fashion. In the forthcoming discussions, *mini hydro* will refer to projects of 1 MW capacity or less, and *micro hydro* to projects of 100 kW capacity or less (see Fig. 1.2).[3] It must be remembered, however, that this terminology is only introduced for the purpose of

FIG. 1.2 Micro hydro plant installed in 1959 in Unduavi, Bolivia, with a capacity of 15 kW.

quantifying the scope of different projects. No particular significance should be attached to defining various classes of projects by capacity.

Hydroelectric sites are divided into three categories by head (vertical difference in elevation between plant intake and discharge): low, medium, or high. Each category requires a different design.

- ***Low head:*** 2 to 20 m (about 6 to 66 ft) head.[4] There had been a tendency in published documents and public statements to confuse small hydro and low-head hydro. These terms are not synonymous although a large number of sites identified for small hydro development in the United States are in the low-head category as in Fig. 1.3.
- ***Medium head:*** 20 to 150 m (66 to 500 ft) head. See Fig. 1.4.
- ***High head:*** 150 m (more than 500 ft) head. Often high-head sites require the greatest capital investment per unit of installed capacity, depending on specific characteristics.

These suggested limits are not rigid but are merely a means of categorizing sites.

FIG. 1.3 Low-head mini hydro plant at Turnip Creek in the Imperial Valley of California with a capacity of approximately 400 kW.

Two additional definitions of small hydro plants may be useful to the reader. A "run-of-river" plant is one whose energy output is subject to the instantaneous flow of the river. On the other hand, a "storage" plant with capacity, i.e., a dam or reservoir, is able to produce continuous energy on the basis of controlled water release.

1.3 HISTORICAL PERSPECTIVE

Waterpower has contributed to the development of humankind since biblical times. References to the use of waterwheels for milling, pumping, and other functions date back to 300 B.C. in Greece, although they were probably in use long before that time. In the years between these early uses of the waterwheel and the advent of the industrial revolution, running water and wind were the only sources of mechanical power, other than animal, available. Improvements in power recovery from flowing water were steadily introduced as exemplified by the sophisticated waterworks designed in the 1600s for the palace of Versailles outside Paris, France. This system had a capacity equivalent to an estimated 56 kW of power. Waterwheels of various types are still used in many countries where they provide economical, low-level mechanical power.[5]

As they evolved, waterpower systems, and ultimately hydroelectric generating plants, were developed from attempts to improve the efficiency of the waterwheel. Early work occurred in France between 1750 and 1850 when various turbine designs were investigated. At the beginning of the industrial rev-

FIG. 1.4 Medium-head mini hydro plant in Cotacachi, Ecuador, with a capacity of approximately 160 kW.

olution, France did not have access to large coal deposits and had to rely on its water resources to generate the energy needed for industrial expansion. (To this day, waterpower is still called *houille blanche,* or white coal, in France.)

Much theoretical work was done during this period by mathematicians and engineers such as Bernard Forest de Bélidor, John Smeaton, Jean Victor Poncelet, Leonhard Euler, Claude Burdin, and Benoît Fourneyron.[6] Their work resulted in significant improvements in turbine efficiency and laid the groundwork for the development of modern turbines of the Francis, Kaplan, and Pelton type. As an example of the progress achieved, the original vertical-axis turbine designed by Bélidor attained an efficiency of 15 to 20 percent. By the mid-1850s this rose to 60 to 70 percent. The modern counterparts of the Bélidor turbine, the Francis and Kaplan turbines, now achieve efficiencies of 90 to 95 percent.[7]

In the early part of the nineteenth century, water provided mechanical power for industrial applications. With the invention of the dynamo or generator in the 1880s, and the popularization of electricity as a source of energy, many water turbines were converted to electricity production. The first hydroelectric unit in the United States is reported to have been a 12.5-kW plant installed in 1882 on the Fox River at Appleton, Wisconsin. Many small hydroelectric projects were installed thereafter, particularly in the western part of the United States. These projects satisfied local demands for electricity, in most cases for lighting and small industry.

The ability of hydroelectric plants to provide electricity for the larger population centers was recognized early. As early as 1890, for example, a hydro project at Niagara Falls was under study to supply electricity to the city of Buffalo, New York. Inadequacies in electrical transmission technology, however, delayed the implementation of this and other projects. With the development of high-voltage transmission lines in the early part of the twentieth century, a shift occurred from small-scale plants serving local electricity markets to large-scale plants feeding into extensive distribution grids. The same trend was observed in thermal plants fired by coal and later by petroleum fuels. In most industrialized countries, therefore, the twentieth century has been characterized by a proliferation of large-scale (hundreds of megawatts) hydroelectric projects serving wide areas.

Independent of their ability to provide reserves of generating capacity and respond to peak demand, large-scale plants, both thermal and hydroelectric, benefit from economies of scale. These economies can be quite significant in terms of capital investment and financing costs. On the basis of a scaling factor of 0.7, generally accepted in many engineering projects, the capital investment in a 200-MW generating plant would only be about 1.4 times that of a 100-MW plant.

Prior to the drastic increases in fossil fuel costs (1973–74), economies of scale were particularly important in justifying hydroelectric projects. (A survey in developing countries quoted (1979) average costs of approximately $1,300 per kilowatt installed for hydroelectric systems as compared to about $860 per kilowatt installed for thermal systems).[8] Large-scale systems were necessary to reduce the unit cost of installed generating capacity. Because hydroelectric plants claim low maintenance and operating costs and do not consume fuel, the major cost component in the early 1980's capital market is financing.

The case for hydro was further damaged by the high cost of power transmission. The location of a hydroelectric plant is determined by water resource availability, site potential, and environmental acceptability. Many such locations are relatively remote from population and industrial centers. Bringing the electricity generated to the end-users and integrating that capacity into a country's distribution grid requires extensive and expensive transmission lines. To maintain the price of distributed hydroelectricity at a level competitive with nonhydro sources, the high capital cost of transmission lines should be amortized over the largest possible number of kilowatthours, contributing again to the logic of large-scale hydro projects. By contrast, thermal plants are more flexible in terms of siting: Fuel can be transported to a convenient location close to markets.

1.4 WORLD HYDROPOWER RESOURCES

Various estimates of hydropower resources have been made over the years: worldwide, regional, and for specific river basins. Since 1929, the World Energy Conference (WEC; founded in 1924 as the World Power Conference) has periodically published a survey of world energy resources, both potential and developed. Energy information supplied by the United Nations and by the 80 member countries of the Conference, is published in a comprehensive volume at 6-year intervals, the first in 1962. In 1974 the Conference established a Commission on Energy Conservation whose responsibility it was to evaluate the world's energy resources in light of recent price escalations. The 1974 volume,[9] excellent in its own right, was updated in 1976 to include this evaluation and provide data for future study.

The results of the Commission studies show that today water provides 23 percent of the world's electric power. In specific regions this figure varies from 100 percent in remote areas to negligible amounts in some oil-rich countries in the Middle East. This was not the case in 1960 when 29 percent of world electricity production was attributed to hydro and geothermal power (see Fig. 1.5). The percentage of fossil fuel inputs in electricity production was 71 and 75 percent in 1960 and 1970 respectively. Projections suggest that by 1990, hydro and geothermal power will account for only about 11 percent of world electricity production compared to 51 percent for fossil fuels.

The number of operating hydro sites worldwide is 16 percent of what is considered to be "reasonably developable" from economic, physical, and environmental standpoints. Exploitable sites are estimated to be only about 12 percent of the total energy that can theoretically be derived from the water in all the world's rivers and streams.

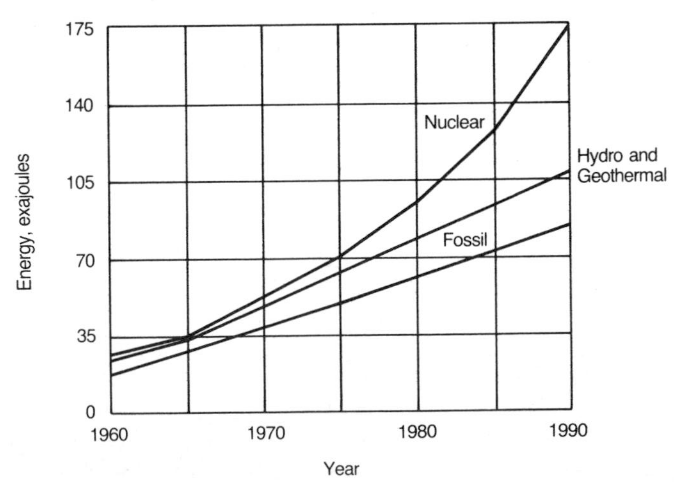

FIG. 1.5 World electricity production.

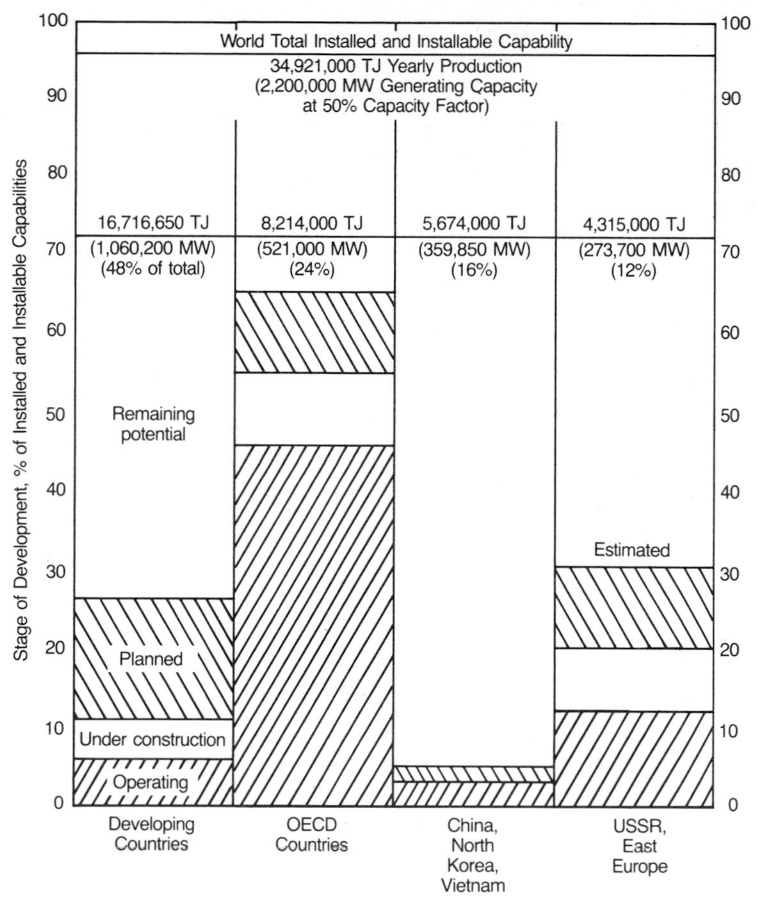

FIG. 1.6 World hydropower resources.

Based on the Commission's study, the total potential considered feasible for development is 2.2 million megawatts of generating capacity with a probable annual production of 9,700 million megawatthours. A 50 percent capacity factor, about average for early 1980s hydropower production, was assumed. These figures were determined from capacities at various current and future installations. To produce the same amount of energy, a thermal plant would burn the equivalent of about 40 million barrels of crude oil per day.

The present annual production of hydropower is about 1,575 million megawatthours, which is approximately 16 percent of the Conservation Commission's estimated expected potential. To produce the same amount of electricity in an oil-fired plant would require burning approximately 6.5 million barrels of crude oil per day.

Figure 1.6 illustrates the percentages of the world's total hydropower potential found in the developing countries, Organization for Economic Cooperation and Development (OECD) countries, and centrally planned countries. It also

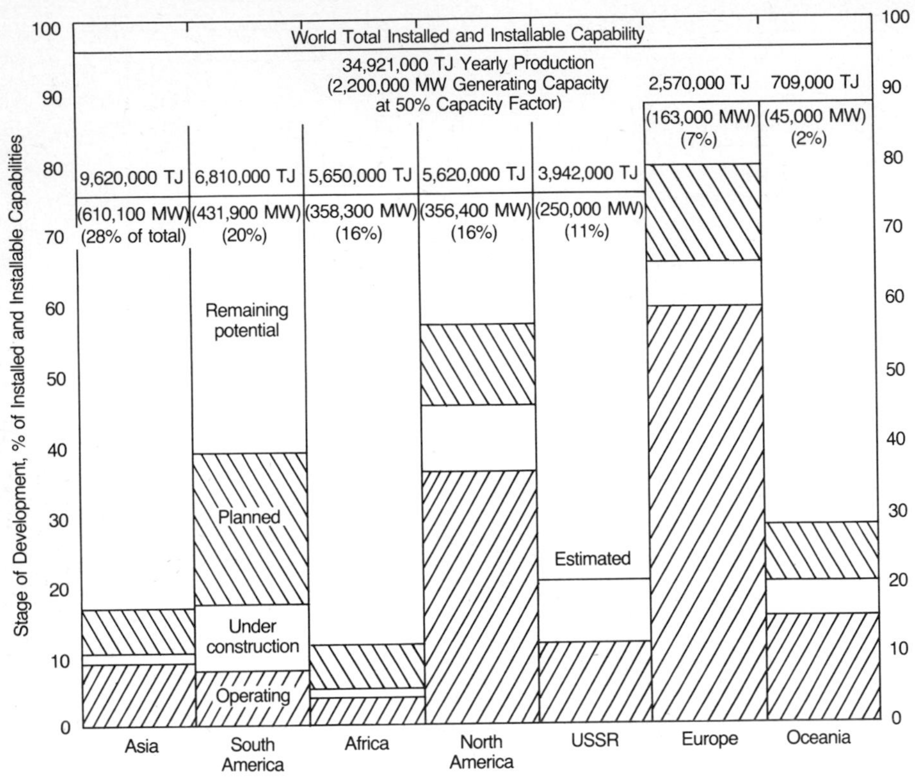

FIG. 1.7 World's installed and installable hydropower capacity.

shows the percentage of potential in each of these areas that is currently in operation, under construction, or in the planning stages. Figure 1.7 presents a geographic breakdown of the same information: Asia, South America, Africa, North America, the USSR, Europe, and Oceania.

These figures are based on the 1979 data bank and show the relative magnitude of hydropower resources in various areas. In 1970, almost 80 percent of installed hydropower capacity was found in North America, Western Europe, the USSR, and Eastern Europe. Most of these systems were large-scale projects. Other areas of the world have made little use of their hydropower resources. In South America, Africa, and Southeast Asia, for example, less than 2 percent of hydropower potential has been developed. This may be attributable to the demand differential between developing and industrialized areas. In the former, the lack of electricity distribution systems and the need to provide electricity on a small scale to a dispersed population has tipped the balance in favor of small and isolated generating plants. For rural areas, small hydropower systems are often in competition with diesel-generator units. In most cases, the higher unit cost of hydropower, contrasted to the flexibility in siting, ease of installation, and availability and low cost of fossil fuels (at least until the mid-1970s) favored internal combustion units.

TABLE 1.1 Estimated Probable Hydroelectric Development*

Divisions	Potential energy, Kwh × 10^9				
	Year 1976	Year 1985	Year 2000	Year 2020	Total developable, 1976 WEC survey
OECD countries	1,050	1,250	1,490	2,166	2,280
Centrally planned economies	200	332	800	2,416	2,775
Developing countries	325	548	1,248	3,278	4,645
World total	1,575	2,130	3,538	7,860	9,700

*The figures shown are the probable annual average energy from installed hydroelectric facilities for the year indicated.

The estimated amount of hydropower development from the WEC Commission on Energy Conservation study is shown in Table 1.1. Based on an approximate cost of 1,000 U.S. dollars per kilowatt installed, the capital investment required is indicated in Table 1.2. It is evident that financial assistance to the developing countries will be required to exploit water resource potential. In an interdependent relationship between power availability and market demand, it is essential that both be developed concurrently in a manner economically beneficial to both.

Between 1973–74 and 1980, the cost of oil has increased by a factor of approximately 10 (from about $4 per barrel to close to $40 per barrel). Assuming a conversion factor of 20 percent or about 17,000 Btu of fuel per kilowatthour of electricity produced, the cost of the fuel required to produce 1 kWh of electricity has increased from $0.012 to $0.12 in less than 10 years (this assumes an energy content of 135,000 Btu per gallon of crude oil). By comparison, a hydro system at $1,300 per kilowatt installed, financed over 40 years at 10 percent (capital recovery factor of 0.1023) and with a load factor of 40 percent could provide electricity at about $0.038 per kilowatthour. The cost of fuel may be even higher in some LDCs where transportation and distribution costs to remote areas under less than ideal terrain and road conditions may be significant. Small hydropower, therefore, appears to be one of the economically most attractive alternative sources of energy for LDCs in the near future.

TABLE 1.2 Estimated Capital Costs of Hydroelectric Installations*

Divisions	Average annual costs between years shown, 1976 U.S. dollars in billions			
	1976–1985	1985–2000	2000–2020	1976–2020
OECD countries	5.1	3.7	7.7	5.8
Centrally planned economies	3.4	7.1	18.4	11.5
Developing countries	5.6	10.6	23.2	15.3
World total	14.1	21.4	49.3	32.6

*Based on Table 1.1.

1.5 DOMESTIC PROGRAMS IN THE UNITED STATES

In the United States, the most desirable sites for large-scale hydropower have been developed. They produce approximately 13 percent of all electricity used domestically. Concern over environmental impacts, long delays in licensing, and increasing construction costs reduce the attractiveness of additional large-scale hydro plants. By contrast, increasing oil prices have brought small hydro systems back into competition with other sources of electricity. Recent improvements in manufacturing and standardizations in the design of small turbines have also increased the economic attractiveness of small hydro systems.

In 1977, the U.S. Army Corps of Engineers estimated the amount of additional generating capacity available for development in the immediate future. About 50,000 existing small dams were identified in the United States of which fewer than 1,000 are currently used to produce electricity. Several hundred of these dams were used to generate electricity in the past. The Corps of Engineers estimated that if all the dams that could produce power were reactivated, and/or if new generating capacity were installed, almost 233 billion kilowatt-hours yearly, equivalent to 10 percent of current U.S. production of electricity in 1978, could be added.[4]

Other sites at which small hydro systems could be installed include existing waterworks such as irrigation canals with drops, municipal water supplies, and some waste management systems. In many of these cases, site development will not raise complex environmental issues, and planning, construction, and licensing phases should be significantly shorter than for larger plants. Small-scale hydro can make a significant and near-term contribution to the electricity needs of the United States and at the same time reduce dependence on fossil fuels. The federal government has recognized this potential. Through its Federal Energy Regulatory Commission (FERC), it has recently simplified the licensing procedure for small hydro (less than 15 MW) to accelerate its penetration into the energy production market.

TABLE 1.3 Peruvian Villages which Possess Hydro Potential, Capacity, and Generation*

Group	Population	No. of villages	Population, thousands	Installed capacity range, kW	Average size, kW	No. of stations	Installed capacity, MWe
A	200–1,999	1,920	1,400	5–50	30	1,728	52
B	2,000–5,999	360	1,300	50–179	120	368	44
C	6,000–10,000	72	550	180–350	270	90	24
Total		2,352	3,250			2,186	120

*From Ref. 11.

1.6 SELECTED DEVELOPING COUNTRY HYDROPOWER PROFILES

In a number of LDCs, significant progress is being made in the implementation of small and mini hydropower systems. Many countries are developing unique approaches suited to their cultures, energy demands, and natural resources. Comparison of these efforts yields valuable information on the opportunities and limitations surrounding this technology.

Peru

Mini hydropower development in Peru is coordinated through the Ministry of Energy and Mining and carried out by ELECTROPERU, the national electric utility. Plans called for the construction of 50 projects between 1980 and 1985 at a cost of $12 million.[10] Since 1980, foreign assistance programs with the United States and Germany have increased these original sums.

Hydropower is one of the most promising options for decentralized electricity production in the villages and secondary towns of Peru. There are a large number of potential sites in the vicinity of inhabited areas in both mountain and jungle regions. Sites in the Andes are characterized by high heads and the low to moderate flows associated with irrigation canals. In the jungle region, the water resource is typically low head and high flow, ideal for run-of-river plants.

The 1972 census data suggest that there are over 3,000 villages in the mountains and jungles of Peru, with a total population of over 4 million. About 100 of these villages are sizable settlements with populations of between 6,000 and 10,000, and 95 percent of them are without electricity. Of the remaining 5 percent, 1 percent have hydropower stations and 4 percent utilize diesel or gasoline generators. Table 1.3 categorizes the size distribution of villages with hydropower resources. To obtain an estimate of the electricity generation requirements, it was assumed that installed capacity per capita would be 25 W in Group A, 30 W in Group B, and 35 W in Group C. The total installed capacity in the year 2000 with such a mini hydro program could be 120 MW.

Electricity requirements in rural areas are generally characterized by low capacity factors, in the range of 15 to 20 percent. Agricultural demand derives largely from irrigation water pumping, a highly seasonal activity, while domestic demand primarily for lighting, particularly in the initial stages of electrification. Assuming an annual capacity factor of 20 percent, the total electricity consumed based on 120 MW installed power will be about 210 GWh in the year 2000, or 0.75×10^{15} J.

Small mining is another sector in which mini hydro plants are potentially very useful. There are currently about 800 such mines in Peru in addition to some 60 concentration plants. On this basis, an additional 67 MW of capacity generating 180 GWh of electricity annually could be installed by the year

FIG. 1.8 Mini hydro plant located in Hoyo, near Lima, Peru, with a capacity of 150 kW.

2000. The total energy contribution from mini hydropower systems could thus reach an installed capacity of 187 MW generating 390 GWh of electricity annually.[11]

Peruvian officials expect the installation cost of mini hydropower systems to average $500 (1978) per installed kilowatt, assuming free labor to complete the necessary civil works. The capital cost, however, will vary considerably, depending on head, size of installation, availability of structures to serve as powerhouses, and plant sophistication. The $500 per kilowatt cost is based on actual experience installing a micro hydro (4 kVA) Michell-Banki turbine at a pilot plant.

The Ministry of Industry and Tourism will coordinate local manufacture of equipment through the Institute of Industrial Technology Investigation and Technical Normalization (ITINTEC) whose job it will be to develop or recommend appropriate design and manufacturing methods. To aid in the dissemination process, ITINTEC has produced and published a handbook for designers. Private industry and the National Engineering University (UNI) are also deeply involved in training and development of this technology.

Indonesia

Hydropower development in Indonesia dates back to the early twentieth century. Some units installed by Dutch engineers before 1940 and ranging in size from 35 kW to 3 MW, are still in operation. By the mid-1940s, hydropower accounted for more than 80 percent of total electricity generation in Indonesia.[12] Since that time, this fraction has steadily declined despite continual increases in capacity. In 1968, total hydropower energy output climbed to 941

TABLE 1.4 Hydroelectric Potential in Indonesia*

Location	Capacity, MW
Sumatra	6,750
Java	2,500
Kalimantan	7,000
Sulawesi	5,600
Irian Jaya	9,000
Other islands	150
Total	31,000

*From Ref. 13.

GWh generated from an installed capacity of 283 MW. By 1976, it had risen to 532 MW. The 552 MW of capacity in the early 1980s represents only 20 percent of the total electricity generating potential utilized by the National Power Agency (PLN). Most of this installed capacity is on Java and is used primarily for irrigation and water supply projects. [It is estimated that currently exploited hydroelectric capacity is only 1.5 percent of Indonesia's total hydropower potential. According to a survey done for the Nuclear Power Planning Commission (NPPC), most of the remaining Indonesian hydroelectric potential is outside Java.[13] Table 1.4 gives an estimate of the total water power potential of major Indonesian islands.]

One of the objectives of the National Energy Plan developed by NPPC is to exploit remaining hydro potential through the installation of mini hydro projects to supply regions far beyond existing distribution grids. Only some 80 kW of small hydro capacity was added in 1976.

Small and mini hydropower systems appear to have an enormous potential, particularly in remote areas of Sumatra, Kalimantan, Sulawesi, and Irian Jaya not served by a central grid. This potential is essentially unexploited and has been only cursorily explored. A rough approximation of the mini hydro potential was obtained from general geographic and rainfall data contained in recent studies.[14] In terms of precipitation and the ratio of mountains to lowlands, Kalimantan and Sulawesi are the most conducive to hydropower development. North Sumatra and parts of Java also appear attractive, but Irian Jaya may prove to be the most economical for development. Use of small-scale hydropower there is, however, limited by the very small population. Table 1.5 provides some approximate figures for the amount of small hydro potential in these areas. The numbers are conservative in that only 1 to 2 percent of potential hydro resources can be effectively utilized. In these areas, a total potential of some 484 MW of capacity is projected. The per capita figures are based on estimates of the population living in areas not currently served by central power sources but in favorable rainfall-topography zones.

Based on the analysis given, it is estimated that approximately 9 of the 20 million people in rural Sumatra, Kalimantan, and Sulawesi could be served by

TABLE 1.5 Estimates of Exploitable Small-Scale Hydropower Potential for Indonesia*

	Region		
	N. Sumatra	Kalimantan	Sulawesi
Area, million hectares	15.0	53.9	18.9
Runoff, mm	1,470	1,590	1,200
Average elevation, m	400	500	1,200
Theoretical power, MW	4,370	17,000	13,500
Practical power, MW	44	170	270
Available population, millions	6	6	8
Practical power per capita, W	7.28	28.3	33.74

*From Ref. 14.

small-scale hydro within 15 years. A per capita capacity of 50 W was assumed, corresponding to 450 MW of connected load. Sizing of the plants is dependent on population density in areas that will be developed. For villages of 500 people with a capacity rate of 0.5 kW per capita, the 25-kW size would be most appropriate. Small towns and areas of approximately 2,000 people require a 100-kW installation. Where 3,000 people are clustered together, a 150-kW plant would be advisable. However, in most locations where 2,000 people are concentrated and diesel power is available, capacity exceeds 50 W per capita. The studies report 100 W per capita as optimal. A population of 1,500 will thus use 150

FIG. 1.9 Mini hydropower laboratory near Jakarta, Indonesia, operated by the National Power Agency (PLN).

kW within 3 years of start-up. If 450 MW is to be installed, even using a 150-kW station, 3,000 units would be required.

Although there is currently no significant utilization of small hydropower in Indonesia, there appears to be widespread recognition of its potential. It is appreciated as a valuable contribution to the government's policy goal of developing rural energy potential to provide cheap energy supplies in poor areas with minimum damage to forests, land, and water resources.

People's Republic of China

With over 5,000 rivers, each with drainage areas of over 100 km^2, as well as some 2,000 lakes and 80,000 reservoirs, the People's Republic of China (PRC) has abundant water resources. The exploitable hydroelectric potential has been estimated at 370 million kilowatts with an existing installed capacity of only 70 million kilowatts.[15] The PRC's program in small and mini hydropower can be traced to 1949, when only some 50 small plants with a combined capacity of 5,000 kW existed. By the end of 1980, some 90,000 plants with individual capacities of less than 6,000 kW had been installed for a total capacity of 7.1 million kilowatts.[16] A 1979 survey reported the average size plant to be 61 kW operating with a plant factor of 20 percent.[17]

The major push for the construction of small hydro plants in the PRC occurred during the 1950s in conjunction with small-scale rural industrialization programs. By the end of the 1960s, installed capacity had reached some 520,000 kW. By 1966, however, many plants fell into disuse when electrical distribution projects were emphasized. During the 1970s, this policy was reversed and widespread construction of small plants was resumed with installed capacity growing from 500 MW in 1969 to 3,000 MW in 1975.[18] Table 1.6 traces this growth.

Early plants were built with local financial, material, and labor resources. Equipment was manufactured in a variety of village shops, at times using wood

TABLE 1.6 Growth of Small Hydropower Capacity in the People's Republic of China*

Year	Total plant capacity, MW	Year	Total plant capacity, MW
1949	5	1973	2000
1960	520	1974	2500
1966	200	1975	3000
1969	500	1976	3500
1970	900	1977	4320
1971	1200	1978	5260
1972	1600	1979	6330

*From Ref. 18.

to make turbine runners. Dams were built of earth and stone and penstocks of cement, wood, and occasionally bamboo. Even though they required frequent repair and maintenance, the readily available supply of labor and large number of rural industries made it possible to build low-cost plants. These activities were initially carried out at the brigade level. As the number of plants grew, they began to be interconnected into regional grids alleviating typical control and outage problems found at isolated plants. The benefits of these projects began to spread to many activities which included irrigation, flood control, and water supply.

With significant progress continuing into the 1980s, the PRC's leaders are currently seeking to expand rural electrification by consolidating planning and construction. The new policy diminishes the local, self-help approach in favor of greater collaboration and system standardization. These steps appear to follow an approach to electrification similar to that which occurred in early twentieth-century America.

In many developing countries, small hydropower has come full circle as result of the energy crisis: from the status of a quickly abandoned source of electricity in the second quarter of the twentieth century to rebirth as a significant potential contributor to meeting energy and development goals.

1.7 SMALL HYDROPOWER AND OTHER SOURCES OF ELECTRICITY

Hydropower is one of several alternative renewable energy sources being considered to replace fossil-fuel-fired electricity generation systems. Other site-specific alternatives include photovoltaic, solar-thermal, geothermal, and biomass-fired generating plants. In many cases, particularly in LDCs, small hydro systems can satisfy basic electricity requirements in small- or medium-sized dispersed communities thereby competing with diesel-fueled generators.

The comparison between small hydro systems and petroleum-fired generators or alternative sources of energy should be made within the context of small-scale dispersed electricity production. System configuration and operating conditions, and thus yearly operating costs, are site specific. Site conditions may include resource availability (hydrology, insolation, geology, land and biomass availability, fossil fuel availability, etc.), equipment and construction material availability, financing, market demand, and local social structure. These conditions vary widely from country to country and often within countries. Generalized comparative assessments of alternative energy sources must be qualitative, whereas only site-specific analyses can result in quantitative contrasts.

The following comments are meant to provide a general indication of the relative attractiveness of small hydro versus other sources of electricity.

Small Hydro Versus Diesel Generators

Figure 1.10 summarizes the results of an analysis performed by the Intermediate Technology Development Group comparing a micro hydro system of 40 kW to a diesel generator of the same capacity installed in Nepal.[19] It illustrates the unit cost electricity characteristics of both systems. The total annual operating cost is basically constant for the hydro system irrespective of the percentage of capacity utilized because recovery of the capital investment is the major component of the cost. The total annual cost for the unused diesel system (no electricity production), i.e., the cost of recovering the capital, is lower than that for the hydro system because of the lower unit cost per kilowatt installed for the diesel. As electricity is produced the annual cost of the diesel unit increases in a linear fashion, reflecting the cost of the fuel consumed.

Annual costs for the two systems intersect at about 32,000 kWh or about 10 percent of capacity utilization. This is noteworthy since in many LDCs, initial demand for electricity may be high during specific periods (lighting at night) and very low during the remainder of the day creating a high peak demand but low capacity factor. As suggested by the data in the figure, under such circumstances the diesel system remains economically attractive. This is an example of site-specific conditions (market demand in this case) which could reverse the validity of general statements suggesting that hydro is more attractive than diesel. Analysis of typical demand patterns in villages in India suggest that capacity factors of about 30 percent might be achieved. Under those conditions, hydro systems appear more attractive than diesel units.

Hydro systems have the obvious disadvantage of being restricted to sites which may be distant from the end user of the energy product. Electricity distribution must then be added to the cost of production, thereby reducing the cost advantage of hydro versus fossil-fuel-fired systems. The data in Fig. 1.10 suggest nevertheless that under favorable conditions, i.e., proximity of source and end-user, sufficient capacity factor, and high world prices for petroleum

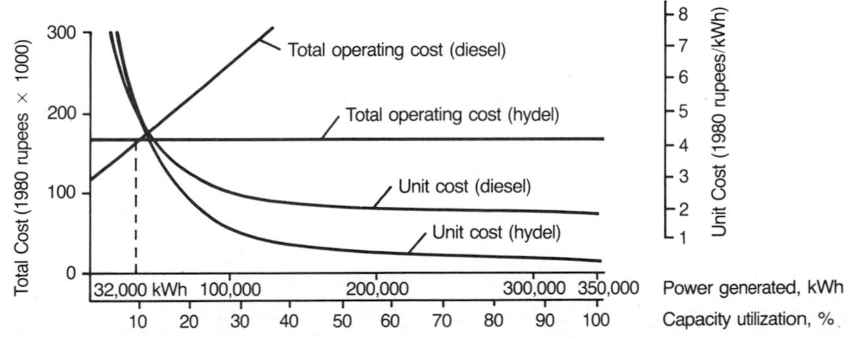

FIG. 1.10 Total annual cost and unit cost versus power generated and capacity utilization for 40-kW diesel and hydroelectric installations.

fuels, small or mini hydropower systems are viable alternatives to diesel systems.

Small Hydropower Versus Other Alternative Sources

Photovoltaic Cells

Table 1.7 shows estimated costs per kilowatt of generating capacity for several electricity producing systems. The estimates for hydropower, geothermal, and thermal systems are average costs for 97 developing countries.[8] Photovoltaics, direct conversion of the visible part of sunlight to electricity, is one of the cleanest and ecologically most attractive methods of converting solar energy to high-grade energy. Solar cells, like hydro systems, require little maintenance and the cost of energy produced is essentially constant, i.e., equal to the cost of financing the system over its lifetime. Present costs of photovoltaic systems are high, in the $10,000 to $20,000 range per peak kilowatt for water pumping systems.

It is projected that the intensive research programs currently under way in the United States, Japan, France, the United Kingdom, and other countries could reduce the capital cost fourfold within the next few years. Even under these conditions, photovoltaic systems will require high investments which may only be justified for special, low-power applications such as telecommunications.[20]

Wind

Wind energy for pumping water or generating electricity is also a well-known, dependable technology. Units require little maintenance and can last 10 to 20 years. Electric generating units cost from $3,000 to $6,000 per kilowatt for small (fraction of a kilowatt) units and from $1,000 to $2,000 for units in the 5- to 20-kW range, the latter being comparable in cost to hydropower systems. Unit prices of $500 per kilowatt are foreseen for new units scheduled to go into production in the near future.[21]

Biomass

There are a variety of ways by which biomass resources can be converted into fuels suitable for thermal generation of electricity. Biomass-fueled steam generators are widely used in the forest products industry and in agricultural processing such as sugar mills (combustion of the bagasse residue). The unit cost of these systems is usually somewhat higher than that of oil-, coal- or gas-fired boilers because of the lower energy density of biomass fuels and more extensive and costly material handling systems (storage, screening, handling). A small

TABLE 1.7 Comparative System Cost per Kilowatt of Generating Capacity

Power source	Cost, $
Hydro*	1,296
Geothermal*	1,564
Thermal*	863
Photovoltaics†	10,000–20,000
Wind‡	
Less than 1 kW	3,000–6,000
5–20 kW	1,000–2,000
Biomass§	
Thermal	1,000
Gasifier/engine-generator	700–1,500
Ethanol	1,550

*From Ref. 8.
†From Ref. 20.
‡From Ref. 21.
§From Ref. 22.

boiler and steam-turbine-generator set can cost approximately $1,000 per installed kilowatt as shown in Table 1.7.

Small biomass gasifiers producing a low-energy combustible gas can be coupled with internal combustion engines to produce electricity. These units were widely used in World War II to replace scarce petroleum products. Today, however, no standard units are being manufactured. Prices can only be estimated and these have been reported to range from $700 up to $1,500 per kilowatt installed for complete gasifier-engine units (10- to 30-kW capacity).[22]

Biomass can also be converted to ethanol through fermentation of sugar crops. This conversion technology is well-established and is the cornerstone of the Brazilian program for independence from imported fuels. Ethanol can be used to replace petroleum in internal combustion engine generator systems. Assuming an overall conversion efficiency of 18 percent (ethanol to electricity), 1 gal of ethanol will produce 4.45 kWh. An ethanol plant capable of supporting 1 kW installed capacity at 50 percent load factor must therefore have a production capacity of about 1,000 gal per year. The investment for a small (farm-size) plant is estimated at $1.55 per gallon capacity for plants with an output of a few thousand gallons per year. The system cost for the biomass-ethanol alternative to electricity production is estimated at about $1,550 per kilowatt in Table 1.7.[23]

The data suggest that hydro, wind, and biomass systems are roughly equivalent in terms of capital investment. The output of hydro and wind systems depends directly on the availability of the solar driven forces, i.e., flowing water and wind of a specific minimum velocity. Providing electric power on demand, that is, matching supply and demand patterns, may require some form of storage to compensate for drought, low water flow, or windless periods. Battery storage is, however, costly and unreliable and is, therefore, only appropriate for micro systems of less than a few kilowatts. A clear benefit is that both wind and hydro systems, particularly those that are small in scale, usually have minimal environmental impacts.

Biomass systems rely on a resource that may be harvested as needed (wood, for example) or stored for variable periods of time (straw, grains, etc.). The electricity generation process may therefore be designed to respond to variable demand. Biomass systems may, however, create significant environmental problems ranging from careless harvesting of the resource to disposal of by-products such as ash or stillage from ethanol production. Biomass is also an inherent part of the ecological cycle and plays an important role as a source of food, feed, and fiber. Disruption of this balance by diverting resources to energy production could lead to serious social problems.

Each of the alternative energy sources considered above must be evaluated in the context of its intended use. Each has advantages and disadvantages that should be assessed in terms of socioeconomic impact as well. These facts suggest that small or mini hydro, where applicable, is among the very attractive alternatives available for dispersed electric power generation.

1.8 PLANNING A SMALL OR MINI HYDROPOWER PROJECT

Implementing a hydro project involves a number of tasks that can be characterized under one of the following headings: planning studies, procurement, construction, or start-up. Chapter 12 will deal only with the planning aspects of hydropower development. Planning studies include all types of investigations performed to determine the desirability of carrying out a hydro project. These studies are initiated when a proposal for a project is considered worthy of interest and are completed with the start of construction. They vary in scope, detail, depth, and intended audience and lead to various decisions and commitments during the preconstruction period. Following business and international practice, planning studies are roughly grouped into three categories: (1) *reconnaissance studies,* also referred to as appraisal or prefeasibility studies; (2) *feasibility studies;* and (3) *engineering design studies.* These will be discussed in greater detail in Chap. 12.

The following is a guide to the contents of this volume as it pertains to project planning. Table 1.8 summarizes the tasks involved in the planning process and indicates the chapters relevant to performing specific tasks.

Chapter 2 Hydrologic Analysis: This discusses methods for estimating the energy production and power potential of a site through the analysis of stream flows. Emphasis is placed on analysis techniques such as flow-duration curve computations, data handling, regional hydrologic assessments, and statistical methods. The chapter also provides a simple approach to making "first cut"

TABLE 1.8 Planning Tasks and Reference Chapters

Tasks	Reference chapters
Power and energy generation	2, 3
Environmental factors and costs	8
Institutional factors	9, 10
Project layout and development	
Overview	4
Dams and reservoirs	5
Turbines	6
Electric equipment	7
Project evaluation/economics	11
System design and case studies	12

estimates of hydropower resources suitable for overall regional assessments or for reconnaissance studies.

Chapter 3 Hydrologic Computations at a Small Hydro Site: This describes a simple method for determining maximum, minimum, and flow-duration information at a small stream with extremely limited hydrologic data. This field procedure was developed specifically for use in rural areas.

Chapter 4 Site Development and Hydraulic Analysis: This presents the tasks involved in developing a site from its original state to that of an electricity producing system. Some of the elements considered are topography, geology, hydraulic design, turbine setting, powerhouse and switchyard, access, and cost.

Chapter 5 Dams and Reservoirs: This describes the types of dams and reservoirs appropriate for small hydro sites. Materials for construction and structural integrity are also discussed.

Chapter 6 Hydraulic Turbines: This traces the history of turbine development and fluid mechanics theory and describes the major types of turbines and cavitation theory. The selection of turbines appropriate for a given application in terms of head and flow is also discussed.

Chapter 7 Generation and Electrical Equipment: This reviews basic theory of ac electricity, operation of electrical machines, equipment selection procedures, controls, and safety considerations.

Chapter 8 Environmental Impact: This discusses potential environmental problems resulting from site development and recommends remedies available to reduce or minimize adverse impacts.

Chapter 9 The Systems of Regulation of Hydroelectric Power in the United States: This discusses legal, licensing, and tariff aspects of small hydro projects in the United States. A clear description of these procedures is given since their effect on the implementation schedule and costs is considerable.

Chapter 10 Institutional and Policy Environment in Developing Countries: This reviews specific aspects of institutional structures likely to be found in developing countries. The organizational arrangements for mini hydro implementation in four countries are contrasted.

Chapter 11 Economic and Financial Feasibility: This reviews techniques which can be applied in the evaluation of the economic and financial feasibility of hydro projects. Various methodologies for economic analysis are compared and illustrated through the use of case studies.

Chapter 12 System Design and Case Studies: This reviews the overall process of project design including the reconnaissance study, feasibility study, planning studies, final design, procurement, construction, and project management. Several cases are used as illustrations.

1.9 REFERENCES

1. Ellis L. Armstrong, *Renewable Energy Resources—Hydraulic Resources,* World Energy Conference, IPC Science and Technology Press, London, 1978.
2. 95th Congress, 2d Session, P.L. 95-617, *Public Utilities Regulatory Policies Act,* November 1978.
3. United Nations, *Report of the Technical Panel on Hydropower,* U.N. Conference on New and Renewable Sources of Energy, New York, 1980.
4. U.S. Army Corps of Engineers, *Feasibility Studies for Small Scale Hydropower Additions,* Hydrologic Engineering Center, Davis, Calif., July 1979.
5. E. Mosonyi, *Water Power Development,* vol. 2, Hungarian Academy of Sciences, Budapest, 1965.
6. N. Smith, "The Origins of the Water Turbine," *Scientific American,* vol. 242, no. 1, January 1980.
7. E. T. Layton, "Scientific Technology, 1845–1900: The Hydraulic Turbine and the Origins of American Industrial Research," *Technology and Culture,* vol. 20, no. 1, 1979.
8. E. A. Moore, *Electricity Supply and Demand Forecast for the Developing Countries,* Energy Department, World Bank, Washington, D.C., January 1979.
9. World Energy Conference, *Survey of Energy Resources—1974.* U.S. National Committee of the World Energy Conference, New York, 1974.

10. UNIDO, *Proceedings Second Seminar—Workshop/Study Tour in the Development and Application of Technology for Mini-Hydropower Generation,* Vienna, Austria, 1981.
11. U.S. Department of Energy, *Joint Peru/U.S. Report on Peru/U.S. Cooperative Energy Assessment,* August 1979.
12. Marjono Natodihardjo, "Hydropower Development in Indonesia," *Proceedings Seventh General Assembly,* World Federation of Engineering Organizations, November 1979.
13. International Atomic Energy Agency, *Nuclear Power Planning Study for Indonesia,* Vienna, Austria, 1976.
14. Energy Development International, *Energy Planning for Development in Indonesia,* Appendix G, Setauket, N.Y., May 1981.
15. J. Hua, "Small Hydropower Stations," *Beijing Review,* no. 32, August 10, 1981.
16. E. El-Hinnawi, *China Study Tour on Energy and Environment,* UNDP, Nairobi, Kenya, 1977.
17. Central Intelligence Agency, *Electric Power for China's Modernization,* May 1980.
18. R. P. Taylor, "Decentralized Renewable Energy Development in the People's Republic of China," World Bank, Washington, D.C., September 1981.
19. R. E. A. Holland, R. J. Armstrong-Evans, and K. Marshall, "Community Load Determination, Survey and System Planning," *Small Hydroelectric Powerplants,* NRECA, Washington, D.C., 1980.
20. J. J. Fritz, "Photovoltaic Water Pumping: A State-of-the-Art Review," *Seminario: Tecnologías Apropiadas Para Elevación de Agua en Areas Rurales,* CEPIS, Lima, Peru, 1982.
21. A. S. Mikhail, *Wind Power for Developing Nations,* SERI/TR-762-966, SERI, Golden, Colo., July 1981.
22. World Bank, *Renewable Energy Resources in the Developing Countries,* Washington, D.C., November 1980.
23. H. R. Bungay, *Energy, The Biomass Options,* Wiley, New York, 1981.

2

Hydrologic Analysis

John S. Gladwell
Leroy F. Heitz
Calvin C. Warnick

The evaluation of hydropower potential requires a knowledge of both the absolute value and the time variability of flow and head. This chapter will familiarize the reader with the data requirement and analysis techniques used in determining these factors.

In general the objectives of the chapter are as follows:

1. To provide an overview of the various analysis techniques available and give guidance as to the applicability of these techniques to different types of hydro studies.
2. To provide the reader with knowledge on what data are required, where the data can be obtained, and how the data can be evaluated and extended.
3. To discuss various methods of synthesizing streamflow data at an ungaged site.
4. To discuss hydrologic considerations other than streamflow that might be needed in the analyses.
5. To discuss various methods of transforming streamflow and head data into energy and power values.
6. To discuss methods of presenting flow, power, and energy data.

2.1 OVERVIEW OF ANALYSIS TECHNIQUES

Types of Studies Used

Before beginning a discussion of hydrologic techniques in hydropower analysis, it is important to determine in what context different techniques will be applied.

Studies that use hydrologic analysis can be classed in two broad groups: (1) *resource inventories* and (2) *site-specific studies.*

Because of the broad nature of a resource inventory, the hydrologic evaluation techniques employed will probably have to be much more simplistic than those employed in site-specific studies.

When a large number of sites need to be evaluated, time and costs dictate using fairly simple procedures. In many cases methods used in the reconnaissance level studies described later would be appropriate for the resource inventory type of study.

In making site-specific studies, there is a systematic sequence of planning studies that has been adopted by those in public and private practice both in the United States and in other countries.[1]

The three primary study types are the *reconnaissance, feasibility,* and *definite plan* (or design) studies. A good set of definitions for these three studies is contained in the U.S. Army Corps of Engineers guide manual *Feasibility Studies for Small Scale Hydropower Additions.*[2] These definitions are as follows:

> ***Reconnaissance Study:*** A Preliminary Feasibility Study Designed to Ascertain Whether a Feasibility Study is Warranted.
>
> ***Feasibility Study:*** An Investigation Performed to Formulate a Hydropower Project and Definitively Assess its Desirability for Implementation (Should an investigation commitment be made?).
>
> ***Definite Plan Studies (design studies):*** The Collective Group of Studies that are Performed Between the Time of an Implementation Commitment and the Initiation of Construction.

The degree of refinement required in different phases of the planning studies varies considerably. In the reconnaissance studies an estimate of average annual flow, average power head, and an estimate of discharges available during the low-flow periods may be all that is required. If the project makes it past the reconnaissance stage, it is justified to spend more time and money on more detailed hydrologic studies in the feasibility stages.

During the feasibility stage, flow-duration curve techniques may be used to describe the time variability of flow. If the project remains viable after the feasibility studies the next step is the actual design studies. At this stage even more detailed hydrologic analysis studies are carried out. These studies might include daily or monthly operation studies which could include detailed simulation of multiturbine operations with fluctuating headwater and tailwater conditions. The degree of sophistication at this stage is dependent on the complexity of the project and availability of data and study funds. The degree of sophistication for the hydrologic analyses in each phase may vary from project to project. In all cases the goal is to make the best estimate of available flows for energy production under existing time and cost constraints.

No matter what phase of analysis is being made, the result of the hydrologic evaluation should strive to predict what the flows will be during the life of the project. The accuracy of this prediction is dependent on the availability of flow data and the time and financial resources available to carry out the hydrologic analyses.

Introduction to Available Hydrologic Analysis Techniques

The purpose of hydrologic evaluations is to provide a value or values of stream discharge that can be used in selecting the size of the power plant units and to determine annual energy production. This implies that site-specific hydrologic data are needed to give the time variation of stream discharge. Rarely is a stream gage located at the desired hydropower site. Thus hydrologic analysis is needed to extrapolate streamflow magnitudes and timing from known gagings to the desired hydropower site.

In some cases and under uniformly varying hydrologic regimes all that may be necessary is to use simple rules of thumb to determine flow values for resource inventories and feasibility level studies. A rule of thumb in the New England states of the United States indicates that for a normal annual precipitation of 500 to 750 mm (20 to 30 in), a flow of 0.022 $m^3/(s \cdot km^2)$ or 2 $ft^3/(s \cdot mi^2)$ can be expected. This flow will correspond to the flow available 20 to 30 percent of the year. Hence, if the normal annual precipitation is known over a drainage basin and the area above the point on the stream is also known, a preliminary estimate can be made of the flow value to be used in determining power capacity. Other such rules of thumb may be available for other parts of the world. Care must be taken not to use these rules where they would not apply.

For crude estimates of potential energy that might be developed at a specific site it is convenient to use the average flow value for the year as the flow value in the power equation, Eq. (2.1). Using an average annual flow value and an average value for hydraulic head in the power equation and multiplying the results by the total number of hours in a year gives the total energy available in kilowatt hours. This may underestimate energy available because it is common practice to specify a runner for which the flow for full utilization is greater than the average flow.

A common method of describing the flow available at a site is through the use of a *flow-duration curve.* A flow-duration curve is a graphical representation of the average time availability of flow. The curve is a plot of flow versus the percent of time that that particular flow is equaled or exceeded. The curve is computed from a sequential list of flows that are representative of flows available for power production at the particular point of interest in a stream. A detailed description of computing flow-duration curves is given in Secs. 2.3 and 2.4.

Another type of analysis technique available is *sequential flow analysis*. This method applies the flows that are available for power production directly in computing the power and energy available. The flows are maintained in their serial form. The flow values may be averaged hourly, daily, weekly, or monthly depending on available data, time, and cost limitations. In its simplest form, outflows from the subject reservoir are recorded and power output is computed using the values of upstream pool and tailwater elevation that correspond to the flow value used. In a more complex situation, a system of multipurpose reservoirs with varying inflow sources and demands can be analyzed to determine power production from all plants in the system. A more complete discussion of sequential flow techniques is given in Sec. 2.5.

Basin flow simulation is also a very popular technique used for modeling flows in hydropower systems. When the stochastic modeling approach is used, the variables that are to be modeled are regarded as being statistical in nature. Sequences of flows are generated using various stochastic techniques.

Deterministic models are generally concerned with real world processes that generate streamflow. These models usually simulate the rainfall runoff process. Again the output from the model is generally a sequence of flows that can be used in the hydropower analyses.

A more complete description of stochastic and deterministic models is covered in Sec. 2.3.

2.2 OBTAINING AND EVALUATING DATA

Basic Data Requirement

The basic data required in determining hydropower potential and capacity at specific sites are hydraulic head and discharge which are related through the general power equation. In USCS units this equation is

$$P_{\text{hp}} = \frac{Q\gamma H\eta}{550} \tag{2.1}$$

where

P_{hp} = power output, hp
Q = plant discharge, ft^3/s
γ = specific weight of water, 62.4 lb/ft^3
H = available head, ft

η = plant efficiency in decimal form
550 = conversion factor, number of (ft·lb)/s in 1 hp

And

$$P_{kW} = P_{hp}(0.746) \tag{2.2}$$

where

P_{kW} = power output, kW
0.746 = conversion factor to change horsepower to kilowatts of power

In SI metric units this equation is

$$P_{kW} = \frac{\rho g Q H \eta}{1{,}000} \tag{2.3}$$

where

ρ = density of water, kg/m^3
g = acceleration of gravity, 9.8 m/s^2
Q = plant discharge, m^3/s
H = available head, m
η = plant efficiency in decimal form

The gross head is the difference between headwater elevation and tailwater elevation. (A refined definition of head and associated losses is given in Chap. 4 as it applies to the turbine runner.) Gross head will vary in streams depending on the flow of the stream and pondage in the water stored behind the dam or diversion.

The plant discharge is generally a fraction of the water flowing in the stream. It should be noted, this is not the maximum stream discharge because it is undesirable to construct turbine openings, runner size, and outlet facilities to accommodate total flow. Thus it is desirable to know the range of highest and lowest flows that occur in the stream. Later discussion of the actual choice of discharge capacity of the hydropower unit will show that this becomes an economic problem.

Data Requirements for Indirect Estimation of Flow

Indirect methods of determining streamflow at an ungaged site depend almost entirely on using streamflows at gaged points in nearby or adjacent streams. The methods of estimating these ungaged flows will be covered later. The following describes the data required in the application of different techniques.

Since almost all the estimation techniques involve using known gaged values, a search of available streamflow data should be made. Thorough documentation of the stream-gage history of each of the gages should be made. Operation period of the gages, amount and location of upstream diversions, and the time and location of new major developments in the gaged basins should also be noted. The gage location and drainage area above each gage must also be determined.

In some cases precipitation and snow-gage data may be helpful in confirming streamflow data. Actual gaged values of snow and rain may be used in deterministic models to predict streamflows. Normal annual precipitation (NAP) maps or plots of NAP at various gages in the basin of interest can be used in estimating average annual runoff. Topographic information from both regular topographic maps and topographic relief maps can be useful in determining basin elevation, aspect, and slopes. Aerial and satellite photography may be helpful in determining ground cover and other useful hydrologic features of the basin.

Sources of Data

To obtain basic data for hydropower studies it is necessary to get preliminary elevation information for hydraulic head determination and planning reservoir design from available maps. In the United States this can be from the Army Map Service, the U.S. Geological Survey, state Geological Survey offices, and county government offices.

In the United States flow data (stream discharge) frequently can be obtained from the U.S. Geological Survey, Water Resources Division. This agency is normally the basic water data gathering entity in the United States. Often the data may have been collected by other government agencies such as

U.S. Forest Service
U.S. Bureau of Reclamation
U.S. Army Corps of Engineers
Environmental Protection Administration
Irrigation Districts
Flood Control Agencies
Water Supply Districts

Other supporting data for correlation studies may require data on precipitation, temperatures, and evaporation. This information is normally obtained from the Weather Service of the U.S. Department of Commerce.

In other countries similar government entities may have data for making studies. In its *Guide to Hydrometeorological Practices,* the World Meteorological Organization (WMO) offers many practical guidelines to subjects that will be of interest to hydropower engineers, with reference to establishing data acquisition networks for precipitation and streamflow.[3] Concerning the development of a minimum network the *Guide* states,

> While a minimum plan should be considered as the first step, it will rapidly become insufficient as countries develop. The establishment of an optimum net-

work would be a much greater undertaking. Also, the gaps which remain even after the establishment of the minimum network would still be sufficiently large to permit the minimum network to become an integral part of the optimum network with very few and only relatively minor changes. Nearly all of the stations of the first network will be principal or base stations in the ultimate network.

The two principal data acquisition networks needed for hydropower studies are those of precipitation and river discharge. The WMO *Guide* suggests minimum densities.[3]

Regardless of the existence or nonexistence of a river discharge measurement network it may be necessary (or at least extremely desirable) to establish a gage in the proposed development area. This will probably be imperative in the case of mini hydro systems.

Considering the level of investment, gaging for the mini hydro systems will tend to be far less sophisticated than those desired for larger investments. In some cases it may be necessary to make measurements on site of both flow and head, set up stream-gaging programs, and take precipitation measurements.

Available data on evaporation will help in making a determination of runoff coefficients for determining expected flow (discharge) values. The sophistication of measurement and calculation of evaporation and other losses to precipitation input to a basin will depend on time allowed for the analysis and the financial resources available for hydrologic analyses.

Evaluating Quality and Applicability of Data

In this section it is assumed that some data series are available for analysis. The following tests are used to determine the adequacy of the data.

Because not all data is invariable in time and space it is desirable to investigate and, where necessary, adjust the series. However, adjustments should be made without violating the basic integrity of the data. As the WMO *Guide* indicates, adjustments are generally made for one or more of three purposes:[3]

1. To make the record homogeneous with a given environment, an example of which is in fitting a uniform period of record for which a "standard period" mean or normal is to be computed.
2. To eliminate, or at least reduce, the effects of changes or otherwise extraneous conditions, for example to correct for changes in gage location or exposure.
3. To selectively summarize data for presentation or examination, an example of which is the smoothing of isohyetal maps.

Adjusting Short-Term Records

A very common problem faced by hydrologists when beginning a regional study is that gaging stations will have differing periods of record. This can be

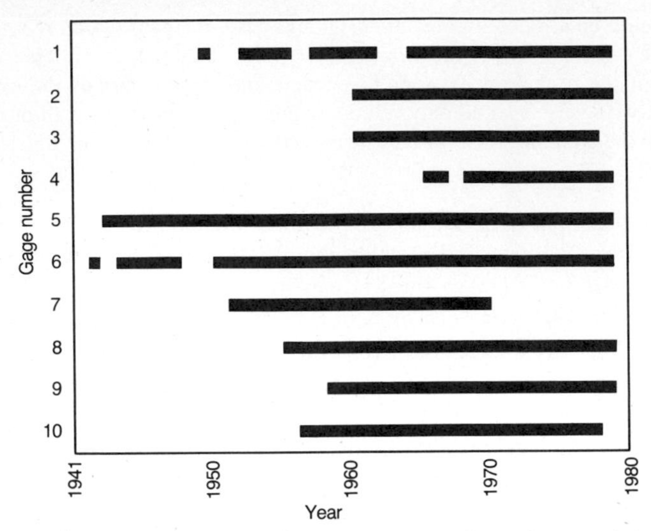

FIG. 2.1 Bar graph of series length to conceptualize optimum period for relating series to short-term records.

particularly important because some gages may have operated during periods of high water availability while others may have records representing only low periods. Some may have overlapped both high and low periods. Attempts to use the records simultaneously could create confusion and misrepresentation of the actual situation. It is therefore important that any procedure used to fill in data gaps not further confuse the situation.

Attempts to fill in missing data can create false security if not done carefully. First, it must be realized that the use of correlation or regression analyses dilutes the value of the data. The added data can only be considered as an approximation of what actually occurred. Furthermore, if the data are to be used in statistical analyses, the use of simple regression analyses will only produce estimates of the means of the missing values. In such a case, the natural variations about the mean will be eliminated and the overall variance thus diminished. It has been suggested that a random element, based on the unexplained variance of the regression equation, be used when more than two or three data points are being filled in. This random element would be added to the mean value determined by the regression.

A valuable way to begin any study in which regional data are to be used is to plot the series length as a bar graph (Fig. 2.1). From this plot it is possible to conceptualize the optimum period to which all data series can be related.

Normal or mean precipitation can be estimated approximately with the general equation

$$N_x = N_r \frac{P_x}{P_r} \tag{2.4}$$

where

N = normal (or mean) precipitation
P = precipitation during a shorter period
x = station with unknown normal (or mean)
r = station with known normal (or mean)

To be used with caution, this equation could result in an estimate that is substantially in error, and thus, in practice, selection of comparable stations should be limited to those in close proximity to that of the short record. It is also better to use several stations to determine the normal (or mean) for the short-record station; for example, the following equation could be used for three stations (a, b, and c) to adjust the short-record station x.

$$N_x = \frac{P_x}{3}\left(\frac{N_a}{P_a} + \frac{N_b}{P_b} + \frac{N_c}{P_c}\right) \tag{2.5}$$

The derived normal value may be questionable where the P values are for periods of 3 years or less.

Where normal (or mean) precipitation values can be established for all stations (such as for the arbitrarily established period of record overlap) Eq. (2.5) can be used to estimate missing precipitation events. Use in this case should be limited to precipitation time periods of no less than a month's duration.

Double-Mass Analysis

The method of applying the technique of double-mass analysis is based on the theory that a graph of the cumulation of one quantity against the cumulation of another quantity during the same period will plot as a straight line, so long as the data are proportional. It is further assumed that a change in the slope represents a change in the proportionality between the variables.

Because of natural variability, in hydrology we usually plot the cumulated series under investigation against the average of several others. The number to be included in the average is limited by the criterion that the area in which the stations are located be small enough to be influenced by the same general conditions. Each of the stations should itself be checked for consistency before it is used in the double-mass analysis, particularly if less than 10 stations are to be used.

The double-mass analysis applies particularly well to precipitation data because precipitation events are in general unaffected by human activities. On the other hand, precipitation gages are too easily moved "slightly," or otherwise not considered when structures are built or altered, or they are not given serious thought when vegetation increases. Thus precipitation data has been shown to be particularly vulnerable to inconsistency and is well served by a double-mass analysis.

Although the double-mass analysis techniques can be used to check streamflow records for inconsistencies in much the same manner as that of precipitations, the assumption that a constant ratio exists between a given time series of streamflow and that of a comparable group of records may not be valid. It has been found that on an annual basis the assumption is sometimes justified, whereas on a shorter basis problems may arise.

As a matter of process, the streamflow is first converted to a comparable basis, such as inches (millimeters) over the basin, discharge per unit area, or percent of mean, so that large rivers will not have more effect than the lesser ones. Of course, reasonably comparable streams should be used.

If a break in the double-mass curve is found, an inconsistency is suggested. As in the precipitation analysis, the first step to be taken is to find a reason for the inconsistency. Unlike that of precipitation, however, the double-mass analysis should seldom be used to adjust streamflow records.

A complete description of the use of double-mass analyses for analyzing both precipitation and streamflow data is contained in Searcy and Hardison.[4]

Length of Record

In general terms it can be stated that the longer the length of a statistical sample the more confidence we can have in calculations made therefrom. As a matter of strict statistical sampling of random data the standard error of the mean and standard deviation of a random series can be shown to be

$$S_{\overline{Q}} = \frac{S_Q}{\sqrt{n}} \tag{2.6}$$

and

$$S_{s_Q} = \frac{S_Q}{\sqrt{2n}} \tag{2.7}$$

where

$S_{\overline{Q}}$ = standard error of the mean
S_Q = standard deviation of the sample
S_{s_Q} = standard error of the standard deviation
n = number of observations

As noted earlier it is possible to extend short records based upon correlation with longer series. For new information to be added (that is, as a practical matter, that the extension be worth the effort), the errors introduced by correlation must be less than the sampling error in the short record.

One test which may be used to estimate the effective improved significance of the mean of the correlation-based extension is[5]

$$N = \frac{N_r + N_e}{1 + \left(\dfrac{N_e}{N_r - 2}\right)(1 - r^2)} \tag{2.8}$$

where

N = effective period of the extended record
N_r = number of years in the short-term record
N_e = number of years in the extension
r = coefficient of correlation between the two records for the comparable short period

The process of extending records is obviously quite ineffective unless N is greater than N_r. Furthermore, the actual amount of effective information may be less than is apparent at first glance, as for example: Assume, $N_r = 5$, $N_e = 15$, and $r = .8$. Then

$$N = \frac{5 + 15}{1 + \left(\frac{15}{5 - 2}\right)(1 - .8^2)} = 7.1$$

The effective length of the extended record is but 7.1 years, not 20, representing but a gain of 2.1 effective years.

The information content of a time series of hydrologic information should be investigated for trend or cyclicity. This was indicated earlier in the reference to adjustment of streamflow data and the double-mass analysis approach. No further sophisticated procedures for analyses are presented here because they require a more in-depth presentation of statistics and stochastic processes. On the other hand, 5- to 10-year moving averages can often detect, subjectively, the existence of trends and cyclicity. The problem remains that on the basis of the small samples usually present in typical hydrological analyses it is impossible to prove, with reasonable assurance, that trends and cycles (in particular) actually exist. It is clear that the addition of long-period climatic analyses would greatly enhance hydrologic studies.

Field Measurement

This section is concerned with describing various field measurements that are required in the course of the hydrologic investigation.

Defining flow variability, as noted earlier, is the major objective of the hydrologic analyses. In most cases, long-term stream-gage records are used to estimate the time variability of flow. However, in some cases it will be necessary to make on-site flow measurements to supplement long-term flow-estimating techniques.

The most common technique of one-time streamflow measurement is the current meter technique, involving the use of a rotating vane meter to measure velocity at various points in the river. The velocity measurements coupled with a knowledge of the cross-sectional dimension of the stream enables the investigator to compute the flow at the time of the measurement. An excellent reference for making streamflow estimates is published by the U.S. Geological

Survey.[6] Many texts on hydrology and hydraulics also contain information on streamflow measurements.[7,8]

A very crude estimate of streamflow may be obtained by simply measuring the stream velocity with the transit time of a neutrally buoyant object between two points a known distance apart. By dividing the distance by transit time, an approximation of stream surface velocity is obtained. The average velocity for the entire cross section is usually taken to be eight-tenths of the stream surface velocity.[9] The cross-sectional area of the stream at the point where the velocity was measured needs to be determined with a measuring tape and sounding rod. After computing the area of the cross section, the estimated flow is found by multiplying the computed area by the average velocity. Care must be taken that the length units used in the area and velocity components are homogeneous.

Field measurements of gross head are usually carried out using surveying techniques. The precision required in the estimate will dictate what methods will be employed. In any case, the difference between headwater and tailwater elevations is what needs to be determined. In the simplest case a surveyor's aneroid barometer might be used to estimate the elevations at the predicted headwater and tailwater. By subtracting the headwater from tailwater elevations an estimate of gross head is obtained.

Several texts on surveying are available that would describe either of the above surveying techniques.[10,11] If ground-controlled aerial photography is available, it may be possible to make the elevation measurements required to estimate the head by using photogrammetric techniques.

If evaporation records are not available at the reservoir site of a proposed power development, it may be necessary to set up an evaporation pan. This should be a standard-type pan so that the data will be acceptable in later calculations. The record should be kept at least for the high-evaporation months of the year. Evaporation-transpiration estimates may be useful to help in determining the value of actual direct runoff from a watershed above the specific site.

A discussion of field data measurement would not be complete without a warning on the reliability of one-time or short-term measurement of time variable data. Measurement of flow, head, and meteorological data fall into this category. It should be noted that estimates of flow and head must be valid for the entire project life. If flow estimates are made from a single- or short-term series of measurements, extreme caution must be taken to be sure that the measured flow is really what can be expected during the life of the project.

Section 2.3 on indirect estimation of flows at a site will give examples of how to estimate flows at a site when long-term measurements of flow are unavailable.

Other Considerations

Often because good published records of streamflow are unavailable it is necessary to obtain information on flow and precipitation data from local sources.

Records are available from irrigation districts, flood control districts, entities responsible for public water supply, soil conservation districts, and local residents who keep private records. Information on upstream development and what uses are being made of water are also necessary. Often in the western United States, irrigation diversions with prior rights have use rights (water rights) that will limit the downstream power development. Information on such water rights can be obtained through local contacts and from contacts with state agencies with water-right jurisdictions.

The following are typical questions that should be answered as planning and feasibility of hydropower development proceeds.

1. What is the highest stage of river flow at the proposed site?
2. Is there a good contour map available from local study?
3. What is the best estimate of maximum annual flow, average flow, and minimum annual flow?
4. What is the approximate length of the flood-flow season?
5. What is the approximate length of the drought or low-flow season?
6. Are there any instream uses and diversions of water that might affect the possible hydropower development at the site?

2.3 INDIRECT ESTIMATION OF FLOWS AT A SITE

General

In some cases the hydro investigator may find a stream gage located in close proximity to the hydro site that is being investigated, making the job of predicting project flows relatively easy. However, this is the exception rather than the rule. In more common cases some means of estimating flows will be required. The techniques that are used will depend on the type of flow to be estimated and also on the availability of data. The remainder of this section deals with computing flow characteristics at these ungaged sites.

Estimating Average Flow

The most basic flow characteristic that is desired is average annual flow or simply the numeric average of the daily average flows. In areas where physical runoff characteristics are common and precipitation is relatively uniform, average discharge will be found to be highly correlated with drainage area. In its simplest form discharge may often be proportioned up- or downstream from an existing gage on a ratio of drainage area. In the New England region of the United States, rules of thumb have been established whereby estimates of average flow can be made.[12] Caution should be used when applying these rules of thumb in areas with large variations in precipitation or runoff.

In areas where precipitation variations are expected, it is better to correlate average flow to both average basin precipitation and drainage area. In studies in the Pacific Northwest region it was found that a simple correlation between average flow and average precipitation times drainage area could be developed.[13] This correlation takes the form

$$\text{Average flow} = K \times NAP \times DA \tag{2.9}$$

where

NAP = basin normal annual precipitation
DA = basin drainage area
K = runoff coefficient

The basin normal annual precipitation could be found using a number of common hydrologic techniques such as Thiessen or isohyetal methods.[7] When NAP maps are available the isohyetal technique would probably yield the best results. The K value relating runoff to the product of drainage area and average annual precipitation can be estimated using known gage points. The K value could be viewed as a runoff coefficient since it relates average flows to average precipitation volumes.

Correlation Techniques to Obtain Series of Flows

Where several physical or hydrologic characteristics modify the runoff significantly, multiple regression techniques have been used to define the flow statistically. A general relationship widely used for this purpose is

$$S_i = a_0 + X_1^{a1} + X_2^{a2} + X_3^{a3} + \cdots + X_n^{an} \tag{2.10}$$

where

S_i = flow parameter of interest, e.g., mean, standard deviation, or mean annual flow
X_i = catchment or hydrologic characteristics
a_i = regional coefficients

Average annual discharge as a function of drainage area is but a simple case of this general model.

It is important that the investigation include a search of the existing literature since regional analyses may have already been completed. Two monumental reports that will provide excellent specific and generalized information that could be of valuable assistance have been produced by the UNESCO Division of Water Sciences.[14,15] The data and maps of worldwide water balances will be particularly useful in areas with minimal data.

A model developed for use where only precipitation and temperature data are available is discussed in the WMO *Guide*.[3] The method is based on a relationship between P/E and R/E, where P is the average annual precipitation, R is the average annual runoff, and E is a temperature factor. Using tables of T and E, and P/E and R/E, the R value (runoff) may be computed. Refinements can be made for regions where most of the precipitation falls within certain seasons.

Stochastic and Deterministic Models

The recognition that the hydrologic cycle is a complex system has led to an increasing awareness that in water resources studies one should maintain a *systems* approach. Models are a basic element of what has come to be known as *systems analysis* or *operations research*. The aim of the process is to assist in identifying those control measures that will tend to ensure that the planning goals are reached.

In general, two broad classes of mathematical models can be identified as being of importance from a planning perspective:[16]

(i) descriptive *simulation* models that relate system inputs to outputs by a direct computation procedure and which are usually re-run a number of times to examine the implications of adopting various alternative designs;

(ii) analytical *optimizing* models, particularly of the mathematical programming variety, which seek to determine the optimum manner of achieving an objective.

Both classes of modeling can be applied to hydropower developments. The optimization models can be particularly valuable where hydropower is to be added to an existing system in which either (1) thermal systems are reasonable alternatives to be considered, or (2) the existing system contains thermal energy production and the hydropower must be properly valued.

In less complex situations, one may be more interested in simulating the hydrologic system than in optimizing operations, at least during the early stages of investigation. For mini hydro investigations, one is probably not interested, except in a theoretical way, in simulation.

The classification of simulation models is not well defined. There are those that claim to model the physical processes, while others are presented as being only approximate (based on empiricisms).

As Diskin indicates[17]

> The most important problem to the potential user [of models] is probably the choice between the comprehensive model versus the specific model. The comprehensive model claims that it reproduces all processes that take place in the watershed. It is thus presented as a tool that can meet the needs of all potential users. The specific model as its name implies is intended to supply only one type of design data. An example of such a model may be one producing monthly runoff volumes. Other examples include a model producing snow melt hydrographs

or a model for converting extreme storm rainfalls into design runoff hydrographs. A specific model also usually produces other data as a by-product but the accuracy and value of these additional data are inferior to those data for which the model is constructed.

He concludes

Practice gained in the analysis and use of various models appears to be the only tool available to the applied hydrologist for assessing the usefulness of a hydrologic model in the process of planning and management of a water resources project in a given watershed [emphasis added].[17]

As a matter of policy it is beneficial to introduce modeling capabilities early in the planning process, since the development of capabilities is not without problems inherent in the learning process. Furthermore, modeling may also assist in guiding the decision to implement data collection networks.

In general two approaches are used in the development of hydrological models: stochastic or deterministic (and, of course, combinations of the two).

Stochastic Models

In the stochastic approach, the variables are regarded as being statistical in character, having probability distributions which may be functions of time. Commonly used (assumed) is a simple lag-one Markov model of the general form

$$X_{i+1} = \overline{X} + r_1(X_i - \overline{X}) + t_{i+1}S_x\sqrt{1 - r_1^2} \tag{2.11}$$

where

X = the variable being stochastically generated
$\overline{X}$ = sample mean
S_x = sample standard deviation
r_1 = first-order serial correlation coefficient
t = random deviate, commonly N (0,1)
i = indexed time interval

It is important to view stochastically generated time series objectively. First, the model is absolutely dependent upon historical data for the estimation of the statistical parameters; e.g., in the model in Eq. (2.11) the statistical estimates of the mean, standard deviation, and first-order serial correlation coefficient must be provided. The validity of those sample statistics is very much a function of the quality and quantity of the data from which they were determined; however, all suffer from *sample error*. Second, the basic assumption is that the "world" to be generated synthetically actually is represented by such a model. Both assumptions can introduce problems regarding the validity (and value) of the results. The main problem is the tool to be used: the computer. Since it is a simple matter to program a computer to generate stochastic data, it is altogether too easy to be misled into believing that, from 10 years of basic data one can generate a 1000-year sequence of more valid events. The hydrologist should always keep in mind the length of the historical record upon which the model is based.

To understand the importance of sample error, the reader is encouraged to review Benson,[18] in which a 1000-year population was postulated and small samples of various sizes were selected randomly. Frequency analyses were then made of the variously sized samples. In general, the study confirms visually what can be shown mathematically: With small samples the potential for error can be quite large. Furthermore, *small samples* is a qualitative term, since even series with which most hydrologists would be overjoyed (say 50–100 years) were shown to have variability. Nevertheless, if done correctly, the sample represents *the best estimate we have* of the population characteristics. We may improve slightly some of the sample information by examining regional characteristics; however, we are always inevitably limited by the samples. Thus, one should always retain at least a mental image of the confidence limits of the basic information.

A final caution in the use of stochastic models: Many series of different lengths can be generated, not one of which will reproduce the historical sequence—in a statistical sense, however, the characteristics of the generated series will converge to those of the original sample from which they were derived. This is not significant, except that the model will reproduce the sample. It should not be used as proof that the extremely long generated series has any great inherent value. Nevertheless, there are at least five important characteristics of correctly applied stochastic models: (1) they suggest other (perhaps more critical) orders of equally likely series which can be evaluated for their impact, (2) it is possible, even if local data are unavailable, that a model can be used with statistics determined by regional analyses, (3) it is possible to generate many sequences of possible occurrences from which levels of confidence in their application could be estimated, (4) they can be used to "fill in" missing data with values that preserve the stochastic nature of the original series, and (5) where, as is most often the case, rainfall data is more available than runoff data, they can be applied to the rainfall series and the generated rainfall sequences used with more deterministic rainfall-runoff models in order to generate runoff sequences.

Deterministic Models

In hydropower studies we are generally concerned with methods by which *streamflow series* can be developed. Of particular interest tend to be the rainfall-runoff process models.

A number of models exist that variously conceptualize the physical processes within the watershed. They may be used with (among others) precipitation data in order to develop the hypothesized streamflow. One well-known example, developed by the U.S. Army Corps of Engineers, is the SSARR model. In this model precipitation is distributed between runoff and soil moisture recharge. A soil moisture index and rainfall intensity are required. Runoff is distinguished between base flow and direct runoff, and the direct runoff is characterized by subsurface and surface components. Storage zones are fed by the runoff components, the sum of which is taken as the streamflow for the watershed. Precipitation and monthly values of evapotranspiration (or weighted-pan evaporation) data are required. Other factors can be established as constants or with tabulated functions. The calibration is executed by trial and error—requiring an existing streamflow series. Obviously, the model's accuracy gives satisfactory results only when sufficient data exist.

A more sophisticated model, with a more complete physical base is the Stanford Watershed Model (and its more highly developed extension and improvement the Hydrocomp Simulation Program). These models require much more input—rainfall, temperatures, radiation, wind speeds, and monthly or daily pan evaporation.

Other sophisticated models such as the Sacramento Model and the SHE (Système Hydrologique Européen) exist as well, including several developed for specific applications. But for generation of mean monthly data these models tend to be much too detailed for the level of data commonly available.

The choice of model is often guided by the size of the watershed. Smaller watersheds will probably be more suitable for representation by the highly detailed physically based models. As the area covered increases there is usually a need to employ larger time units in the computations; thus the coefficients and parameters tend to depart from or lose their original meaning.

Finally, there is the natural desire to use models for which the coefficients and parameter values could be easily transferred from a region of known values to another with insufficient data. Such presumed sophistication would be very desirable. However, as Body[19] states, such is not yet the case:

> Many models have been developed which attempt to describe the form of relationships which exist in a basin. In some instances particular models will provide excellent results for specific purposes, such as time series extrapolation. However, it seems that no such model has been successful in providing a framework into which basin characteristics can be inserted with any confidence that the streamflow time series produced will provide parameters any more accurate than those derived from the regression approach.

Flow-Duration Curve Analysis Techniques

Another method of describing the time availability of flow at a certain point in a river is the flow-duration curve technique. As was mentioned earlier, the flow-duration curve is simply a plot of flow versus the percent of time that that particular flow can be expected to be exceeded (called the "percent exceedance"). A sample flow-duration curve is shown in Fig. 2.2. The flow-duration curve is a very useful tool in hydrologic analysis in general and especially useful in hydropower studies. Its compact form makes it very easy to estimate magnitudes of high and low flows and the time availability of flows between these

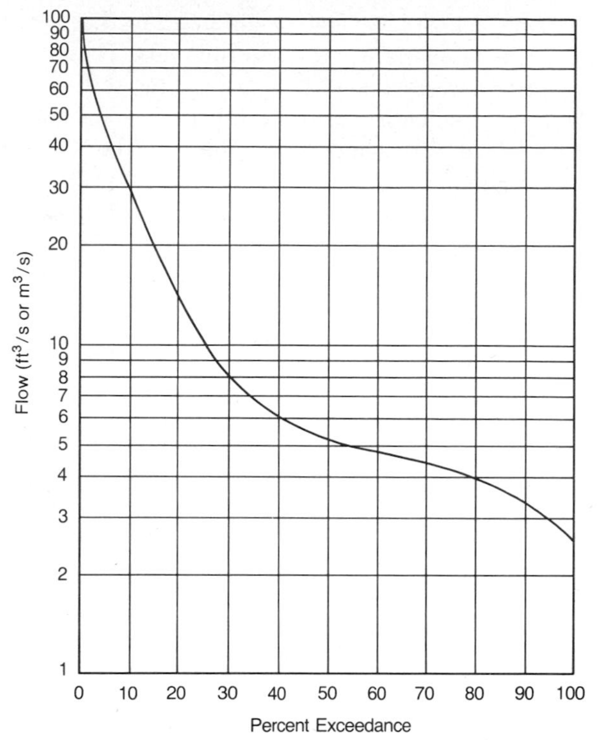

FIG. 2.2 Typical flow-duration curve.

two flow levels. In hydropower analyses the flow-duration curve can be used to determine estimated power and energy from a proposed hydropower installation. A discussion of using flow-duration curves to compute power and energy output from a hydropower site is contained in Sec. 2.5. The flow-duration curve is also useful in other hydrologic investigations where knowing the average time availability of flow is required, such as for irrigation projects.

The two basic methods of computing flow-duration curves at gaged points are the *ranked flow* and *class interval* techniques. In the ranked flow-duration technique the time series of flows is rank-ordered according to magnitude of flow. Mean annual, monthly, and weekly or daily flows may be used. The use to which the information is to be put determines the choice of what type of average flow is used. The rank-ordered values are then assigned order numbers, the largest beginning with order 1. The order numbers are then divided by the total number in the record and multiplied by 100—representing the percent of time intervals (days, weeks, etc.) that a particular mean flow has been equaled or exceeded during the period of record analyzed. The flow value is then plotted versus the respective "percent exceedance." As in any statistical analysis, the value of the information contained is a function of the length of record. References to flow-duration curves are usually made as Q_{50}, Q_{30}, Q_{10}, etc., indicating the flow values at the percentage point subscripted.

The class interval technique is slightly different, in that each of the time series flow values are categorized into class intervals. These classes of flows range from highest to lowest value of flow in the time series. A tally is made of number of flows in each class and the number of values greater than each class can be determined. The number of values greater than each class is divided by the total number of flows to get the percent exceedance. This percent exceedance is plotted versus the upper class interval to get the flow-duration curve values. This technique is usually faster than the rank-ordering technique, especially where the time series of flow is lengthy. A more thorough description of both methods and listings of computer programs to do the required computations is available in a University of Idaho Ph.D. dissertation by L. F. Heitz.[20]

When flow-duration curves are developed using other than average daily flows, it must be remembered that any flow variation within the averaging period is camouflaged by using just the average value. This problem should always be considered when deciding whether to use average daily, weekly, monthly, or annual flows in the flow-duration analysis.

Because flows at specific sites generally follow cyclical variations as a function of within-year periods, greater utility can be derived if the analysis is based on monthly flow durations. This may be done in at least two manners. In one, all the January means (for example) are listed as a data series of N values, and the analysis made. The monthly averages used, however, will mask the within-month variations. Thus, an analysis of all the daily January flows (in this example) will provide a better basis for design consideration. Depending upon the purpose for the analysis, it may only be necessary to evaluate the critical monthly periods (which for small hydro should include the high-flow as well as the obvious low-flow months).

Another procedure provides "index" years by preparing the yearly average flow-duration curve first. From the curve the Kth percentile index year may be identified, and by using the historic monthly and daily flows occurring during the selected index years, the capacity and energy characteristic can be determined. Although this procedure has been called "probabilistic," it is only the index year that has any true probabilistic inference. There is nothing certain about the probability of that year's within-year distribution of flow.

It is thus very important to inspect that year for any perceived anomalies, and since the acceptance of an index year concept is a subjective decision, there may be some advantage to purposely "normalizing" the within-year distribution. By ordering the index year daily flows, a more realistic and useful flow-duration curve for determining capacity and annual energy will be available for that selected year. It has been suggested that the Q_{50} index year can offer a good estimate of primary energy, anything above that value being secondary. Figures 2.3 through 2.5 show some of the various flow-duration techniques by example. Table 2.1 shows the calculated values. Experience has shown that Q_{20} or Q_{30} values are good starting flows for sizing equipment.

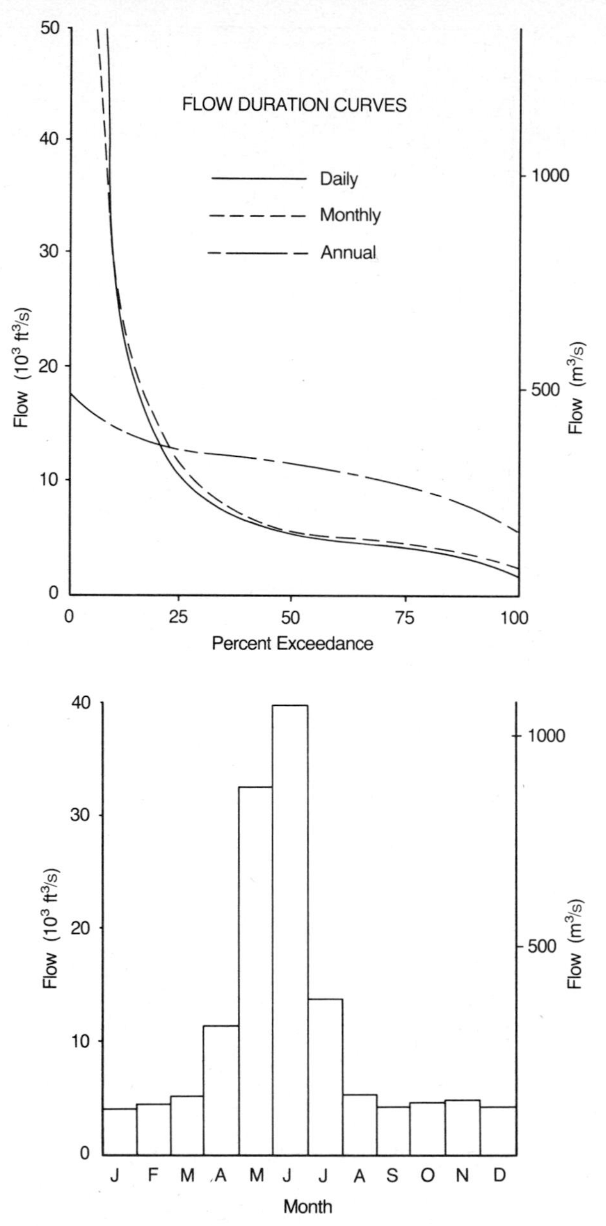

FIG. 2.3 Daily, monthly, and annual flow analyses, Salmon River at Whitebird, Idaho.

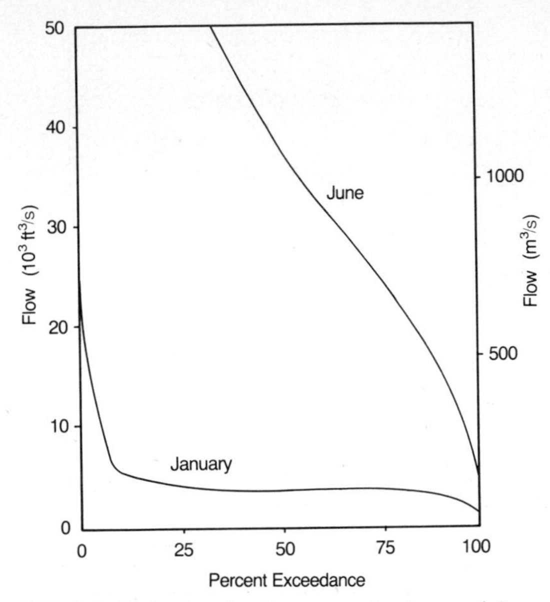

FIG. 2.4 Daily flow-duration curves for June and January, Salmon River at Whitebird, Idaho.

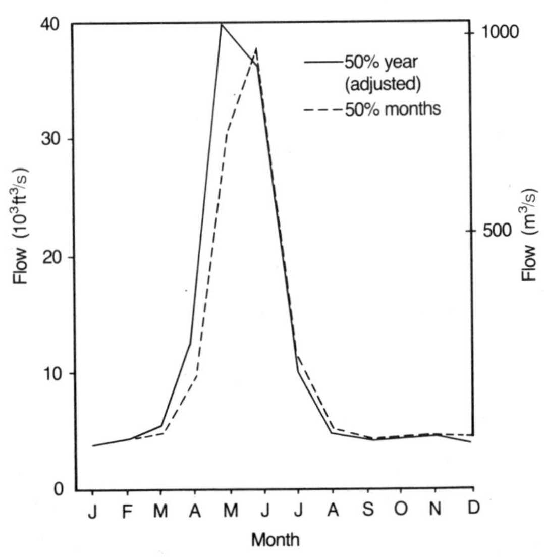

FIG. 2.5 Monthly flow values from 50 percent index year and monthly flow-duration curve analysis.

TABLE 2.1 Flow Analyses, Salmon River at Whitebird, Idaho (ft^3/s)

Period	Percent exceedance 100	90	75	50	25	10	0	$\overline{Q}$
Daily*	1,500	3,300	4,100	5,300	10,300	30,300	129,000	11,255
Monthly†	2,500	3,500	4,200	5,300	11,400	30,000	82,600	11,247
Annual‡	5,800	7,700	9,200	11,400	12,900	14,700	17,480	
Daily flow duration analyses by months								
Jan.	1,800	3,000	3,400	3,900	4,600	5,800	27,800	4,166
Feb.	2,000	3,200	3,600	4,200	4,800	5,700	14,100	4,365
Mar.	2,500	3,600	4,100	4,800	5,800	7,500	18,200	5,258
Apr.	3,000	5,300	6,700	9,300	14,800	21,700	46,100	11,555
May	5,000	13,800	20,000	29,200	42,900	70,200	104,000	32,687
June	5,000	15,700	25,400	38,100	60,900	101,800	129,000	39,963
July	2,000	5,300	7,600	11,300	17,100	26,300	62,000	13,955
Aug.	2,000	3,100	4,100	5,200	6,400	8,000	13,000	5,438
Sept.	2,000	3,100	3,800	4,400	5,000	5,900	10,500	4,441
Oct.	2,500	3,300	4,000	4,600	5,400	6,200	20,400	4,810
Nov.	1,800	3,400	4,100	4,700	5,500	6,600	17,100	4,917
Dec.	1,500	3,000	3,600	4,200	4,900	6,200	25,900	4,519

Period	50% Index year monthly averages 1925	1949	1922	Adjusted§
Oct.	3,280	5,172	4,610	4,354
Nov.	3,950	5,057	4,810	4,606
Dec.	3,500	4,042	4,600	4,047
Jan.	3,400	3,684	3,740	3,608
Feb.	4,780	4,094	3,890	4,255
Mar.	4,870	6,232	4,760	5,287
Apr.	15,900	14,370	7,990	12,753
May	42,700	47,860	29,500	40,020
June	31,600	28,780	48,800	36,393
July	11,900	8,316	11,100	10,439
Aug.	5,000	4,234	5,590	4,941
Sept.	4,530	3,677	4,090	4,099
Mean	11,300	11,330	11,100	
Min. day	3,020	2,760	2,900	
Max. day	58,600	73,900	67,200	

*Daily records WY 1911–1974.
†Monthly records WY 1911–1917, 1920–1977.
‡Annual records WY 1911–1974.
§Adjusted is average of 3 central years.

A number of limitations must be recognized in the flow-duration analyses. Some of these limitations are inherent in the method itself, some may be due to the basic flow data.

The first type of problems are those that simply result from method. While the flow-duration curve does indicate the time availability of flow, the serial relationship between flows is lost. Another limitation is that the flow-duration curve only reflects the regulation that was present in the flow data. If a flow-duration curve for a site with storage capability is to properly reflect the storage, the flow record used to compute the duration must be modified to reflect that storage. The flow data used must reflect conditions as expected at the site of interest during the economic life of the site for which the flow-duration curve is being used.

Indirect Methods of Computing Flow-Duration Curves

The flow-duration curve discussion thus far has centered on computing these curves at a known gage location. The availability of streamflow gage data at the site of interest is probably the exception rather than the rule in most areas. Therefore, methods have been developed to predict flow-duration curves at ungaged sites. The methods that will be discussed were used in a large scale hydroelectric power inventory in the Pacific Northwest region of the United States.[21] Because of the nature of this survey, it was necessary to predict flow-duration curves at ungaged points on streams. A regionalized approach was developed that depended on the availability of an estimate of mean annual precipitation values for the drainage basin of interest. The procedures used permitted the development of synthetic flow-duration curves at any point in any stream in the region, within the constraints of the process.

Following is a short discussion of one of the regionalization methods. A complete discussion of this method is available in several publications.[13,20,21]

The first method that will be described involves developing a set of parametric flow-duration curves based on flow values at gaged points in the region and average annual flow at these gaged points. A sample of a completely developed set of parametric curves is shown in Fig. 2.6. This method allows prediction of only certain exceedance values, not all values on the flow-duration curves. Which values are used are at the discretion of the investigator. To develop the parametric flow-duration curves, a separate data set of values is constructed for each exceedance value to be predicted. This data set includes the values of Q for that particular exceedance value versus average runoff for all gages in the region.

Each data set is plotted and a curve fitting procedure is applied to that data to develop a "best fit" curve for that exceedance value. This procedure is repeated for each of the particular exceedance values. To predict a complete estimate of flow values for the particular exceedance values at an ungaged

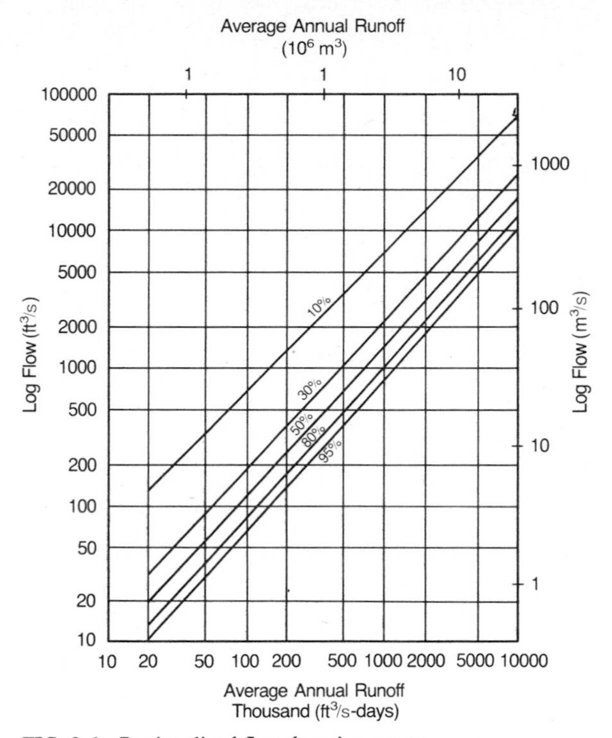

FIG. 2.6 Regionalized flow-duration curves.

point, all that is required is to have an estimate of average annual runoff at that point.

Another method of predicting flow-duration curves at ungaged points is to normalize the curve at a gage point by dividing each flow value by the average annual flow. The values of flow at the point of interest are found by multiplying the normalized values at the known gage location by the estimated average flow at the point of interest. An example of the normalized flow-duration curve technique is shown in Fig. 2.7.

Both the regionalized method and the normalized gage method depend on estimating average flows at the ungaged point of interest. In some areas average flow could be correlated very well with drainage area. In areas where there are large fluctuations in rainfall, a better method would be to use normal annual precipitation (NAP) or average runoff maps. Using this method, the average flow is estimated by integrating the areas on the NAP or average runoff maps. If NAP maps are used runoff coefficients must be estimated. The runoff coefficients can be estimated from runoff coefficients at gage locations in the area of interest. A brief description of this method is contained at the beginning of this section.

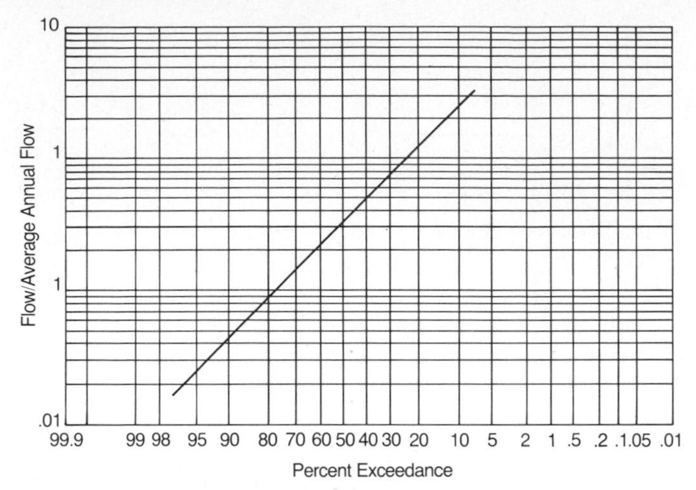

FIG. 2.7 Normalized flow-duration curve.

2.4 OTHER HYDROLOGIC AND HYDRAULIC CONSIDERATIONS

Determining Reservoir Volume-Area-Elevation Relationships

Generally, at hydro sites there is a dam with impounded water. The water behind the dam is frequently pondage for short-term storage to meet fluctuation in daily power load. In other cases a large reservoir may be required to meet seasonal variation in flow. The stored water is used to make the firm water flow more consistent and provide a higher discharge over a longer period of time. This necessitates the preparation of volume of storage or pondage versus elevation curves or tables. At the same time it is usually necessary to have a water surface versus elevation curve. This is obtained by planimetering good topographic maps and making the necessary calculation. Those two curves are typically combined into what is called an area capacity curve. A typical area capacity curve for a hydropower reservoir is shown in Fig. 2.8.

Operating Rule Curves

In many cases the operation of the releases of water from the impoundment is dictated by needs for flood control, demands to supply water for domestic use, or to supply water for downstream irrigation projects. It is necessary to develop rule curves for power and reservoir operating personnel to guide them as to when water is to be released. Rule curves may also be developed for operating power water releases to meet certain electrical load requirements. These often require rather careful reservoir operation studies using historical flow data and estimates of demands for water that are likely to occur in the future.

Tailwater Relationships

As streamflow changes and if water is being spilled past the hydro plant, it is important to recognize that the discharge from the outlet of the hydro plant will cause fluctuation in the level of the tailwater. It is necessary then to develop a discharge versus elevation curve over the complete range of flows expected to occur in the channel below or downstream from the draft tube or turbine outlet. Figure 2.9 is a typical example of such a curve. This will require contour maps of the stream channel area and estimates of velocity in the channel at various stages of flow. This tailwater information for normal tailwater, maximum tailwater, and minimum tailwater elevations is essential for making estimates of operating hydraulic head and to study turbine setting elevation. Rough estimates of stream channel velocity can be made by making slope-area calculations using the conventional Manning open channel flow equation.

Evaporation Considerations

In cases where the impoundment area is considerable, it should be recognized that surface evaporation of water from the reservoir may cause a sizable amount of loss. Thus, estimates of at least monthly evaporation values should

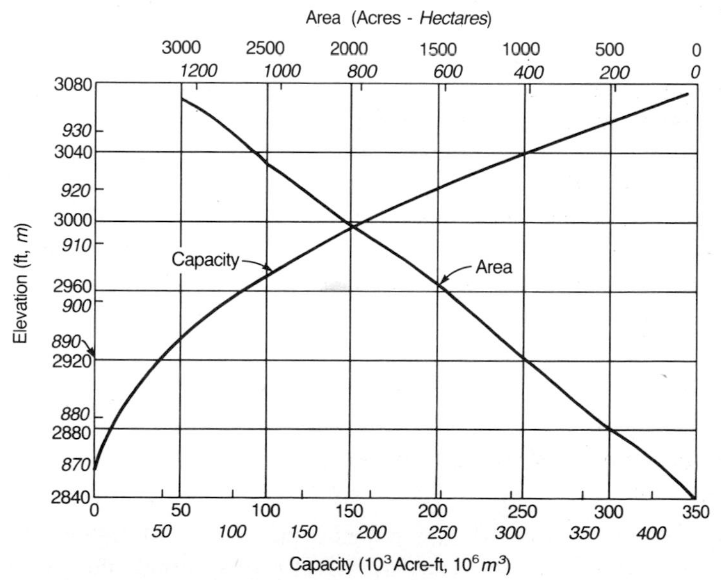

FIG. 2.8 Lucky Peak Reservoir area-capacity curve.

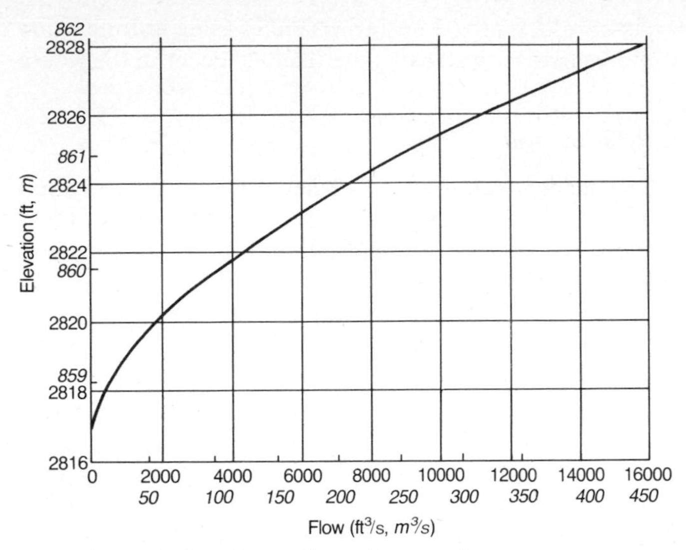

FIG. 2.9 Lucky Peak Power Plant tailwater rating curve.

be generated to determine the effect of water loss from reservoir surfaces. In the warmer, higher wind velocity areas this should be carefully checked since losses of as much as 4 ft (1.2 m) of water have been reported during a summer operating period.

Other Meteorological Considerations

In studies of turbine setting and possible cavitation, it is important to recognize that the normal barometric pressure at the site should be known as well as the highest water temperature. The latter datum is necessary for determining the vapor pressure of water that might prevail at critical period of use of the hydropower plant. This critical period occurs when there is the greatest negative pressure existing at the runner outlet. This condition occurs when there is very high headwater and very low tailwater. Useful tables for determining values for this are available in most fluid mechanics and hydraulics textbooks.[22,23]

2.5 PROCESSING SITE FLOW DATA

Determining Power and Energy from Flow and Head Data

Reference to Eq. (2.1) shows that power is the product of several constants, estimated efficiency, and two basic variables of Q (discharge) and H (hydraulic head). Once the capacity of the hydropower unit has been determined the maximum flow of water that will pass through the turbine is established. If a flow-duration curve study is being used, it is recognized that the runner discharge capacity is defined when the size of the runner is specified as shown in Fig. 2.10. The discharge capacity is labeled as Q_c, where Q_c is the discharge at full gate opening of the runner under design head. Although to the left of that point on the flow-duration curve the stream discharge is greater, it is not possible to pass the higher discharges through the unit. If the reservoir or pondage is full, the extra flow must be bypassed over a spillway. It should be noted that to the right of the runner discharge capacity point, all the water that can pass through the hydroplant is flowing in the stream at that percent of time. This indicates that the full rated power production will not be produced. If hydraulic head at the particular flow in the stream and stage of reservoir is known then it is possible to generate a power-duration curve from the flow-duration curve. Figure 2.11 gives an example of the power-duration curve. The P_c value is the full gate discharge value from the flow-duration curve, multiplied by the simultaneous value of hydraulic head, the estimated efficiency, and the conversion constants. Energy production for a year or the time period of study is merely the product of the power ordinate and time, thereby representing the area under the power-duration curve, multiplied by an appropriate time conversion factor. It is conventional practice to use as the unit of energy measurement the kilo-

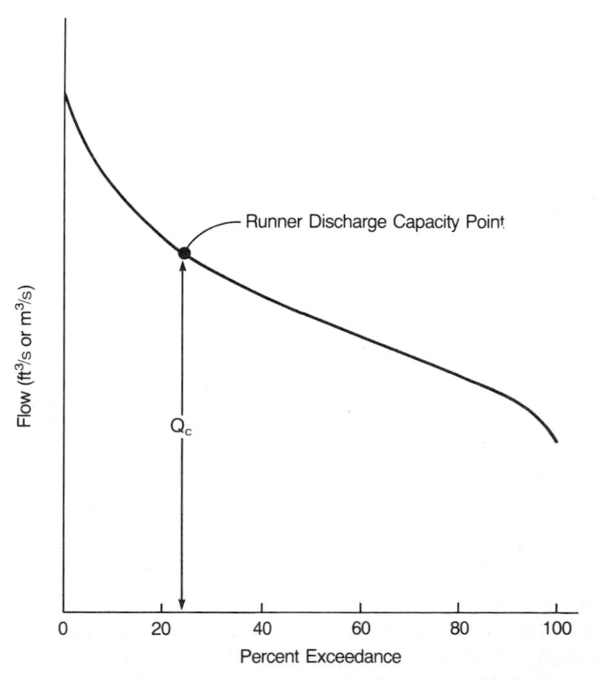

FIG. 2.10 Flow-duration curve indicating runner discharge capacity.

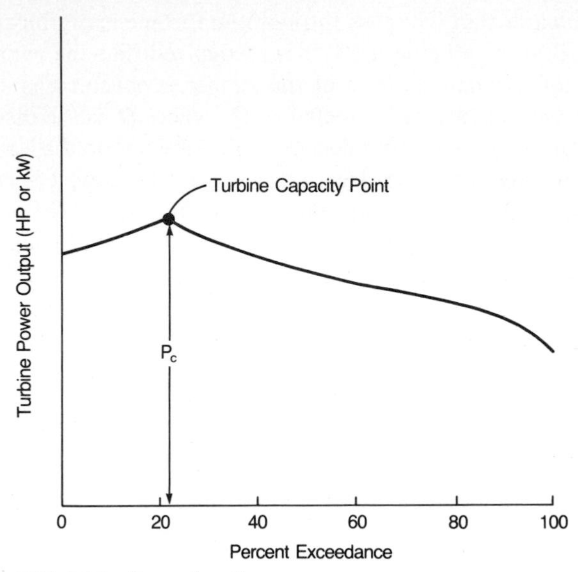

FIG. 2.11 Power-duration curve.

watthour. An example tabulation of how the calculations might be done is shown in Table 2.2.

For crude estimates of the power available one might use the average flow and the average head available in the power equation [Eq. (2.1)]. The energy available is average power in kilowatts multiplied by the time in hours for the time period being investigated.

Sequential Flow Techniques

The flow-duration approach to estimating energy production at a site is not always applicable when storage of water alters the flow and provides for later release in a river systems. A flow-duration approach could be used if an operational study of reservoir releases had been made and the data used in preparing the flow-duration curve reflects the requested releases. It is customary in most cases to resort to a tabular operations study using the true time sequential flows as modified by the reservoir regulation. This is often referred to as a sequential-flow type of study. Such a study entails obtaining records of flow into a reservoir as they naturally occur, and the use of appropriate hydraulic head and operational decisions of how the water will be released either through the turbine or bypassed over a spillway. A typical tabular study analysis form or table can be developed.[24] A sequential flow approach can likewise be used on run-of-river developments if desired. These can be done on an hourly, weekly, or monthly time interval basis. Generally this analysis is done by com-

TABLE 2.2 Annual Energy Calculations Using Daily Flow-Duration Analysis, Plant Installed Capacity Designed for Q_{20}

Percent exceedance	0	10	20	30	40	50	60	70	80	90	100	(1)
Q, 1,000 ft^3/s	130	30	15	8	7	6	5	4	3	2	1.5	(1)
Net head, ft	48	50	52	54	56	58	60	65	70	75	80	(2)
Turb. disch., Q_t, ft^3/s	14.4	14.7	15	8	7	6	5	4	3	2	1.5	(3)
Assumed eff., e	87	87	87	87	87	87	87	87	87	87	87	(4)
	51,150	54,390	57,720	31,970	29,000	25,750	22,200	19,240	15,540	11,100	8,880	(5)
Percent of time	10	10	10	10	10	10	10	10	10	10		
Dev. energy, MWh	46,230	49,100	39,280	26,700	23,980	21,000	18,150	15,230	11,670	8,750		(6)
Annual energy, MWh						260,090						(7)

NOTES: (1) From flow-duration analyses.

(2) From hydraulic analyses of site and development.

(3) In this example, the turbine is designed for Q_{20} = 15,000 ft^3/s. Flow through turbines for lesser discharges is river flow. Flow for higher flows (Q_{10}, Q_0) is assumed to be orifice flow controlled, or proportional to the square root of relative net heads: flow at 20% is $15\sqrt{50/52}$ = 14,700 ft^3/s.

(4) Efficiency assumed to be constant for this example only. The value combines turbine(s) and generator(s) efficiencies.

(5) Use the power equation $P_{kW} = \frac{HQ_te}{11.81} = .074HQ_t$

(6) Energy = $\left[\frac{P_1 + P_2}{2}\right]\left[\frac{\text{Percent of time}}{100}\right]\left[8760\,\frac{\text{hours}}{\text{year}}\right]\left[\frac{1}{1000}\right] = 0.438(P_1 + P_2)$ MWh

(7) Sum of the incremental developed energies.

puter. Many special computer programs have been developed for this purpose. A good example of this is HEC 5C.[2]

Mini Hydro Considerations

Mini hydro development has been defined as utilizing hydroelectric units of smaller than 1000 kW capacity. Often, this type of development will involve remote locations where the units will not be connected to a national transmission grid. Most plants use impulse-type turbines under rather high heads with small flows. Recent developments indicate conventional centrifugal pumps can be operated as turbines and result in considerable savings. Evaluations and estimates of flow duration and head variations with time are still very necessary to make feasibility and design studies.

The magnitude of the local load may dictate what head and what discharge to use in the capacity sizing of the specific installation. For some locations it may mean just proceeding higher with the penstock to obtain a higher head. In other cases, it may mean diverting a higher proportion of the streamflow that is readily available. In these cases the two variables head and flow may be interchanged in many different combinations to meet a desirable load.

2.6 PRESENTATION OF HYDROLOGIC AND HYDRAULIC DATA

In earlier sections, methods of hydrologic, power, and energy analyses were discussed. In the following paragraphs methods of presenting the information developed in these analyses is discussed, specifically those that have been used in resource inventories in the Pacific Northwest region of the United States. An outline of the hydrologic data and format required by the U.S. Federal Energy Regulatory Commission in obtaining a U.S. Federal Power Permit will be presented as an example of the type of hydrologic information that may be required by governmental organizations.

In making resource inventories of hydro potential it is important to be able to condense the desired hydrologic information into an easily understood format. Information such as physical location of the stream under study, its location within political boundaries (city, county, state, etc.) are very important to hydropower planners and should be included with the hydrologic and energy potential data wherever possible.

An example of a data presentation format is shown in Refs. 13 and 20. It was used in a resource inventory study of the Pacific Northwest region of the United States presenting the theoretical hydropower potential on particular segments of streams called reaches. Included were location information such as state, county, geographical location, river mile, and map of the region. Hydrologic and hydraulic information included upstream and downstream river elevations, reach length, slope, and upstream drainage area. A flow-duration curve in tabular form with plant capacity, a normalized typical annual hydrograph, and energy values were also provided.

In the United States, the Federal Energy Regulatory Commission requires adequate coverage on all hydrologic information that is necessary for determining the size of the installation and the economic and political acceptability of a given development. The following is a brief checklist of the data that is normally reported in a license application:

1. Tailwater rating curve
2. Head loss versus discharge relationship
3. Reservoir outflow on a monthly basis
4. Reservoir elevations on a monthly basis
5. Net head (normal or design)
6. Turbine discharge (normal or rated Q)
7. Minimum turbine discharge for generation
8. Minimum net head for generation
9. Maximum net head for generation
10. Plan output versus turbine discharge capacity
11. Reservoir elevation versus storage capacity
12. Reservoir elevation versus surface area
13. Projected spillway releases
14. Generated output versus net head
15. Flow releases and demands for competing uses of water
16. Operating rule curve for reservoir release
17. Flow-duration curve with period of record
18. Gaging stations used in streamflow analysis
19. Specification of critical streamflow period
20. Location of point of diversion

2.7 REFERENCES

1. United Nations, *Manual of Standards and Criteria for Planning Water Resources Projects,* Water Resources Series No. 26, 1964.
2. U.S. Army Corps of Engineers, *Feasibility Studies for Small Scale Hydropower Additions,* The Hydrologic Engineering Center, Davis, Calif., 1979.
3. World Meteorological Organization, *Guide to Hydrometeorological Practices,* WMO-No. 168. Tp. 82, 2d ed., 1970.

4. J. K. Searcy and C. H. Hardison, "Double-Mass Curves," *Manual of Hydrology: Part I. General Surface-Water Techniques, U.S. Geological Survey Water Techniques,* U.S. Geological Survey Water Supply Paper 1541-B, 1960, 66 pp.
5. G. W. Kite, *Frequency and Risk Analyses in Hydrology,* Water Resources Publication, Fort Collins, Colo., 1977.
6. R. J. Buchanan and W. P. Somers, "Discharge Measurement at Gaging Stations," Chapter A8, "Techniques of Water-Resources Investigations of the U.S. Geological Survey," Book 3, *Applications of Hydraulics,* 1969, 65 pp.
7. R. K. Linsley et al., *Hydrology for Engineers,* 2d ed., McGraw-Hill, New York, 1975.
8. A. L. Simon, *Practical Hydraulics,* 2d ed., Wiley, New York, 1981.
9. O. T. Lind, *Handbook of Common Methods in Limnology,* 2d ed., C. V. Mosby, St. Louis, Mo., 1979.
10. R. E. Davis et al., *Surveying Theory and Practice,* 6th ed., McGraw-Hill, New York, 1980.
11. R. C. Brinke and P. R. Wolf, *Elementary Surveying,* 6th ed., Harper, New York, 1980.
12. H. A. Mayo, Jr., *Low Head Hydroelectric Fundamentals,* Allis-Chalmers, Hydro-Turbine Division, York, Penn., n.d.
13. J. S. Gladwell et al., "A Resource Survey of Low-Head Hydroelectric Potential in the Pacific Northwest Region—Phase I and II," project completion reports to U.S. Department of Energy, Idaho Water Resources Research Institute, University of Idaho, 1979.
14. UNESCO, Division of Water Sciences, *Atlas of the World Water Balance,* explanatory text (35 pp.) and 64 maps, produced by USSR National Committee for the International Hydrological Decade, Paris, 1977.
15. UNESCO, Division of Water Sciences, "World Water Balance and Water Resources of the Earth," *Studies and Reports in Hydrology 25,* produced by the USSR National Committee for the International Hydrological Decade, Paris, 1978.
16. Food and Agriculture Organization, *Approaches to Water Development Planning in Developed and Developing Economies,* W/K4859/c, n.d. 16 pp.
17. M. H. Diskin, "General Report, Topic 2, Section 2.3—Use of models derived on the basis of representative basins for water planning and management" (separate paper), Symposium on the Influence of Man on the Hydrological Regime with Special Reference to Representative and Experimental Basins, Helsinki, Finland, June 1980, 9 pp.
18. M. A. Benson, "Characteristics of Frequency Curves Based on a Theoretical 1000-year Record," in T. Dalrymple, *Flood Frequency Analysis.* U.S. Geological Survey Water Supply Paper 1543-A, 1960, pp. 51–74.
19. D. N. Body, "Further Research Needs in Representative and Experimental Basins Development and Understanding," (separate paper), background paper Topic 3, Symposium on the Influence of Man on the Hydrological Regime with

Special Reference to Representative and Experimental Basins, Helsinki, Finland, June 1980, 6 pp.

20. L. F. Heitz, "Hydrologic Evaluation Methods for Hydropower Studies," Ph.D. dissertation, College of Engineering, University of Idaho, Moscow, Idaho, 1981.
21. J. S. Gladwell and C. C. Warnick, *Low Head Hydro,* Idaho Water Resources Research Institute, University of Idaho, Moscow, Idaho, 1978, 206 pp.
22. J. A. Roberson and C. T. Crowe, *Engineering Fluid Mechanics,* 1st ed., Houghton Mifflin, Boston, 1980.
23. R. K. Linsley and J. B. Franzini, *Water Resources Engineering,* 3d ed., McGraw-Hill, New York, 1979.
24. U.S. Army Corps of Engineers, *Hydrologic Engineering Methods for Water Resources Development,* vol. 1: *Requirements and General Procedures,* The Hydrologic Engineering Center, Davis, Calif., 1971.

3

Hydrologic Computations at a Small Hydro Site

Norman H. Crawford

3.1 INTRODUCTION

The methods included in this chapter are for hydroelectric site planning on small watersheds that do not have streamflow records.[1] The calculation methods use minimum field and meteorologic data, since very little data are available for small watersheds in developing countries. The methods will give results of moderate accuracy based on a minimum of field information.

This chapter describes methods and includes sample calculations for estimating:

1. Peak streamflow at a site
2. Flow-duration curve at a site

Collection of field data on historic flows for peak flow and flow-duration estimates, and calculation techniques for peak flow and flow duration from related meteorologic data and watershed characteristics are discussed. The methods described can be applied to watersheds up to 1,000 km^2 (386 mi^2) in area where snow accumulation and melt is minimal and where streamflows are not regulated by large lakes or reservoirs.

Hydrologic Estimates

The hydrologic cycle operates continuously in all watersheds. The hydrologic results needed for a small hydroelectric project are a peak flow and the flow duration at the site. The basic hydrologic processes that produce peak flow and flow duration at a site are as follows:

Peak Flows: Peak flows are caused by high intensity storm rainfall. Rainfall that moves as surface runoff into stream channels produces a *flood hydrograph.* The balance of the rainfall infiltrates into the soil and is "lost." The infiltrating water will later evaporate or transpire or will provide groundwater flow in the stream.

The factors that must be estimated or calculated for peak flows are

1. Rainfall intensity and duration on the watershed
2. Amount of "losses" or infiltration during the storm
3. Flow time and storage in stream channels

Flow-Duration Curves: The flow-duration characteristics of a watershed give the percent of time that flows exceed specific levels. The factors that control flow duration are

1. Annual cycle of precipitation, potential evapotranspiration, and actual evapotranspiration in the watershed.
2. Amount of rainfall that infiltrates and moves on subsurface flow paths into stream channels. Infiltration rates depend on the permeability and depth of the watershed soils.
3. Subsurface flow velocities and the storage capacity of subsurface aquifers.

Accuracy

The important factor in the accuracy of a hydrologic estimate is the accuracy of the meteorologic data on which it is based.

An additional factor in accuracy is the adequacy of the calculation technique for application to the watershed being studied. The calculation techniques in this chapter are based on sound hydrologic principles and they represent key processes in flood flows and continuous monthly streamflows. They do not include processes like snowmelt or detailed hydraulic routing and should not be used where snowmelt is a key process or where there are large lakes or reservoirs.

3.2 HYDROLOGIC PROCESSES

The hydrologic cycle is sketched in Fig. 3.1. Precipitation provides the water to the system. Precipitation will either infiltrate through the land surface into

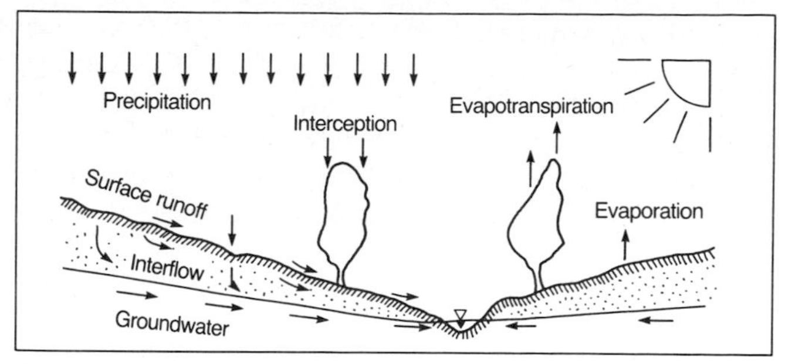

FIG. 3.1 The hydrologic cycle.

the soil or move toward stream channels as surface runoff. Water that infiltrates may move as subsurface flow or may be evaporated or transpired by vegetation. The amount of water lost by evapotranspiration is larger than the amount of water that becomes runoff in most watersheds. Water that becomes runoff moves over the surface as direct flow to streams or moves subsurface as delayed or groundwater flow. Direct runoff causes floods, while the delayed subsurface flows provide continuous or low flows to the rivers.

For a hydroelectric site, two types of information are needed. First, the flood flow or expected maximum water level is needed to size a spillway (if any), to locate turbines and generators above the highest expected water level, and to design diversion structures or canals. Second, the statistical distribution of monthly streamflow volumes is needed to estimate the reliability of the site for the production of a given amount of electrical power and to size the turbine.

Peak Discharge

Peak flows result from a combination of heavy rainfall and high soil moisture levels. Heavy rainfall provides the water and high soil moistures prevent the water from moving into the soil. Peak flows on small watersheds are frequently caused by thermal or thunderstorm rainfall. Peak flows on larger watersheds are caused by a series of storms or by snowmelt.

Estimating Peak Flow

Flood flows are plotted on special semilog scales. If flood flows are measured for a period of years, this historic data can be used to statistically estimate the magnitude of floods. Hydrologists refer to 50-year or 100-year floods, which are the flood levels that would be equaled or exceeded once in 50 or 100 years. To estimate peak discharges, a hydrologist must estimate the maximum rainfall rates that will occur in a watershed for the duration or length of storms that would be expected to cause a maximum flow. The runoff from this rainfall must then be estimated. The difference between the rainfall and the runoff will be the infiltration through the soil surface during the storm event.

When rainfall occurs and exceeds the capacity of the soil to absorb water, surface runoff occurs and enters the river channels. This surface runoff will take place during and immediately after the rainstorm. It flows through the river channels and is measured at a streamgage as a flood hydrograph and a peak flow. When rainfall on a watershed is plotted together with streamflow from the watershed, the close relationship between rainfall and peak flow is clear.

Factors Determining Peak Flow

The factors that determine the peak flow, or maximum flow, in a flood hydrograph are (1) the intensity of the rainfall that causes the runoff; (2) the amount of water that infiltrates and follows the slower, subsurface paths to the stream channel; and (3) the amount of attenuation or subsidence that the peak flow experiences as it moves through the river channels to the streamgage.

The first factor, the intensity of rainfall over the watershed, is a characteristic of the climate of the region. For example, in the United States, a rainfall of 30 mm (1.18 in) per hour is a "heavy intense storm" in Seattle, Washington, but it is a "light rain storm" in Houston, Texas. Data on the intensity of rain that is to be expected in the watershed is needed.

The second factor that determines flood magnitude is the amount of water that is absorbed through the soil to follow the slower subsurface flow paths. Infiltration rates or losses depend on soil properties and on soil moistures. Soils may range from tight, low-permeability clays, to high-permeability silt and sand. Soils with high permeability like sandy loam soils absorb water quickly. Forest soils that have thick organic layers of decaying vegetation absorb water quickly. In watersheds that have high infiltration capacities, surface runoff may be unknown. Watersheds that have soils of low permeability such as clay will absorb very little water and some rainfall will become surface runoff.

The third factor that determines peak flow is attenuation. The runoff that enters stream channels from the land surface may be modified substantially as water moves through stream channels toward a gaging site. Flood waters that enter a natural lake or reservoir are "routed" or attenuated as they move through the reservoir. The peak flow leaving the storage may be much less than the peak flow entering the storage. This is illustrated in Fig. 3.2.

Flow along natural channels in a river basin also tends to attenuate. The maximum rate of discharge tends to decrease as the flood moves downstream when the peak is expressed in units of flow per unit of watershed area. Hydrologists account for this attenuation process by using *flood routing* or *flow-routing procedures*. The amount of attenuation that a flood will experience in a watershed depends on the length, shape, and roughness of the channels. If the channels have broad floodplains with vegetation and if the flood moves out of the incised channels onto the floodplains, the attenuation of peak flows will be dramatic. If the channels in the watershed are narrow and steep, and if flow

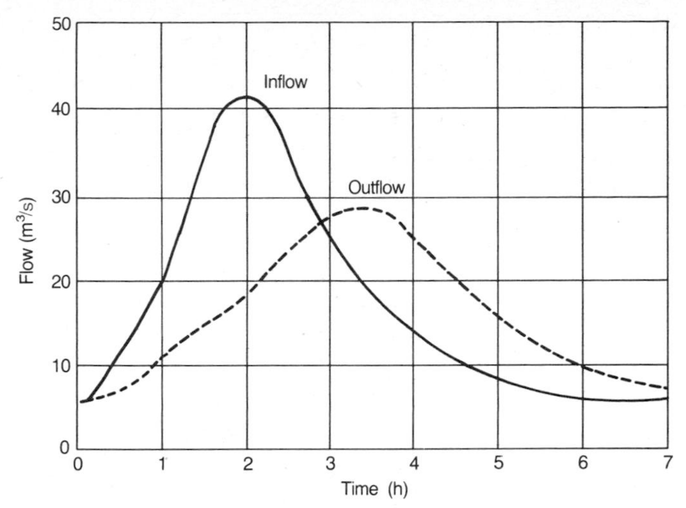

FIG. 3.2 Flow routing through storage in channels or reservoirs.

velocities are high, the attenuation of flood peaks will be much less. If flows move into natural lakes, swamps, marshes, or constructed reservoirs, reduction in peak flow due to "reservoir routing" can be expected, as shown in Fig. 3.2.

In summary, the problem of estimating peak flow on natural watershed requires:

1. Data on rain storms that will occur in the region, and on the maximum rainfall intensities that will occur.
2. Estimates of the infiltration losses that will occur during the storm event. This infiltration will depend on typical soil moisture levels and on the characteristics of soils in the watershed.
3. Attenuation that peak flows will experience as they move from the headwaters of the watershed through the basin.

Minimum Discharge

Low-flow periods in streamflow can be seen in Fig. 3.3. The lowest flows are reached during droughts when no rainfall or very little rainfall has occurred for an extended period of time. Just as the peak flows, or maximum flows, are reduced by water that is absorbed through the soil surface, the low flows are increased by this infiltrating water. Low flow is supplied to the stream channels by the subsurface or groundwater flow paths shown in Fig. 3.1. Thus, the factors that control the low flow in a stream are the length of the no rain or minimum rainfall periods, the amount of water that is absorbed through the soil surface during rainstorms, and the time needed for water to flow along the subsurface or groundwater flow paths.

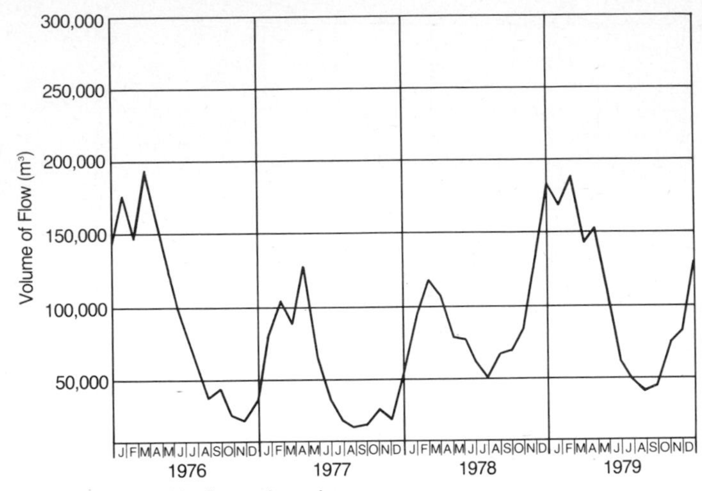

FIG. 3.3 Monthly flow volume in a stream.

The climate in the watershed controls the length of drought periods. In areas that experience seasonal rainfall, there may be a drought each year for several months. In these climates, the lowest flow will occur toward the end of the annual dry season and will be much lower than the minimum flow in a climate where rainfall occurs throughout the year. In arid climates where the potential evapotranspiration exceeds the rainfall, minimum flows will be less than in humid climates where the rainfall is greater than the potential evapotranspiration. The subsurface flow paths in a watershed are stable and predictable. If a watershed is known to have a minimum flow over a 10-year period of 10 m^3/s (353.1 ft^3/s) it is unlikely that the minimum flow will suddenly drop to 2 m^3/s (70.6 ft^3/s). Low flows respond to cumulative climate changes. A period of 1 or 2 weeks of dry weather will not cause unusually low flows in a watershed, but low rainfall for a period of 6 to 36 months may result in the minimum observed low flow. A large subsurface groundwater storage will take periods of months rather than periods of weeks to react to a change in climate. Observations of low flows are very helpful to document the low-flow regime in a watershed, even when the observations are limited.

The hydrologic processes important to low flow in streams occur at and below the land surface. Water that is infiltrated at the land surface may move along the subsurface flow paths into the stream channel. Time delays for water moving on the subsurface groundwater flow paths range from weeks to years. These time delays are much longer than the time delays in channel flow. When water enters a river channel, the typical time delay to move to a downstream measurement point is hours to days.

In summary, to estimate the minimum discharge in a stream, the important factors are

1. Climate of the watershed and the duration of drought periods
2. Infiltration of rainfall through the soil surface and water movement into the subsurface groundwater flow paths
3. Typical time delays on the subsurface flow paths
4. Marshes, lakes, or reservoirs (if any) that are present in the watershed

Frequency of High and Low Flows

The continuous measurements of streamflow plotted in Fig. 3.3 show both the high and the low flows from a watershed. It is convenient to reorganize these data to show the frequency of high or low flows separately on semilog graphs like Fig. 3.4. For example, for high flows, the peak flow measured each year for a series of years can be plotted. If 20 years of data are available, the flood flows are listed in order of magnitude and the highest flow is assumed to recur once in 20 years, the second highest flow, once in 10 years, the third highest flow once in 6.67 years, and so on. This allows the points to be plotted on the graph, as shown in Fig. 3.4.

Peak flows for many watersheds produce linear or near linear curves when plotted on semilog paper. These flood frequency curves are used to estimate

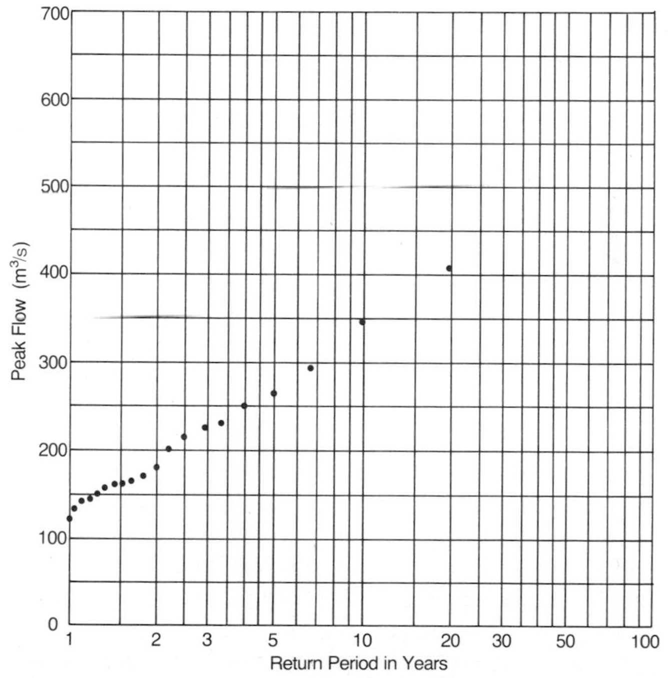

FIG. 3.4 Flood frequency: peak flow vs. return period.

TABLE 3.1 Monthly Runoff Volumes

Month	Runoff volume, 10^6 m^3
January	102.3
February	189.0
March	291.9
April	211.6
May	98.7
June	32.0
July	12.6
August	8.7
September	14.5
October	68.7
November	96.1
December	113.4

TABLE 3.2 Monthly Runoff Volumes in Order of Magnitude

Rank	Runoff volume, 10^6 m^3
1 (March)	291.9
2 (April)	211.6
3 (February)	189.0
4 (December)	113.4
5 (January)	102.3
6 (May)	98.7
7 (November)	96.1
8 (October)	68.7
9 (June)	32.0
10 (September)	14.5
11 (July)	12.6
12 (August)	8.7

floods for project design. A designer might choose to design a structure for a once in 100 years flood, a flood that would be equaled or exceeded an average of once every 100 years.

Minimum flows can be treated like maximum flows. A frequency curve of the minimum discharge each year could be prepared and plotted using the same procedure that was described for peak flows.

A very useful technique for graphically representing the continuous flow measured in a watershed is the flow-duration curve as outlined in Chap. 2. To construct a flow-duration curve, assume that Table 3.1 is the monthly volume of runoff for a year in a watershed. These runoff volumes are rearranged in order of magnitude in Table 3.2. The data in Table 3.2 are plotted in Fig. 3.5. It can be seen in Fig. 3.5 that the runoff equals or exceeds 100 million cubic meters (131 million cubic yards) 40 percent of the time. Similarly, the runoff exceeds 175 million cubic meters (229 million cubic yards) 20 percent of the time.

The flow-duration curve is very useful for study of projects that divert water from a river for water supply, irrigation or hydroelectric power production. Flow-duration curves can be constructed from any period of stream-flow data but are more reliable if they are constructed for several years of streamflow data. Flow-duration curves use all of the streamflow information that is available at a site, while peak-flow and low-flow frequency studies select out maximum or minimum flows for analysis. In locations where very little streamflow data is available, the flow-duration plot may be feasible to construct while peak-flow and low-flow frequency curves may be infeasible. Of course, both peak-flow and low-flow frequency curves and flow-duration curves increase in accuracy as the length of the basic data observations increases.

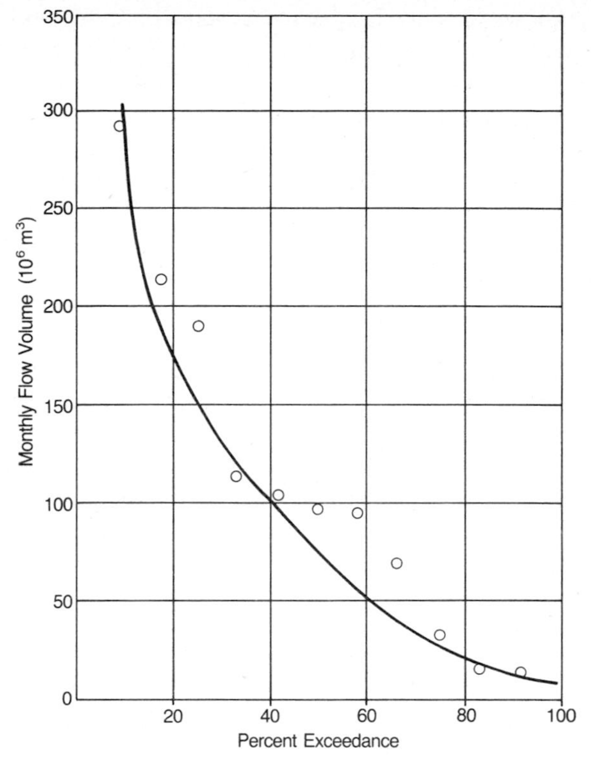

FIG. 3.5 Flow-duration curve for monthly flow volumes.

3.3 ESTIMATING PEAK FLOW AT A SITE

To estimate the peak flow which could occur at a potential hydroelectric site, one should (1) study the watershed to find out what high flows have occurred in the past, and (2) study the data on the watershed to calculate what high flows might occur in the future.

Field Study

The field study of the watershed will answer the question, "How large have past floods been?" Two methods are introduced to determine out how high the water has been. These methods are

1. Study the physical features near the stream to find signs of high water.
2. Talk to people living near the stream and to officials in the area about past floods.

Once the high water mark has been determined, we will use some calculations to find the flow rate for that water level.

Physical Signs of High Water

Flowing water moves many objects that it touches. Grass, branches, and other light objects float along near the water surface. Earth and sand are eroded by fast-moving water and carried downstream or are left on beaches or in pools. To begin your survey of the stream channel, stand on the bank of the stream and watch the edge of the stream as it flows against the plants and rocks or sand. Imagine that the water level is higher by 1 or 2 m (3.28 or 6.56 ft) and see what would be under water. Look for signs, like eroded tree roots or sand or gravel beaches, which indicate that the water has been at that level.

If water is not moving very fast or is flowing through bushes and trees, the light objects which are floating will often become caught in the branches or left on floodplain lands when the water level goes down. Look for clumps of grass and sticks caught in the trees or on beaches as you move away from the stream.

Look at the general slope of the land toward the stream. Water will always form a flat surface. If you find a high deposit of old floating material, imagine that the water is at that level everywhere, and determine what area would be under water. Look at the ground at the same elevation as your high water level. Are there signs that the ground has been under water? Look for large branches and logs which might have floated to the edge of the stream during a flood. Once you have found what appears to be the highest deposit of flotsam, look for other signs that the water was once at or near that level. Look for marks on trees or buildings or more branches or logs at the same height.

After getting some idea of how high the recent water levels have been, talk with local residents about high flows, as outlined in the next Section. If you have some information from the plants and soil deposits near the stream, it will be easier to evaluate the recollections of residents about high flows.

Interviews with Residents and Historical Accounts

People who have lived for a long time near a stream may remember large floods quite well. If a historic flood was large and if water marks remained visible for some time, residents will know the highest water level and when the flood occurred. By talking with several people and by comparing their recollections, you may get accurate information on the largest floods in the last 50 years.

Ask people being interviewed to point out marks where the high water has been recorded or where they remember high water. Also, ask for any available written documentation or photographs to confirm their reports. Diaries when the floods occurred and a search for high water levels at a number of nearby points may also help confirm reports of floods. Other ways of verifying residents' reports of high water levels include researching old local newspapers (if any exist), church records, or local government records of road or bridge repairs.

Calculating Discharge from Water Level

The final step in the field study for peak flow is to transform the flood level we have determined into an estimated peak discharge. To calculate the flood discharge, we need to know the cross-sectional area and the velocity of the stream at the peak water surface elevation. The cross-sectional area of the stream may be calculated by first plotting the height above the ground of the peak water surface along a line perpendicular to the stream channel. Plot peak water depth measurements at regular intervals from the point where the high water line touches the ground on one side of the stream, all the way over to the same point on the other side. Figure 3.6 shows an example of a stream cross-sectional plot at flood stage.

The next step is to count the number of boxes between the ground surface profile and the high water line. This number is then multiplied by the scale factors on your plot to yield the peak flood's cross-sectional area. Figure 3.6 shows how this calculation is done.

After you have determined the area of the peak flood, you must determine the velocity of the flow. This can best be done by measuring off a stretch of 50 m (164 ft) along the channel, then timing a floating object as the stream carries it through the 50-m (164-ft) stretch. The velocity of the flow is then:

$$v = 0.85 \left(\frac{50}{t} \right) \tag{3.1}$$

where

v = velocity of flow, m/s

t = average time to float through stretch, s

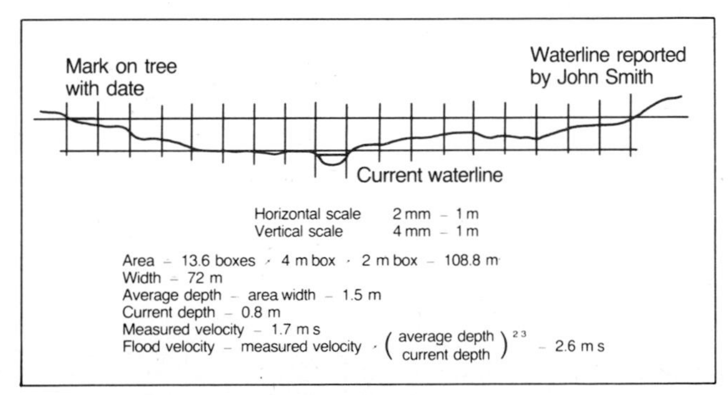

FIG. 3.6 Stream cross section at flood stage.

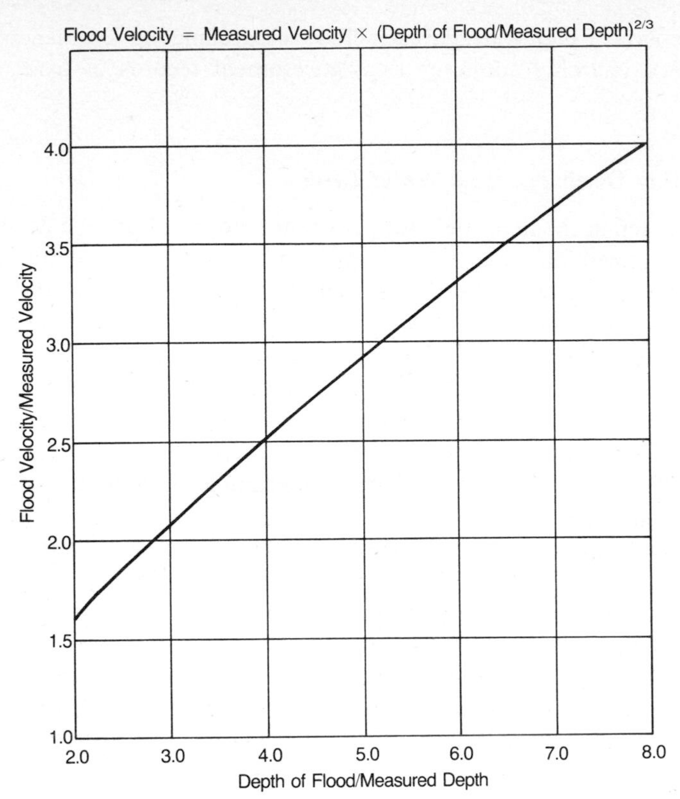

FIG. 3.7 Flood velocity factor.

Throw the object into the center of the stream and make sure it doesn't strike anything as it moves through the reach. Time the object through the reach at least 5 times.

This velocity needs to be adjusted to take into account the higher velocities of flood flows. To do this, find or estimate the water depth near the center of the channel and the average water depth at flood levels from your cross-sectional plot. Use these numbers as shown in Fig. 3.7 to find the factor which you must multiply your measured velocity by in order to get flood velocity. Once you know the flood velocity and the cross-sectional area A, just multiply the two numbers together to obtain the peak flood discharge Q.

$$Q = vA \tag{3.2}$$

Calculation of Peak Flows from Watershed Data

This section describes how to calculate the peak flood flow at a possible hydroelectric site using the available data on the watershed and the tables and figures presented here. Requirements include a contour map of the watershed and regional data on rainfall intensity for different storm or rainfall durations. Rainfall intensity is given in units of millimeters per hour (mm/h). If 45 mm (1.77 in) of rain is measured in 2 h, the rainfall intensity is 22.5 mm/h (0.79 in/h). Rainfall intensity will decrease as storm duration increases. In Boston, Massachusetts, for example, a rainfall intensity of 150 mm/h (5.91 in/h) will occur for storms of 10-min duration, but a rainfall intensity of 30 mm/h (1.18 in/h) will occur for storms of 4-h duration.

The steps involved in calculating the peak discharge are

1. Using channel and watershed characterisitics, determine the flow time for water to move through the watershed.
2. From regional meteorologic data, determine rain intensity for the design storm. The duration of the design storm is assumed to be equal to the flow time for water to move through the watershed.
3. Determine watershed losses, the "excess rain" or the amount of rainfall that becomes surface runoff, and the peak discharge.

This calculation method assumes that a peak discharge will occur when the storm duration is equal to the flow time for water to move through the watershed. This is an idealization of the actual interactions between storm rainfall duration and peak flow, but it gives peak flow estimates of reasonable accuracy. Each step in the calculation is explained in the following sections and easy to follow examples have been included.

Calculating the Flow Time for Water to Move Through the Watershed

To compute the flow time for water to move through the watershed you will need to determine L, h_e, and A:

L = channel length, km (or mi)
h_e = change in elevation between the highest point in the watershed and the site, m (or ft)
A = watershed area, upstream of the site, km^2 (or mi^2)

These data can be found from a contour map of the watershed. L and h_e are used in Eq. (3.3) to calculate the flowtime t_f through the watershed. This flowtime t_f is assumed to equal the duration of the storm rainfall that will cause a peak flow. The duration of the storm rainfall is used to find the rainfall intensity I during the storm, using local data similar to Fig. 3.9. The storm rainfall intensity I less a loss rate L_r is used to find the rate of runoff or rain excess X_r during the storm in Eq. (3.4). Finally, the excess rain is converted to flood peak discharge by multiplying by a necessary constant and the watershed area in Eq. (3.5).

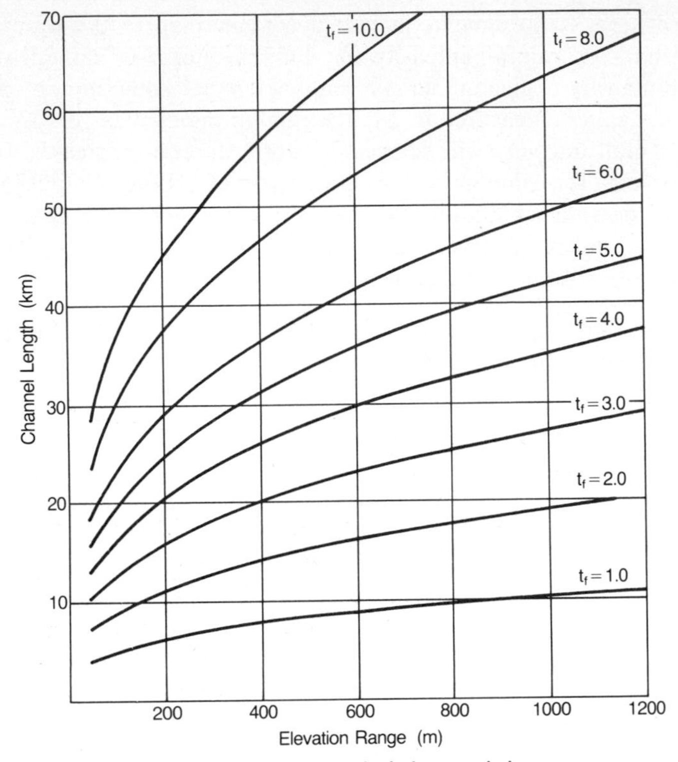

FIG. 3.8 Time of rainfall for watershed characteristics.

The flow time in hours t_f can be calculated from L and h_e using Eq. (3.3) or it can be read from Fig. 3.8.

$$t_f = 0.95(L^3/h_e)^{0.385} \tag{3.3}$$

In Eq. (3.3), L is in kilometers, h_e is in meters, and t_f is in hours.

Estimating Rain Intensity

Once t_f is calculated from Eq. (3.3), data on duration of rainfall versus rain intensity may be used to find the rainfall intensity on the watershed. The relationship between rainfall duration and rainfall intensity is usually displayed graphically, as shown in Fig. 3.9. It may be possible to make a graph similar to Fig. 3.9 with your own data. Graphical plots of rainfall intensity versus duration of rain are often made for different statistical return periods. A rainfall intensity of 60 mm/h (2.36 in/h) for 4-h duration would be equaled or

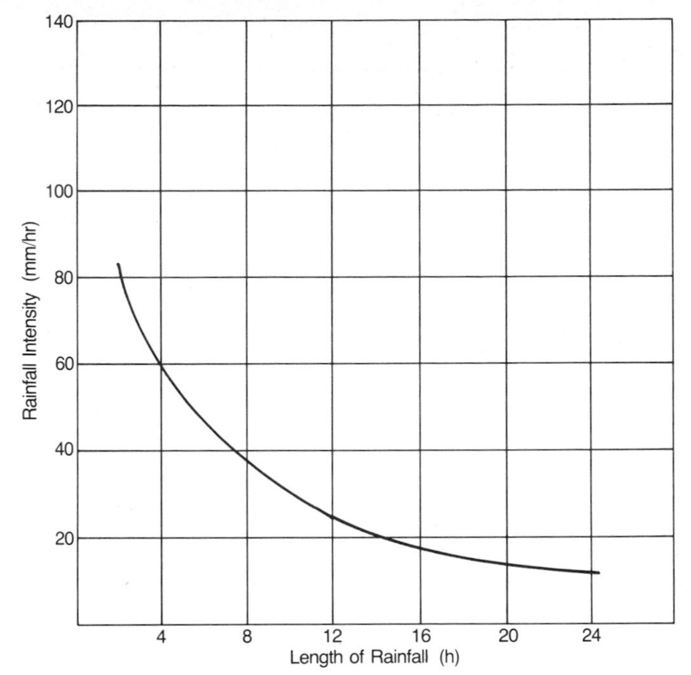

FIG. 3.9 Example of rainfall intensity vs. length of rainfall for 50-year return period.

exceeded once every 50 years, as shown in Fig. 3.9. To use the figure, find the duration of rainfall in hours that equals t_f on the bottom scale, intersect the line, and read the rainfall intensity I from the vertical scale.

Estimating Excess Rain and Determining Peak Flow

Rainfall will be lost during a storm through infiltration. Table 3.3 describes various soil types and gives a loss rate for each. Losses are greater in watersheds which have very heavy vegetation. A correction factor for vegetation density can be selected that is multiplied by the loss rate for the watershed soils to give the total loss rate for the watershed L_r.

After determining the rate of rainfall loss, excess rain may be computed by subtracting your losses from the rainfall intensity. The total excess rainfall rate is then

$$X_r = I - L_r \tag{3.4}$$

where X_r, I, and L_r are all in mm/h (in/h).

TABLE 3.3 Watershed Loss Rates and Correction Factor for Predominant Vegetation

Predominant soil type	Loss rate, mm/h
Impervious rock	1
Tight clay	1
Clay and silt	3
Silt and sand	5
Sand and gravel	10

Predominant vegetation	Correction to loss rate (multiply by)
Sparse: little vegetation, bare soil, scrub brush	0.5
Moderate: grassland, cropland, mixed forest	1.0
Heavy: dense forest, tropical forest	2.0

EXAMPLE: A watershed with clay silt soils and dense forest cover would have a loss rate of 3.0 mm/h and a correction factor of 2.0 so its total loss rate would be 3.0 × 2.0 = 6.0 mm/h.

The final step is to determine the peak discharge, Q_p in m^3/s (ft^3/s) at the site from Eq. (3.5).

$$Q_p = 0.28X_rA \tag{3.5}$$

where X_r is excess rain, mm/h (in/h), and A is watershed area, km^2 (mi^2).

Determining Final Peak Flow Value

From the foregoing analysis one may have two values for peak flow which may be quite different. This is to be expected. The remaining task is to evaluate the two values to arrive at one best value. To do this, you need to review the process by which you calculated each number and determine where the largest uncertainty lies.

Which method yields a higher value? Was one value more than twice as much as the other? If the calculation method value was much larger, what was the return period for the rainfall intensity versus duration of rainfall data that you used? If it was more than 50 years, it should have yielded a larger peak flow, because the field study interviews will document floods which have occurred within the last 50 years.

If the field study gave you a lower peak flow value, compare the velocity figure you used in the field study to the average flood velocity which is channel length divided by the flow time through the watershed t_f. The two velocities should be fairly similar, although the field-study velocity is likely to be lower. Try calculating the field-study peak flow using the Chezy-Manning equation introduced in Chap. 4.

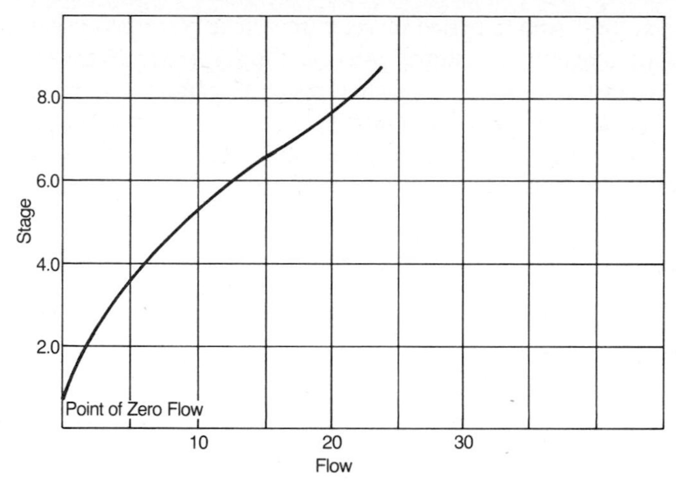

FIG. 3.10 Stage discharge rating curve.

3.4 ESTIMATING THE FLOW-DURATION CURVE AT A SITE

The flow-duration curve is a most useful tool for evaluating low flows at a hydroelectric site. These data can be used to estimate the flow that can be used for power production 90 to 95 percent of the time. This gives an immediate indication of the reliability of power production at a site. For example, if a 20-m (65.6-ft) head were available, the flow-duration curve in Fig. 3.5 would indicate that a 1,200 kW power source could be developed at 85 percent reliability, assuming an 80 percent overall efficiency for the project.

Field Investigation for Estimation of Flow-Duration Curves

High and low stages as well as normal stages throughout the year will be known to local residents, even though they will not know the corresponding discharge. Knowledge of stage can be converted to discharge using a stage-discharge-rating curve. A stage-discharge-rating curve can be constructed using measured cross sections at two or more stages and calculating flow velocities at each stage, as was described in the previous section.

Discharge at a given stage is the flow velocity times the cross-sectional area of the flow at that stage. When stage-discharge ratings are plotted on log scales, linear or near-linear curves result. Fig. 3.10 is an example of a stage-discharge-rating curve.

To construct a flow-duration curve, the following sources of information might be used:

1. Stage records over a period of years at a lake or reservoir can be used to calculate streamflow volumes entering the reservoir. When monthly stage records exist at a reservoir, locate a reservoir stage versus reservoir volume chart for the reservoir. The streamflow volume entering the reservoir each month will be the reservoir volume at the end of the month, less the reservoir volume at the first of the month, plus the volume of water released from the reservoir during the month.
2. Correlation of low flows on a regional basis is possible. The U.S. Geological Survey has summarized commonly used methods for regional correlation of flows in Water Supply Paper No. 1975, "Generalization of Streamflow Characteristics from Drainage-Basin Characteristics," by D. M. Thomas and M. A. Benson (55pp.).
3. Any existing data on flow duration compiled in the region in previous studies should be reviewed.
4. A survey of local residents, particularly people who are using the stream in some way; ferry operators, irrigation farmers, or people who divert water for water supplies may have additional information.

Where monthly streamflow can be calculated from lake or reservoir levels, as in item 1, or if regional studies to relate flow at a site to nearby gaged streams are done as in item 2, the data can be used to construct a flow duration as was done in Sec. 3.2. If regional studies of flow duration already exist from item 3, flow-duration curves will be available.

When interviews are used to establish seasonal stages in item 4, use Manning's equation or float measurements to establish a stage-discharge relationship, as was described earlier.

Convert stages to discharge to estimate monthly flows for a typical year. These mean monthly flows can be used to construct a flow-duration curve as in Sec. 3.2. If the field interviews give clear evidence of occasional low flows that are substantially less than the minimum expected annual flow, a multiyear flow-duration plot might be made. To make a multiyear flow-duration plot, estimate monthly flows for additional years and include the historic low flows. If your field information shows the minimum flow in an 8-year period, 8 full years of estimated monthly flows should be used to construct the flow-duration curve. These additional monthly flow estimates ensure that the lowest historic flow correctly influences the flow-duration curve: The lowest monthly flow in 8 years is a 99 percent condition; in 95 months out of 96, this minimum flow would be equaled or exceeded.

Calculations of a Flow-Duration Curve from Meteorologic Data

Flows of water from the land surface during and immediately following precipitation create flood hydrographs and peak flows. Water that is absorbed by the soil during rainstorms moves as subsurface flow into stream channels and provides low flows in periods when rain does not occur. Flow-duration curves are based on continuous streamflow data. On ungaged streams where streamflow measurements are not available, precipitation and potential evapotranspiration records can be used to calculate continuous flows. The calculations mimic key hydrologic processes: infiltration of water into the soil profile, surface runoff, and flow along subsurface flow paths into the stream.

The calculation method described in this section uses monthly precipitation and potential evapotranspiration data to calculate monthly streamflow.

When the calculated monthly streamflows are found, they are used like observed flows to calculate the flow-duration curve. A sketch of the calculations is shown in Fig. 3.11. The time of flow in stream channels in small watersheds is usually less than 1 day and this time of flow can be neglected when monthly interval runoff volumes are calculated. Calculations of monthly flows from meteorologic data are based on the water balance in the watershed. The water balance equation is:

$$\text{Precipitation} - \text{actual evapotranspiration} + \text{storage} = \text{runoff}$$

The water balance equation applies to the watershed over any time interval. Precipitation, actual evapotranspiration, and runoff are the volumes of water entering and leaving the watershed in the time interval, and storage is the

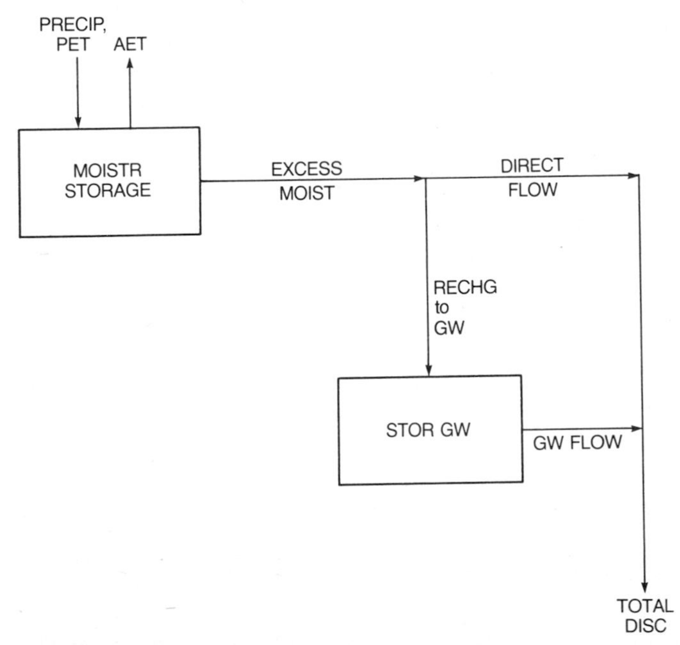

FIG. 3.11 Monthly runoff calculations from rainfall and potential evapotranspiration data.

change in soil moisture and groundwater storage in the time interval, the initial storages less the final storages. Water is held in storage in the soil, in groundwater aquifers, and in lakes and snowpacks. All water flows into or out of the watershed are assumed to be included in the runoff.

The following steps are used to calculate monthly runoff and a flow-duration curve from meteorologic data:

1. Assemble five or more years of concurrent rainfall and potential evapotranspiration data.
2. Estimate the watershed characteristics of the basin.
3. Calculate, using a tabular form, the monthly streamflows for five or more years based on these rainfall and potential evapotranspiration data.
4. Calculate the flow duration curve using the calculated monthly streamflows, as would be done if observed data were available.

A description of these tasks is in the following sections.

Rainfall and Potential Evapotranspiration Data

Streamflow calculations require data on monthly rainfall on the watershed and monthly potential evapotranspiration. Rainfall data that is observed in or near the watershed must be found. National meteorologic services or agricultural agencies maintain records of meteorologic stations. Records at individual stations are often incomplete. Records from two or three stations are needed to fill in missing records. Adjustments are usually needed to estimate watershed rainfall from precipitation-gage records. Rainfalls often increase with elevation, and gages are usually placed in villages along river valleys. Meteorologic and agricultural services prepare maps of mean annual rainfall. These maps can be used to adjust gaged rainfall and to estimate rainfall on the watershed.

Potential evapotranspiration (PET) is the amount of water that would evaporate from the watershed if water supply is ample. The actual water loss, called *actual evapotranspiration* (AET), is less than or equal to the potential evapotranspiration. The potential evapotranspiration may be estimated and published by a national meteorologic services office. Potential evapotranspiration is quite uniform from year to year. A mean monthly distribution of potential evapotranspiration will be sufficient for purposes of calculating monthly runoff.

Estimating Watershed Characteristics

The tabular calculations of monthly flows use three coefficients that represent watershed characteristics:

NOMINAL = an index to the soil moisture storage capacity in the watershed, mm (in).

PSUB = the fraction of runoff that moves out of the watershed as baseflow or groundwater flow, dimensionless.

GWF = an index to the rate of discharge from the groundwater storage to the stream, dimensionless.

NOMINAL, PSUB, and GWF are watershed characteristics that will change from one watershed to another. NOMINAL is the soil moisture storage level that permits half of any positive monthly water balance to leave the watershed as *excess moisture,* where excess moisture is either direct runoff or groundwater flow. The soil moisture storage level varies and may be less than or greater than NOMINAL. When the soil moisture storage is less than NOMINAL, the majority of any positive monthly water balance is retained in the soil moisture. When the soil moisture storage is greater than NOMINAL, the majority of any positive monthly water balance becomes direct runoff or an addition to the groundwater storage.

PSUB is the fraction of runoff that moves out of the watershed on subsurface flow paths rather than as direct or surface runoff. The total flow that a watershed provides consists of surface or direct runoff that creates the peak flows and subsurface flows that provide the low flows. Low-permeability soils that have low-infiltration capacities yield large amounts of surface or direct runoff and low sustained discharges. Soils that have high-infiltration capacities yield higher sustained discharges. Therefore streams that have high minimum discharges are those with highly permeable sandy soils and fractured or permeable subsurface geology.

GWF is an index to the time of flow along subsurface flow paths that enter the stream. It is the fraction of the total volume of water on groundwater flow paths that will enter the stream in the current month.

These watershed characteristics can be estimated for ungaged streams using the following guidelines:

$$\text{NOMINAL} = 100 + C \times \text{mean annual precipitation}$$

where C is approximately 0.2 in watersheds where precipitation occurs throughout the year and 0.25 in watersheds with seasonal rainfall. The value of NOMINAL can be reduced by up to 25 percent in watersheds with limited vegetation and thin soil cover.

$$\text{PSUB} = 0.6 \qquad \text{median value}$$

PSUB would increase to 0.8 in watersheds known to have highly permeable soils and would decrease to 0.3 in watersheds with low permeability or thin soils.

$$\text{GWF} = 0.5 \qquad \text{median value}$$

GWF would increase to 0.9 in watersheds that have little sustained flow and would decrease to 0.2 in watersheds known to have reliable sustained flows.

Watershed characteristics can be estimated from limited field data based on historic streamflows. The characteristic that controls runoff volumes is NOMINAL. Increasing NOMINAL will decrease runoff and decreasing NOMINAL will increase runoff. The actual evapotranspiration (AET) increases when NOMINAL is increased, and the AET/PET ratio in Table 3.4 will change. When some historic flow data is available, the calculations in Table 3.4 can be repeated for different values of NOMINAL until the calculated flow volume and the historic flow volumes agree. In some cases, the rainfall amounts that are assumed to occur on the watershed may need to be adjusted. A change in rainfall on a watershed of ± 10 percent will change watershed runoff by at least 10 percent.

The fraction of excess moisture that moves as subsurface flow (PSUB) and the rate of outflow of subsurface or groundwater flow (GWF) control the low streamflows between storms. A value for GWF can be found when little rainfall occurs for 2 or more months. When rainfall does not occur, GWF = 1 − (streamflow volume in current month/streamflow volume last month). Alternatively, if monthly measurements of streamflow are made during a period when rainfall does not occur, GWF = 1 − (streamflow today/streamflow 1 month ago). Calculated values for GWF will vary, so GWF should be calculated for different months when observed flows are available. The lowest calculated value of GWF is usually the most reliable estimate.

When NOMINAL and GWF are established, the value of PSUB can be considered. PSUB increases or decreases the volume of water moving on the subsurface flow paths. If, for example, the following monthly streamflows are available from field measurements and calculations, PSUB needs to be increased.

Streamflow	June	July	August	September
Observed	1,070	642	385	231
Calculated	720	432	259	155

In this example, the calculated flows are all too low and more subsurface flow volume is needed. Note that GWF for both the calculated and the observed monthly flows is 0.4, so changing GWF would not improve the calculated streamflows.

The trial calculations to establish watershed characteristics that are outlined above can be carried out for nearby streams and the watershed characteristics that are found may be assumed to apply to the ungaged stream that is being studied. Watershed characteristics usually change only moderately in a region and changes will correlate with changes in vegetation, soils, and subsurface geology.

Calculations of Monthly Runoff

Calculations of monthly runoff are made using a tabular sheet, Table 3.4. A step-by-step procedure for computing data in the table follows. These calculations can be done with a hand calculator or adding machine, but a flowchart for a FORTRAN program for handling the calculation is given in Figure 3.12.

Initial Steps

Select the values of the coefficients representing watershed characteristics and enter them in the column headings: NOMINAL above column (5), PSUB above column (13), and GWF above column (16).

Initial or starting conditions are needed for the soil moisture storage [column (4)] and the groundwater storage [column (14)]. If rainfall is seasonal and the tabular calculation begins in the dry season, the initial storages will be low. Typical values would be an initial soil storage of 10 percent of

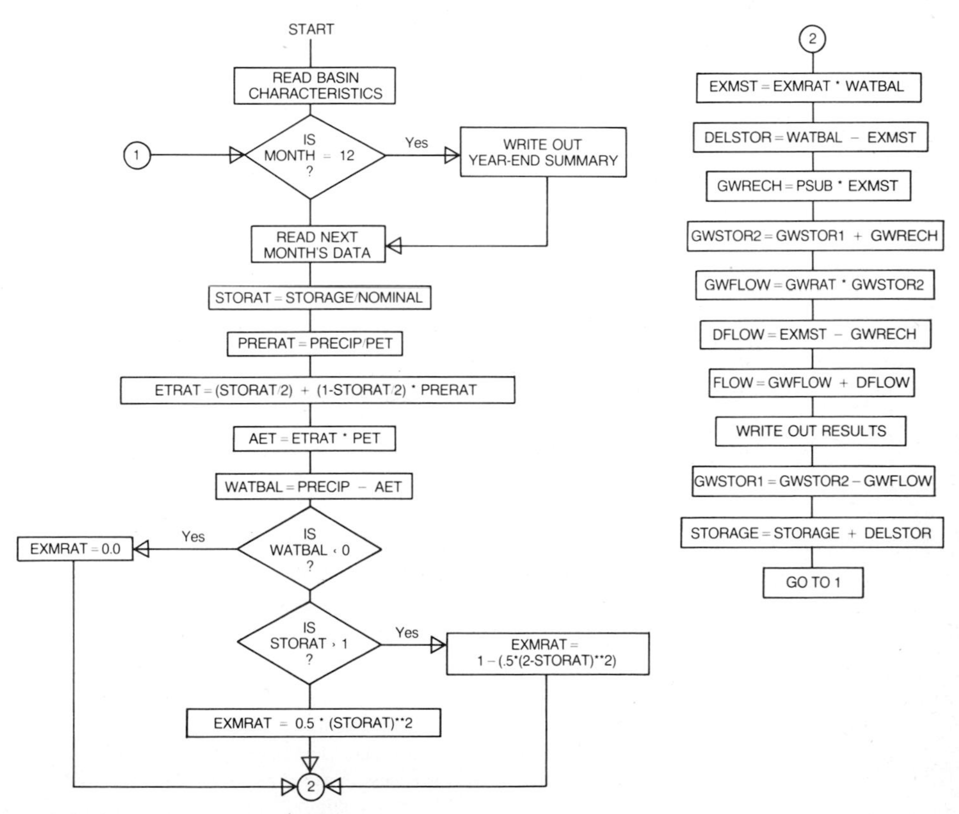

FIG. 3.12 Flowchart for flow-duration curve.

TABLE 3.4 Flow-Duration Model

(1) DATE, mo/yr	(2) PRECIP, mm	(3) PET, mm	(4)* MOIST STORAGE, mm	(5) STOR RATIO	(6) PRECIP/ PET	(7) AET/ PET	(8) AET, mm	(9) WATER BALANCE, mm
1/80	356.3	21.7	500.0†	1.22	16.42	1.00	21.7	334.6
2/80	196.4	38.4	601.9	1.47	5.12	1.00	38.4	158.1
3/80	145.6	79.1	624.3	1.52	1.84	1.00	79.1	66.5
4/80	59.8	118.3	631.9	1.54	.51	.89	104.9	−45.1
5/80	26.9	155.4	586.7	1.43	.17	.76	118.8	−92.0
6/80	11.6	171.5	494.8	1.21	.07	.63	108.1	−96.5
7/80	19.9	191.6	398.3	.97	.10	.54	103.3	−83.4
8/80	33.7	154.7	314.9	.77	.22	.52	80.2	−46.5
9/80	17.2	137.9	268.5	.65	.12	.41	56.7	−39.5
10/80	299.6	88.9	228.9	.56	3.37	1.00	88.9	210.7
11/80	275.8	41.7	406.8	.99	6.61	1.00	41.7	234.1
12/80	350.0	29.8	525.7	1.28	11.74	1.00	29.8	320.2

*NOMINAL = 410.
†Initial conditions MOIST STORAGE = 500, BEGIN STOR GW = 25.0.
‡PSUB = 0.61.
§GWF = 0.64.
Note: EXCESS MOIST RATIO in col. (10) was entered two significant figures; EXCESS MOIST in col. (11) was calculated on a small calculator that retained more than two significant figures for the EXCESS MOIST RATIO

Column	Comment
(1)	Enter the month and year of the data.
(2)	Enter precipitation (PRECIP) on the watershed for the month (gage rainfall must be adjusted to represent watershed rainfall).
(3)	Enter potential evapotranspiration (PET) on the watershed.
(4)	MOIST STORAGE was entered previously as an initial condition or is MOIST STORAGE plus DELTASTORAGE, column (4) plus column (12) from the prior month calculation.
(5)	Calculate the soil storage ratio (STOR RATIO), the value in column (4) divided by NOMINAL.
(6)	Calculate the ratio PRECIP/PET, column (2) divided by column (3). Enter the result in column (6).
(7)	Enter Fig. 3.14 with PRECIP/PET and the STOR RATIO [column (5)], and find the value of the ratio of actual to potential evapotranspiration, AET/PET. Enter the value of AET/PET in column (7).
(8)	Calculate AET as PET multiplied by the AET/PET ratio, column (3) times column (7), and enter AET in column (8).
(9)	Calculate the WATER BALANCE for the month where WATER BALANCE = PRECIP − AET [column (2) less column (8)]. Enter the result in column (9).
(10)	If the WATER BALANCE is positive, enter Fig. 3.13 with the STOR RATIO [column (5)], and find the excess moisture ratio (EXCESS MOIST RATIO). If the WATER BALANCE in column (9) is negative, the EXCESS MOIST RATIO is zero. Enter the EXCESS MOIST RATIO in column (10).

(10) EXCESS MOIST RATIO	(11) EXCESS MOIST, mm	(12) DELTA STORAGE, mm	(13)‡ RECHG TO GW, mm	(14)§ BEGIN STOR GW, mm	(15) END STOR GW, mm	(16) GW FLOW, mm	(17) DIRECT FLOW, mm	(18) TOTAL DISC, mm
.70	232.7	101.9	141.9	25.0†	166.9	106.8	90.7	197.6
.86	135.7	22.4	82.8	60.1	142.9	91.4	52.9	144.4
.89	58.9	7.6	35.9	51.4	87.4	55.9	23.0	78.9
.00	.0	−45.1	.0	31.5	31.5	20.1	.0	20.1
.00	.0	−92.0	.0	11.3	11.3	7.2	.0	7.2
.00	.0	−96.5	.0	4.1	4.1	2.6	.0	2.6
.00	.0	−83.4	.0	1.5	1.5	.9	.0	.9
.00	.0	−46.5	.0	.5	.5	.3	.0	.3
.00	.0	−39.5	.0	.2	.2	.1	.0	.1
.16	32.8	177.9	20.0	.1	20.1	12.9	12.8	25.7
.49	115.2	118.9	70.3	7.2	77.5	49.6	44.9	94.5
.74	237.7	82.5	145.0	27.9	172.9	110.6	92.7	203.3

Column	Comment
(11)	Calculate the excess moisture (EXCESS MOIST) where EXCESS MOIST is the EXCESS MOIST RATIO times the WATER BALANCE, column (10) times column (9). Enter the result in column (11).
(12)	Calculate the change in soil storage (DELTA STORAGE) where DELTA STORAGE is WATER BALANCE minus EXCESS MOIST, column (9) minus column (11). Enter the result in column (12).
(13)	Calculate the recharge to groundwater storage (RECHG TO GW) as PSUB times EXCESS MOIST, PSUB times column (11). Enter the result in column (13).
(14)	BEGIN STOR GW was entered previously as an initial condition or is END STOR GW less GW FLOW, column (15) less column (16), from the prior month calculation.
(15)	Calculate the end of month groundwater storage (END STOR GW) by adding the RECHG TO GW in column (13) to the BEGIN STOR GW in column (14). Enter the result in column (15).
(16)	Calculate the groundwater discharge to the stream (GW FLOW) where GW FLOW is GWF times END STOR GW or GWF times column (15). Enter the result in column (16).
(17)	Calculate the direct runoff to the stream (DIRECT FLOW), where DIRECT FLOW is EXCESS MOIST less RECHG TO GW, column (11) minus column (13). Enter the result in column (17).
(18)	Calculate the streamflow as the DIRECT FLOW plus GW FLOW, column (17) plus column (16). Enter the result in column (18). This result is in millimeters and can be converted to cubic meters for the month by multiplying by the watershed area in square kilometers times 1000.

NOMINAL, and groundwater storage of 5 percent of NOMINAL. In watersheds with uniform rainfall throughout the year, an initial soil storage equal to the NOMINAL, and an initial groundwater storage of 20 percent of NOMINAL would be expected.

Calculations that begin in the wet season would have an initial soil storage of 125 percent of NOMINAL and an initial groundwater storage of 40 percent of NOMINAL.

The effects of initial storages in a water-balance calculation "damp-out" over a period of 6 to 12 months. There are two ways of avoiding bias caused by unknown initial storages: The first year of calculated monthly flows could be ignored and not included in the flow-duration curve. Alternatively, the values of soil and groundwater storage that are found at the end of the first year of calculations could be accepted as representative of typical storages at that time of year. These storages could then be entered as the best estimate of initial conditions for the watershed and calculations could be redone using these storages.

Enter the initial soil moisture in column (4), and the initial groundwater storage in column (14).

Reset Initial Conditions for the Next Month

The BEGIN STOR GW for the next month is END STOR GW less GW FLOW, column (15) minus column (16). Enter this result in column (14) for the next month.

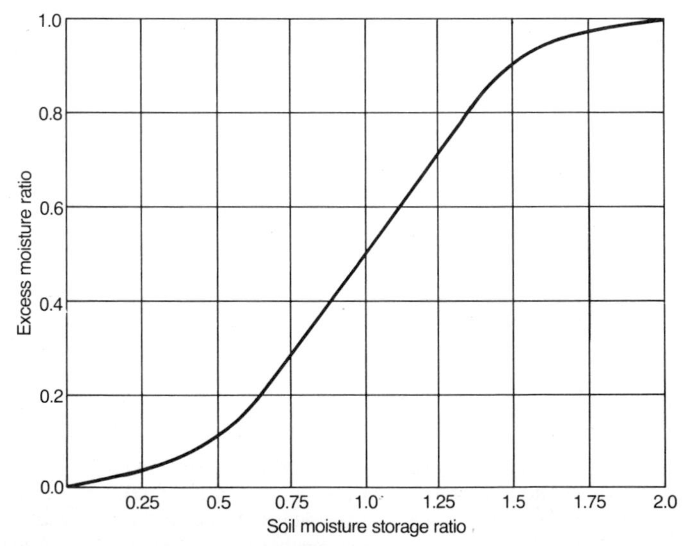

FIG. 3.13 Excess moisture ratio.

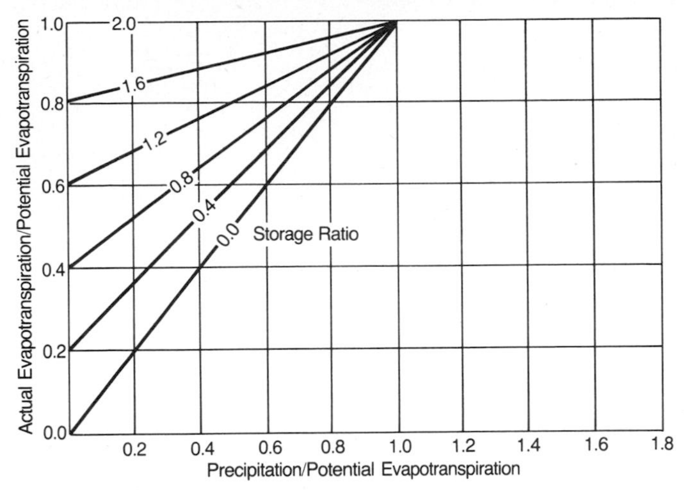

FIG. 3.14 Ratio of AET/PET.

The beginning MOIST STORAGE for the next month is the current MOIST STORAGE plus DELTA STORAGE, column (4) plus column (12). Enter this result in column (4) for the next month.

The tabulation steps are repeated until all the months of precipitation and potential evapotranspiration have been considered. An example of a completed table for monthly flows is given in Table 3.4. Note that in Table 3.4 the EXCESS MOIST RATIO in column (10) has been entered as two significant figures. The EXCESS MOIST in column (11) was calculated on a small calculator that retained more than two significant figures for the EXCESS MOIST RATIO.

Determination of the Flow-Duration Curve

The procedure to draw a flow-duration curve was described in the previous section. To make a flow-duration curve from calculated monthly flows, list all the monthly flows which were computed in column (18) of Table 3.4. The flows should be listed in order of magnitude. Next, calculate the percent of months each flow volume is equaled or exceeded. Plot the flow-duration curve as shown in Fig. 3.5.

3.5 REFERENCE

1. N. H. Crawford and S. M. Thurin, *Hydrologic Estimates for Small Hydroelectric Projects,* Hydrocomp, Inc., Mountain View, Calif., May 1981.

4

Site Development and Hydraulic Analysis

John J. Cassidy

A project site, because of its topography and geology, indirectly dictates the overall project configuration. Each will have physical characteristics that are most economically and, thus, most easily utilized in a particular design. This chapter considers the investigations, analyses, and design determinations which must be completed in formulating a design concept for a particular site. The necessary considerations are numerous, but as in other chapters, only those pertaining to the development of small hydropower systems are included. Since economic considerations of site development are covered in Chap. 11 only the physical aspects are considered here. Nevertheless, planners and designers must work hand in hand to assure that the site potential is developed as economically as possible while maintaining the necessary operational reliability.

Once the final project configuration and size has been chosen, the necessary technical analyses follow directly. However, the ingenuity and imagination exercised during conceptual design will ultimately determine the project's economic advantages. A wise project manager should aggressively seek review and evaluation during this conceptual design period in order to ensure that the simplest and most economical configuration is identified and developed.

4.1 SITE CONFIGURATIONS

Small hydropower projects can be either high-head or low-head depending on power requirements and physical characteristics of the available site. In general, high-head sites will be less expensive to develop than low-head sites because, for a given capacity, the turbine and required structures will be smaller.

High-Head Sites

Small high-head hydropower projects will almost always utilize a penstock or tunnel to develop the necessary head. Fig. 4.1 shows a typical configuration where a canal is utilized to convey water diverted from the river to the upper end of the penstock. Since the canal would be expensive and difficult to construct on a steep canyon wall, its use is best suited where the valley wall has a natural bench or at least a moderate slope toward the river. Concrete or wooden flumes have been constructed on very steep canyon walls. However, the construction as well as maintenance will be prohibitive unless the site has other features which provide for extraordinary economy. A small diversion dam is generally required unless the flow rate to be diverted is a very small part of the low flow of the river. The diversion dam may be constructed of concrete or masonry and will be designed for overflow. Where a significant part of the low flow of the stream must be diverted, stop logs may be placed temporarily along the top of the dam.

Proper location of the diversion dam is critical. In order to minimize problems created by sediment conveyed by the river, the dam should preferably be located on a straight stretch of the river. If necessary, the dam may be located

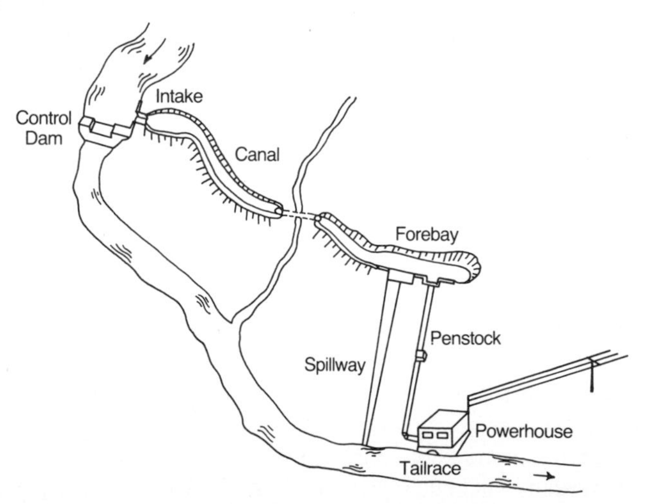

FIG. 4.1 Typical configuration for a high-head site.

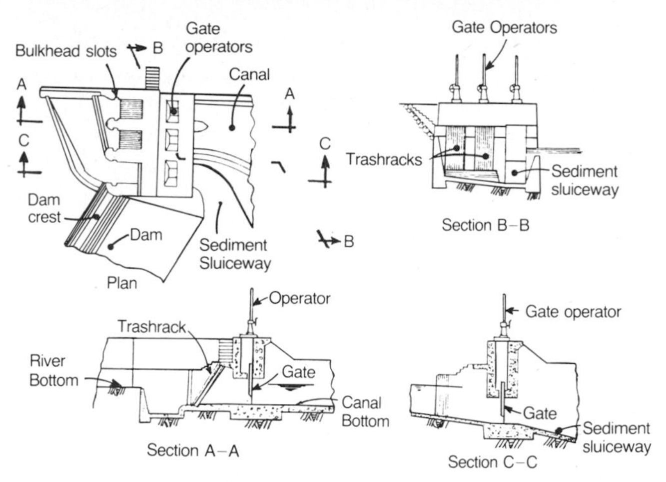

FIG. 4.2 Typical arrangement for a diversion dam and power canal intake.

on a bend. Because the secondary currents in a river bend tend to deposit sediment at the inside of the bend, the canal intake should be located on the outside of the bend. The intake should never be located on the inside of a bend, since sediment carried by the stream (particularly bedload) will be deposited against the intake resulting in continuous blocking problems and continuous cleaning operations.

The outside of a river bend generally has a steep bank produced by continuous erosion. A thorough study of the river's stability should be conducted prior to choosing a diversion site. If there is evidence that the river's location is unstable, expensive stabilization measures may be required to prevent the river from creating serious erosion problems around the intake, or more importantly, from developing a new channel during a flood and rendering the intake and diversion dam useless. Evidence of instability of rivers can be found in the form of old abandoned channels which are readily observed from the air. The ideal diversion site will be one where the river channel is stabilized by rock. A rock foundation is also highly desirable for a stable foundation for the diversion dam.

If a concrete diversion dam is used, the power-canal intake structure can be integrated with the dam. Figure 4.2 shows a typical structure and diversion dam. A depressed channel with a controlling gate should be provided to evacuate sediment deposited in front of the intake gates. Reduced velocities in the river, produced by the diversion dam, will always result in some sediment deposition in front of the intake structure, reducing the discharge capacity of the intake gates. Incorporation of a sediment sluicing structure will minimize operational problems.

Ideally the power canal should be oriented at nearly 90° to the river to minimize the need for structural measures to ensure separation of the canal and river downstream from the diversion dam. In a narrow valley, however, it may be necessary to direct the canal nearly parallel to the river. In such cases, stability of the soils in the vicinity of the river is very important. It may be necessary to construct a retaining wall downstream between the canal and river to provide for integrity of the canal. It may also be necessary to line the canal with concrete for some distance to eliminate seepage with resulting soil saturation and resulting slope instability.

Gates should be provided at the intake to the power canal to control flood-induced flows entering the canal. If the canal is short, downstream control is provided, and sufficient freeboard has been allowed to minimize the effects of the flood flows, mechanical gates may not be required. It will still be necessary to occasionally stop flow into the canal for maintenance and inspection. Thus, provision should be made for closure at the intake. Wooden, steel, or concrete stop logs can be used for this closure if the canal is small. However, it is very difficult, and sometimes impossible to remove or install stop logs under an unbalanced head. In addition, because of their infrequent use, stop logs are often appropriated for other uses with the result that they are not available in an emergency.

The power canal may be lined or unlined depending upon the native soil. If soils are quite porous, such as gravels and sands, a lining may be required to prevent excess seepage and possible piping of the canal banks.

Immediately downstream from the gate structure, the canal should be widened and deepened to provide a settling basin for suspended sediment entering the canal. In addition, the enlarged section allows for dissipation of energy in the high-velocity flow leaving the gates during partial opening.

Slope and cross-sectional area of the canal must be jointly established to convey the necessary maximum flow rate to the turbines. During steady-state operation of the canal, the water surface within the canal will be parallel to the canal bottom. However, during start-up or shutdown of the turbines, unsteady flow will develop in the canal. If the turbines are shut down before the gates at the upstream end of the canal are closed, a positive surge will travel upstream and the canal will continue to fill to overflowing. When the turbines are started up there is a tendency to unwater the canal until steady-state flow is established. Thus, there is need to provide for a reasonable freeboard in canal operation. An enlarged section should be constructed at the downstream end of the canal to provide extra flow volume during start-up. Figure 4.3 shows a typical canal section and a typical expanded section at the downstream end of the canal.

Cross drainage must be provided throughout the length of the canal. Storm flow from very small cross-drainage channels can be directed into the canal provided large quantities of debris will not also be brought into the canal. Spillway sections must be provided at strategic locations on the canal to provide for flood discharge. The spillways also function to keep the canal from overflowing if the turbines are shut down, but flow into the canal continues. These spillways

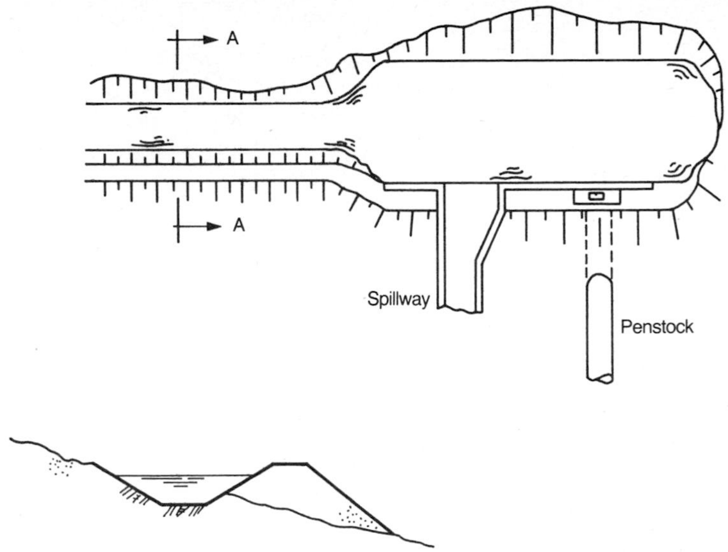

FIG. 4.3 Typical canal section and penstock forebay. Plan (top) and section A-A (bottom).

should be located where downstream erosion will not endanger the canal or other structures which may be along the river below.

Diversion to the penstock should be made through a gate structure constructed in the side of the canal, as in Fig. 4.3. From a strictly hydraulic standpoint, having the gate structure at the end of the canal is more desirable since there is less tendency to form an air-entraining vortex at the gate. Locating the gate structure on the canal bank provides easier access for equipment used in cleaning debris out of the canal. In addition, large floating objects, such as logs, cannot strike the gates directly when gate structures are located on the canal bank. Vortices at the gate entrance can be prevented by proper submergence.

Penstocks will most likely be constructed from steel. For very small plants consideration should be given to modern plastic pipe. Wooden penstocks with steel reinforcing rings have been used extensively in older plants, and where available are entirely satisfactory. The penstock should be constructed in as straight a line as possible. Wherever changes in direction are required, concrete thrust blocks will be needed to overcome the unbalanced forces produced by the bend.

Structural strength of the penstock must allow for both static pressure and dynamic pressure. In addition an allowance must be made for corrosion with time. The fewest problems arise in the steepest penstocks. If the penstock is quite steep, a surge tank can be provided near the powerhouse to protect the penstock from rupturing during sudden shutdown or from collapse during sudden start-up. An air vent should be provided at the upper end of the penstock to prevent collapse of the penstock if the gates at the end of the penstock are

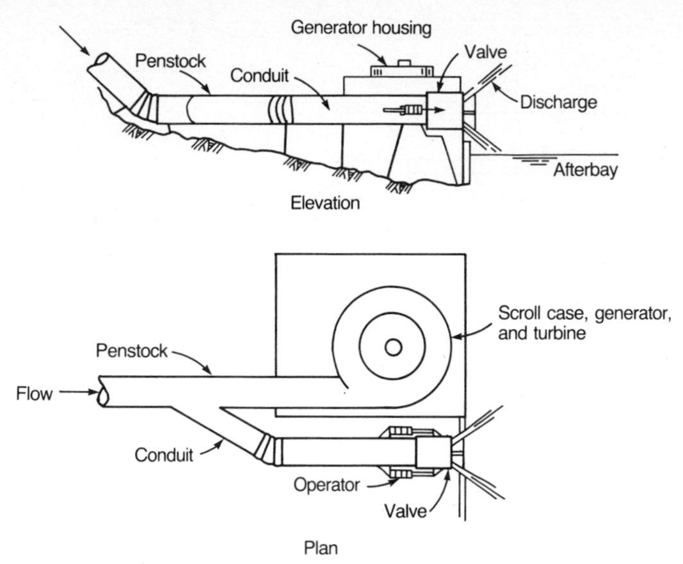

FIG. 4.4 Arrangement of a synchronous bypass valve.

closed and the control valve at the turbine is opened. An alternative is to make the penstock strong enough to be safe under these emergency stresses. However, that will generally be uneconomical unless the penstock is very short or very small in diameter.

Protection against both overpressure and underpressure in a penstock can be provided by frangible disks located so that the penstock is vented when the disk ruptures. However, if a ruptured disk is inadvertently replaced by one which is too strong, the penstock could be seriously damaged during the next unexpected shutdown.[1]

Instantaneous bypass valves can also be arranged to pass flow around the turbine. They are arranged and controlled so that flow is slowly established in the penstock by opening the valve prior to starting up the turbine. As the turbine is brought up to speed, the valve closes at a synchronized rate to maintain constant flow in the penstock. For shutdown the process is reversed. Figure 4.4 shows a typical arrangement for the bypass valve. In theory the synchronous bypass valve can control the penstock pressures perfectly. In practice, however, the controls may be electronically complex calling for very specialized maintenance.

Tunnels

Some site configurations can be more readily adapted to a tunnel than to a supply canal. However, tunneling is always much more expensive and the decision should be carefully analyzed. In general for a tunnel to be economical the designer must establish that:

1. The geologic formation through which the tunnel will pass is stable.
2. The tunnel would be much shorter than an alternative canal.
3. The construction of the canal would be prohibitively expensive because of nearly vertical canyon walls, unstable soils, or other natural barriers.

Modern tunneling methods utilize tunnel boring machines in almost any kind of material. However, diameters of at least 8 ft (2.44 m) are required. In some regions where labor is not expensive, excavation of the tunnel by drilling and blasting may be economically feasible.

If the native materials are such that a structural lining of the tunnel is required, the cost of the tunnel will increase greatly. If no lining is used, provision must be made for access and inspection of the tunnel on at least an annual basis to be certain that the tunnel walls remain stable and do not spall or tend to collapse.

Needless to say, good geologic exploration is mandatory when planning for a tunnel. However, some uncertainty will always remain about geologic conditions along the tunnel centerline until excavation is complete. Thus, even after design has been completed it is important to continue the geologic assessment throughout construction.

Low-Head Sites

Small low-head hydroelectric sites generally can be classified into two types. The first type uses a diversion dam and power canal and is very similar to the configuration shown in Fig. 4.1. However, the canal will generally be short and the penstock either nonexistent or very short.

The second type of low-head configuration involves a dam with an integral powerhouse. For equal power capacity, this second configuration is usually more expensive because all structures will need to be more substantial for safety during flooding. However, with increasing energy costs, more dams, originally constructed without generating facilities, are being retrofitted for power generation.

4.2 TOPOGRAPHIC SURVEYS

Topographic survey requirements vary significantly with the phase of investigation. For reconnaisance-level studies precise topographic detail is not required. In general, elevations and horizontal distances need to be accurate to only ±5 ft (±1.5 m). For feasibility studies elevations and horizontal distances should be known to at least ±1 ft (±300 mm). If right-of-way must be purchased, a precise survey, accurate to ±0.1 ft (±30 mm), may be required prior to construction. Any survey should show property boundaries, true elevations, bearings, and distances. Surveys for canal alignments should show the location of major drainage channels and provide sufficient information for calculation of their drainage areas. A topographic map is a necessity for all levels of planning. If an existing topographic map is used, it should be checked against known features in the field to verify that bearings, distances, and elevations have been accurately portrayed.

4.3 REQUIRED GEOLOGIC INVESTIGATIONS

A small hydroelectric development will generally have the same design considerations inherent in a large project. However, it can rarely afford a large budget for geologic exploration and mapping. Structural loadings for a small project will be relatively small, and the required structural strength of the native geological formations and soil formations will also be relatively small. Stability of the soils and geologic features cannot be compromised however. Thus, it is necessary to develop a geologic map in as complete detail as possible. Observations, recording of the observations, and interpretation must be done carefully under the supervision of an experienced engineering geologist. The geologic map produced will ultimately be used in layout and design of the project. Thus, the map must include *all* identifiable geologic characteristics and distinguish carefully between features that have actually been observed and those that are interpretations. Layout of the project will identify those areas where more detailed field study is required.

Structural integrity will be most important for the turbine and dam foundations and beneath the penstock anchorages. However, stability of slopes will be important throughout the project. Significant geological information can be obtained from surface observations. All rock outcroppings should be examined and rock type, strike and dip of the formation, and the presence or lack of joints should be noted. Ravines and streambeds tend to expose the subsurface features and should be carefully studied throughout the project area.

Features such as rodent holes, mine shafts, or tunnels, which could contribute to leakage at the dam or along the canal route, should be carefully noted and studied for additional subsurface information. Water wells in the area will also provide information on the subsurface geology as well as on groundwater levels and foundation permeability. Such data may be of particular use in interpreting geology and in assessing the potential for seepage and drainage problems.

Vegetation types in the area should also be carefully observed: They may provide information on soil depth, or since some species have a preference for particular soils, they may help identify the presence of a particular geologic unit. Soil depth and type will be most important along the canal route but must be carefully checked wherever excavation is considered.

If a power tunnel is considered, geologic cross sections along the tunnel alignment must be developed to decide whether the tunnel must be lined, what excavation methods can be used, and, most important, whether the tunnel is feasible from a technical standpoint. Studies must include a careful examination to locate any faults in the area, their extent, and history. Even with very careful surface exploration and interpretation, the subsurface structure will only be fully revealed through tunnel excavation. Hence, design changes may be required as tunnel excavation progresses and this possibility should be recognized during planning and design.

Geologic and soils exploration will also identify potential areas for rock quarries for concrete aggregate or riprap and borrow areas for soil to be used in embankments. Preliminary estimates of quantities available and of potential maximum rock sizes obtainable from the quarry will also be valuable for project planning. Estimates should also be made of maximum safe-cut slopes for both temporary and permanent excavations. Since most landslides occur when the bedding is undercut or when some exposed geologic feature, such as a fault or a joint, is adverse, these features must be identified as completely as possible during the geologic exploration.

4.4 HYDRAULIC DESIGN AND ANALYSIS

The hydraulic features of a small hydroelectric development are as numerous as those of any large one. In general, spectacular failures due to cavitation, such as those which have occurred on the Boulder Dam and Yellowtail Dam spillways,[2] are very unlikely. However, the loss of energy due to poorly designed flow passages may be relatively more important in a small hydro development. Design philosophy for small hydro facilities is not different from that in any other engineering-design situation: An adequate design can be formulated provided all possible failure modes can be anticipated. In developing this section on hydraulic design and analysis, that philosophy has been followed in selecting the specific items to be covered.

Dams and spillways, diversion structures, gates, canals, tunnels, penstocks, and surge tanks all involve elements of hydraulic design and analysis. Calculation of power potential is repeated here as

$$P_{\mathrm{kW}} = \frac{(0.746)\,Q\gamma H\eta}{550}$$

in the U.S. Customary System or as

$$P_{\mathrm{kW}} = \frac{\rho g Q H \eta}{1{,}000}$$

in the SI metric system

where

P = power, kW
Q = flow rate through turbine, ft^3/s (m^3/s)
H = head drop through turbine, ft (m)
η = plant efficiency
γ = specific weight of water, 62.4 lb/ft^3
ρ = density of water, 1,000 kg/m^3
g = acceleration due to gravity, 9.81 m/s^2

If electrical power output is desired, η must include the combined efficiency of turbine, generator, and transformer. Typical values of those individual efficiencies are 0.8, 0.9, and 0.9 respectively. The product of the three, or the overall efficiency will be approximately 0.6 to 0.7 in most cases.

Calculating the net head H will involve determination of all system losses. Figure 4.5 shows schematically the energy gradient through the system. As shown, the net head H is equal to the difference between the static water-surface elevations upstream and downstream minus the sum of all losses. Hydraulic losses must include that due to resistance to flow in uniform straight sections as well as additional turbulence losses contributed by pipe bends or other appurtenances which cause flow disturbances.

Steady Flow in Open Channels and Canals

Selection of the slope of the power canal (or tunnel flowing as an open channel) and the dimensions of the cross section are closely related. The hydraulically most efficient cross section will be one which minimizes the hydraulic radius R (ratio of cross-sectional area to wetted perimeter). However, practical consid-

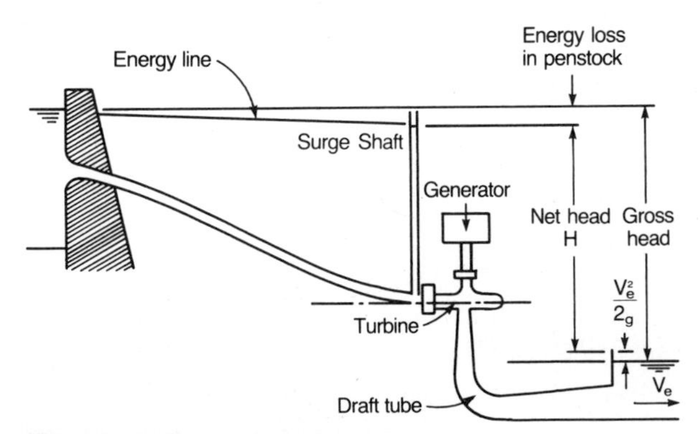

FIG. 4.5 Definition sketch for net head.

TABLE 4.1 Values of n for Use in Equations (4.1) and (4.2)

Value of n	Condition of channel
0.011–0.012	Smooth clean wood, metal, or concrete surfaces, without projections, and with straight alignment
0.013	Smooth wood, metal, or concrete surfaces without projections, free from algae or insect growth, mostly straight alignment
0.014	Good wood, metal, or concrete surfaces with only small projections, with some curvature, slight algae or insect growths, and minimal gravel or sand deposition. Troweled gunnite surfaces
0.015	Wood with algae and moss growth, concrete with smooth sides and rough bottom, metal with shallow projections
0.016	Metal flumes with large projections. Wood or concrete with well-developed moss or algae growths
0.017	Rough concrete or untroweled gunnite surfaces
0.018	Smooth natural earth channels, free from growths with straight alignment
0.019–0.020	Smooth natural earth, free from growths with some curvature. Large canals in good condition
0.020–0.025	Small canals in good condition, larger canals with some growth
0.025–0.035	Canals with dense aquatic growths

erations, such as necessary width to operate excavation equipment, may dictate the use of a wider section.

The flow rate Q (m^3/s or ft^3/s) which will pass through a canal with cross-sectional area A (m^2 or ft^2), hydraulic radius R (m or ft), and slope S is given by Eq. (4.1).

$$Q = \frac{1.5}{n} AR^{2/3}S^{1/2} \qquad \text{USCS units} \tag{4.1}$$

$$Q = \frac{1}{n} AR^{2/3}S^{1/2} \qquad \text{SI units} \tag{4.2}$$

where n is the Manning roughness coefficient. Values of n vary from 0.010 for very smooth surfaces to 0.05 for natural streams with rocky beds.

Table 4.1 gives accepted values of n for lined and unlined channels. Solutions of Eqs. (4.1) and (4.2) involve trial and error if water depth is required and channel slope, shape, width and roughness, and flow rate are known. Figure 4-6 shows typical canal cross sections and the relationships for hydraulic radius, area, and wetted perimeter.

Considerable convenience in solving Eqs. (4.1) or (4.2) is achieved for particular cross sections if the equations are rearranged. For the rectangular channel of Figure 4.6

$$\frac{Qn}{S^{1/2}} = 1.5bb^{2/3}D\left(\frac{D}{b + 2D}\right)^{2/3}$$

$$= 1.5b^2b^{2/3}\frac{D}{b}\left(\frac{D/b}{1+2D/b}\right)^{2/3}$$

$$\frac{Qn}{S^{1/2}b^{8/3}} = 1.5\frac{D}{b}\left(\frac{D/b\cdot}{1+2D/b}\right)^{2/3} \quad \text{(USCS units)} \tag{4.3}$$

$$\frac{Qn}{S^{1/2}b^{8/3}} = \frac{D}{b}\left(\frac{D/b}{1+2D/b}\right)^{2/3} \quad \text{(SI units)} \tag{4.4}$$

Since the right-hand side of Eqs. (4.3) and (4.4) are functions only of D/b, a table of values of $Qn/S^{1/2}b^{8/3}$ may be formed using a range of values of D/b. If a channel width b is assumed, the depth can be calculated using that table. This procedure can be followed for other cross-sectional shapes as well. Table 4.2 is such a tabulation for trapezoidal sections with side slopes ranging from $Z = 0$ (rectangular section) to $Z = 2$ (side slope about 27° from horizontal).

Energy Losses in Open Channels

Calculation of the total net head available for energy production will require careful considerations of energy losses through the hydraulic system. Losses due to friction in uniform flow can be calculated indirectly using Eqs. (4.1) and (4.2) since for steady uniform flow the slope of the energy gradient will be parallel to the canal bottom. Thus, the total energy loss h_l equals the total fall of the canal or

$$h_l = LS \tag{4.5}$$

where L is the canal length and S is the bottom slope.

However, consideration should be given to other so-called minor losses. Following is a selection of equations for calculating energy losses which arise as a result of flow through bends, expansions, contractions, and other special structures or geometric changes. Values reported by various investigators vary somewhat, and point to the fact that the exact values depend on the exact geometry of the canal.[3] The values given below represent reasonably average values.

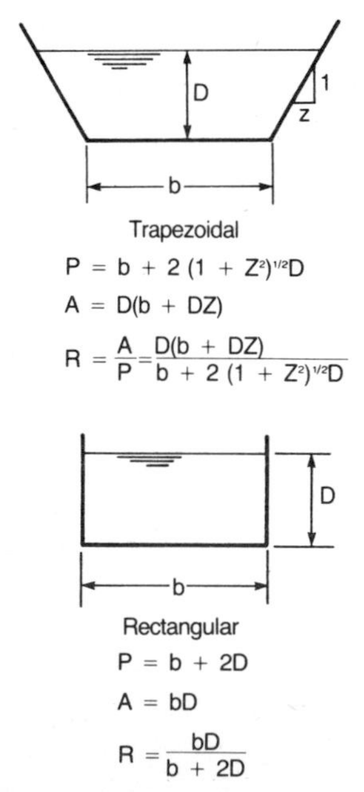

FIG. 4.6 Typical canal sections and hydraulic properties.

Losses at Bends:

$$h_B = SL_B + 2\frac{b}{r}\frac{v^2}{2g} \tag{4.6}$$

where

S = longitudinal slope of canal
b = bottom width of canal
r = radius of curvature of bend
L_B = centerline length of bend
v = average velocity of flow

Losses at Canal Entrances:

$$h_e = 0.05 \frac{v_2^2}{2g} + \frac{v_2^2 - v_1^2}{2g} \tag{4.7}$$

where

v_2 = velocity in canal
v_1 = velocity upstream from canal entrance

Losses Due to an Expansion:

$$h_{ex} = \left(1 - \frac{A_1}{A_2}\right)^2 \frac{v_1^2}{2g} \tag{4.8}$$

where

A_1 = upstream flow area
A_2 = downstream flow area
v_1 = velocity upstream from expansion

Losses in a Contraction:

$$h_c = 0.2 \frac{v_2^2}{2g} \tag{4.9}$$

where v_2 is the downstream velocity.

TABLE 4.2 Values of $Qn/b^{8/3}S^{1/2}$ for Determining Uniform Flow Depth in Trapezoidal Channels*

	Values of $\frac{Qn}{b^{8/3}S^{1/2}}$				
D/b	$z = 0$	$z = ½$	$z = 1$	$z = 1½$	$z = 2$
.05	.00947	.00980	.0100	.0102	.0103
.10	.0283	.0305	.0318	.0329	.0339
.15	.0528	.0585	.0628	.0662	.0692
.20	.0813	.0932	.102	.110	.116
.25	.113	.133	.150	.163	.176
.30	.146	.179	.205	.227	.248
.35	.181	.230	.270	.303	.334
.40	.218	.286	.341	.389	.433
.45	.256	.346	.422	.487	.548
.50	.295	.411	.512	.599	.679
.60	.375	.556	.717	.858	.988

*Adapted from Ref. 9.

Losses through a Trashrack: For rectangular bars free of debris

$$h_t = 2.3 \left(\frac{t}{b}\right) \sin \theta \left[\frac{v^2}{2g}\right] \tag{4.10}$$

where

t = thickness of bars
b = spacing of bars
θ = angle of inclination of rack from horizontal
v = velocity of flow approaching rack

Unsteady Flow in Open Channels

As pointed out earlier, flow in the power canal will take time to respond to a change in turbine flow rate. Thus, if the canal is operating at a steady flow rate and the flow to the turbines is stopped suddenly, a positive surge will move upstream in the canal. Figure 4.7 shows this surging action schematically. Momentum and continuity considerations can be used to evaluate the height of surge y and its velocity c, in accordance with Eqs. (4.11) and (4.12) for the particular case of a rectangular channel.

$$y_2 = \frac{y_1}{2} [\sqrt{1 + 8(v_1^2/gy_1)} - 1] \tag{4.11}$$

$$C = \sqrt{gy_1} \left[\left(\frac{y_2}{2y_1}\right)\left(1 + \frac{y_2}{y_1}\right)\right]^{1/2} \tag{4.12}$$

where g = acceleration due to gravity and C, y_1, and y_2 are shown in Fig. 4.7.

For a trapezoidal canal, the equations of momentum and continuity are somewhat more complicated, but Table 4.3 gives values of depth y_2 in terms of the upstream depth y_1, the channel side slopes z, and the total upstream specific energy $H_1 = y_1 + v_1^2/2g$.

Sufficient freeboard should be provided to allow for the formation of this surge at the downstream end of the canal. However, additional freeboard may be required to allow for wind-generated waves, sedimentation, or excess storm flows. This freeboard should not be less than 0.5 ft (0.2 m) for a canal carrying 10 ft^3/s (0.3 m^3/s) or 1.0 ft (0.3 m) for a canal carrying 100 ft^3/s (2.8 m^3/s).

Eventually as the surge progresses upstream in the canal, the growing down-

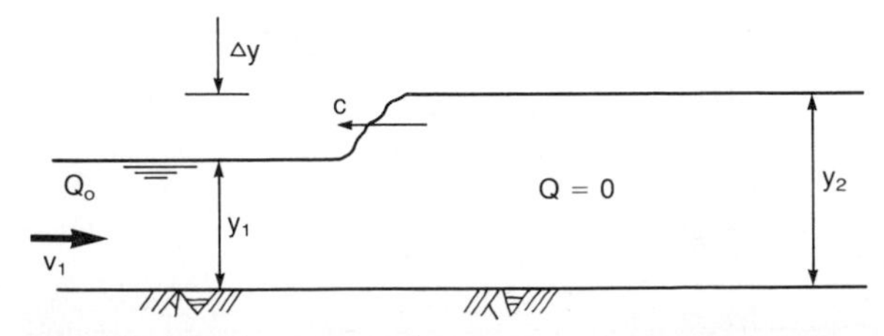

FIG. 4.7 Positive surge moving upstream in a power canal.

TABLE 4.3 Downstream Depths for a Surge in a Trapezoidal Channel*

	Values of y_2/H_1				
y_1/H_1	Triangular sections, $\frac{b}{zy_1} = 0$	$\frac{b}{zy_1} = 1$	$\frac{b}{zy_1} = 4$	$\frac{b}{zy_1} = 20$, $\frac{zy_1}{b} = 0.05$	Rectangular sections, $\frac{zy_1}{b} = 0$
0.10	.370	.418	.476	.528	.552
0.20	.562	.616	.664	.696	.706
0.30	.694	.738	.766	.778	.780
0.40	.790	.814	.820	.808	.800
0.50	.852	.852	.828	.796	.780
0.60	.877	.846	.792	.742	.724
0.70	.866	.796	.712		
0.80	.800	.750	.706	.677	.667

*Adapted from Ref. 9.

stream volume of water will overtop the enbankment unless an uncontrolled spillway is provided. The capacity of the spillway must be equal to the canal capacity. Equation (4.12) can also be used to estimate the velocity at which a negative surge will move upstream if flow into the penstocks is started suddenly. An increase in flow cannot be introduced in less than approximately twice the time a surge takes to travel from the penstock entrance to the canal intake gates. A forebay volume at the penstock intake must be large enough to provide flow to the penstock until an increase in flow through the canal can be achieved.

If unexpected load changes will not occur, the system operation can be simplified. Flow in the canal can be increased to the necessary final flow rate before any turbine settings are changed. Once flow has been established, which may include flow over the spillway, the turbine flow rate can be increased. If the turbine is to be shut down or operated at a reduced flow rate the process would be reversed. This type of operation is simple but will result in significant quantities of water being wasted, an acceptable practice if water supply is not critical.

Allowable Velocities

Canals must be designed to convey water at velocities below that at which objectionable erosion of the channel will occur. For lined channels, velocities will always be below erosive levels. However, for unlined channels the guidelines in Table 4.4 should be observed.

Flow over Spillways and Weirs

Flow over the overflow portion of a dam and over the spillway is constructed to provide relief of the power canal from overtopping. The physical laws orig-

TABLE 4.4 Maximum Recommended Velocities for Unlined Open Channels

Material	Maximum recommended velocity	
	ft/s	m/s
Fine sand	1.50	0.46
Silt loam	2.00	0.61
Volcanic ash	2.50	0.76
Stiff clay	3.75	1.14
Shale	6.00	1.83
Fine gravel	2.50	0.76
Coarse gravel	4.00	1.22

inally developed for weirs apply. In general the overflow section on a dam will be designed as a smooth section to develop the greatest flow rate with the smallest rise in headwater elevation. Spillways located on canals however will usually be constructed using a relatively thin concrete or timber wall which will have hydraulic characteristics similar to weirs.

The shape of a smooth spillway section can be developed from the curves given in Fig. 4.8. The definitions in Fig. 4.8 are illustrated in Fig. 4.9. The head on the spillway h should be chosen properly for the design flood (probably a 100-year peak flow). The discharge for an overflow spillway proportioned in accordance with Fig. 4.8 can be computed as

$$Q = \frac{2}{3}\sqrt{2g}C_D B(h)^{3/2} \tag{4.13}$$

where C_D is given by Fig. 4.10. Thus, using Eq. (4.13), the designer can calculate the width of spillway channel B required to pass the design flow Q, without exceeding the upstream water surface in agreement with the head h.

Broad Crested Weir: Flow over a broad crested weir is illustrated in Fig. 4.11. Discharge of such a weir is given by

$$Q = 5.67\left(\frac{w + h}{w + 2h}\right)^{1/2} h^{3/2} \qquad \text{(USCS units)} \tag{4.14}$$

where Q is in ft^3/s, and w and h are in feet, or by

$$Q = 1.36\left(\frac{w + h}{w + 2h}\right)^{1/2} h^{3/2} \qquad \text{(SI units)} \tag{4.15}$$

where Q is in m^3/s and w and h are in meters.

Where weirs or spillways are utilized it is necessary to consider the possibility of erosion downstream. If the natural material downstream is highly

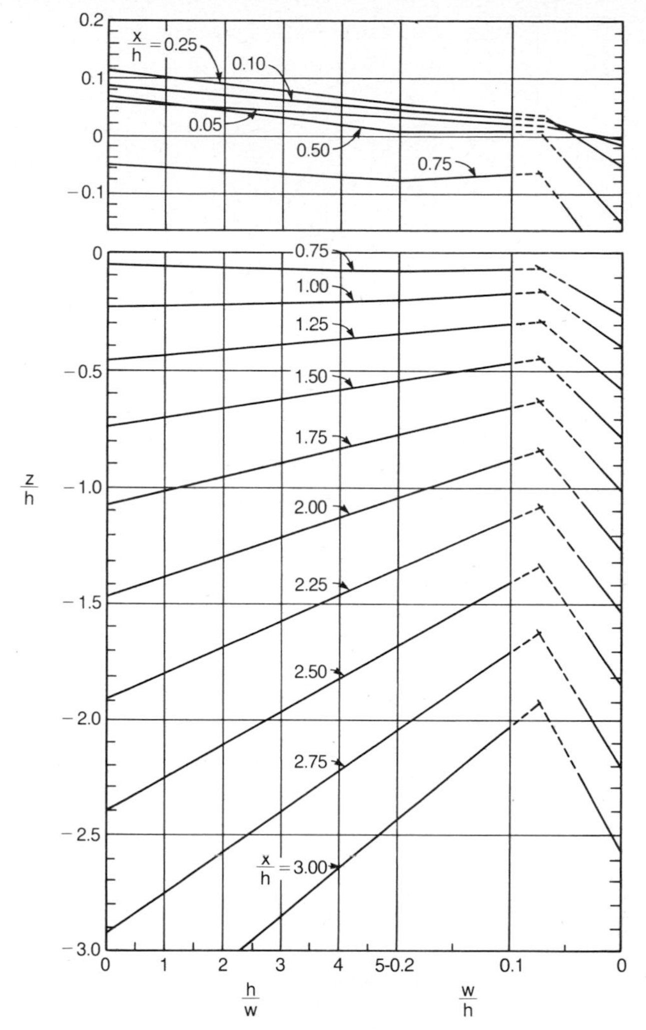

FIG. 4.8 Dimensionless coordinates for proportioning an overflow spillway crest. (Adapted from Ref. 10.)

erodible, it will be necessary to provide lining or riprap to prevent erosion from undermining the structure and eventually causing its failure. Details on some typical structures of this type are given in Chap. 5.

Flow in Closed Conduits

Flow through penstocks, pressure tunnels, and conduits follows the physical laws of flow in closed conduits. Thus, in the design of these components of small hydroelectric systems it is necessary to properly consider resistance to flow in

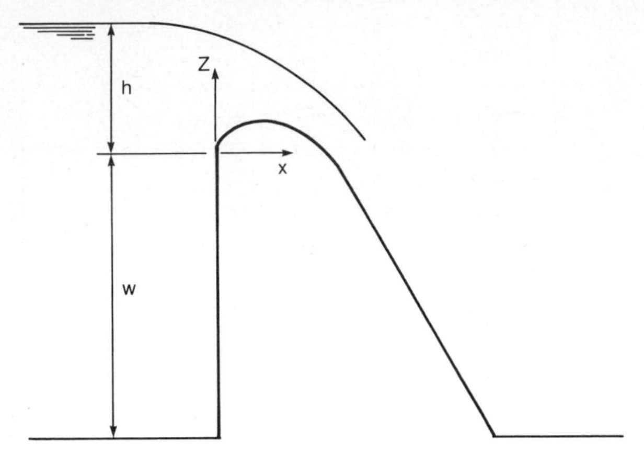

FIG. 4.9 Definition sketch for flow over a spillway.

sizing the conduit. Very frequently the Manning equation is used to calculate resistance to flow in conduits.

$$Q = \frac{1.5}{n} AR^{2/3}S_f^{1/2} \qquad \text{(USCS units)} \tag{4.16}$$

$$Q = \frac{1}{n} AR^{2/3}S_f^{1/2} \qquad \text{(SI units)} \tag{4.17}$$

For closed conduits, S_f is the slope of the energy gradient and not necessarily the slope of the pipe. The coefficient n is a function of absolute roughness of the conduit, and particularly for large smooth conduits with low velocities of flow, its use may cause an overestimation of the head loss.

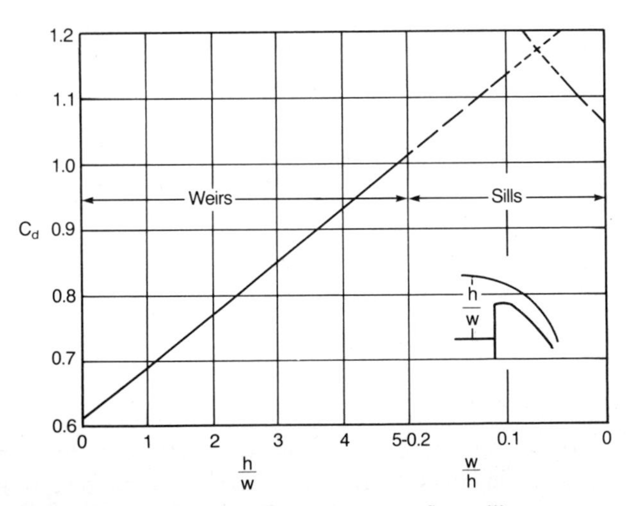

FIG. 4.10 Discharge coefficient for an overflow spillway.

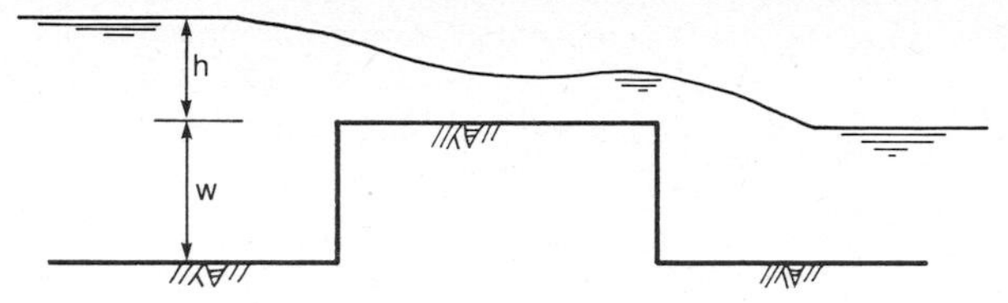

FIG. 4.11 Flow over a broad-crested weir.

It is preferable to use the well-known Darcy-Weisbach equation

$$h_L = f\frac{L}{4R}\frac{v^2}{2g} \tag{4.18}$$

where h_L is the energy loss expressed in ft (m) of water, L is the conduit length, v is the average velocity in the conduit, and R is the hydraulic radius.

Since Eq. (4.18) is dimensionally homogeneous, g may be used in either SI or USCS systems with the proper value of g (32.2 ft/s^2 in USCS or 9.80 m/s^2 in SI units); f is a dimensionless function coefficient which depends on the relative roughness $k/4R$ of the conduit walls and the Reynolds number:

$$Re = \frac{4vR}{\nu} \tag{4.19}$$

where v and R are as previously defined, ν is the kinematic viscosity of water, and k is the absolute roughness of the conduit walls.

Kinematic viscosity of water at various temperatures is given in Table 4.5. The resistance coefficient has been found to vary in accordance with the curves shown in Fig. 4.12.

Values of k vary with the conduit material. Table 4.6 shows values of k for various materials. Conduits are manufactured and thus k has been quantified in terms of material. For tunnels, however, the value of k depends not only on the nature of rock encountered but on the tunneling method. Bored tunnels in mudstone have a wall roughness height equivalent to relatively smooth concrete

TABLE 4.5 Values of Kinematic Viscosity of Water

Temperature		Kinematic viscosity, ν	
°F	°C	ft^2/s	m^2/s
32	0	1.93×10^{-5}	0.179×10^{-5}
50	10	1.41×10^{-5}	0.131×10^{-5}
60	15.6	1.21×10^{-5}	0.112×10^{-5}
70	21.1	1.05×10^{-5}	0.0974×10^{-5}
80	26.7	0.930×10^{-5}	0.0862×10^{-5}
90	32.2	0.823×10^{-5}	0.0763×10^{-5}

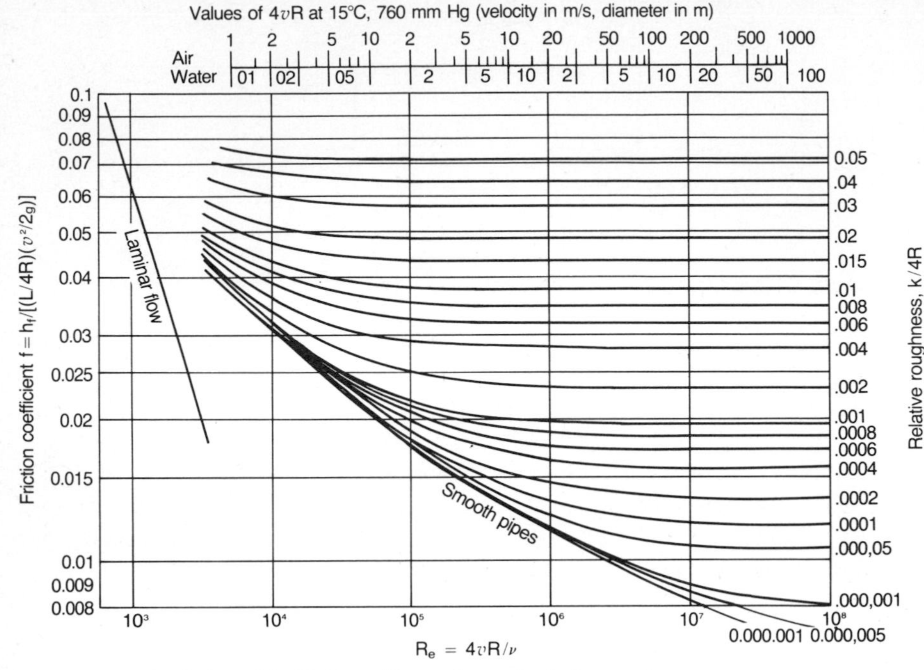

FIG. 4.12 Variation of f with Reynolds number and relative roughness.

conduits. However, tunnels blasted in hard rock will have absolute roughness heights k as great as 6 in (15 mm). If the tunnel is concrete lined, the value of k will equal that of a concrete conduit provided smooth forms are used.

Energy losses in closed conduits due to bends and other obstructions to flow can be calculated in accordance with Eqs. (4.20) through (4.25).

Entrance Losses:

$$h_e = K_1 \frac{v^2}{2g} \tag{4.20}$$

TABLE 4.6 Values of Absolute Roughness Height k for Conduits and Tunnels

Material	k	
	in	mm
Unlined rock tunnels:		
Drilled and blasted	2.0–6.0	5.0–15.0
Bored in shale or soapstone	0.12–0.25	0.30–0.60
Bored in limestone or igneous rock	0.25–0.50	0.60–1.20
Concrete lined	0.10–0.25	0.25–0.60
Steel pipe with welded joints	0.08–0.20	0.20–0.50

TABLE 4.7 Values of Entrance Loss Coefficient for Various Types of Entrances

Entrance type	K_1
Protruding	0.75
Sharp-edged	0.5
Well rounded	0.01

where v is the entrance velocity, and values of the entrance loss coefficient K_1 are given in Table 4.7.

Abrupt Contractions:

$$h_c = K_2 \frac{v_2^2}{2g} \tag{4.21}$$

where v_2 is the downstream velocity, and values of the contraction coefficient K_2 are given in Table 4.8 for various ratios of upstream and downstream flow areas.

Abrupt Expansions:

$$h_{ex} = [1 - (A_1/A_2)^2] \frac{v_1^2}{2g} \tag{4.22}$$

where A_1 and A_2 are the upstream and downstream flow areas, v_1 is the upstream velocity.

Bends:

$$h_B = K_B \frac{v^2}{2g} \tag{4.23}$$

TABLE 4.8 Values of Abrupt Contraction Loss Coefficient for Various Ratios of Downstream to Upstream Flow Areas*

Ratio of downstream to upstream flow area, A_2/A_1	Contraction loss coefficient, K_2
0.60	0.13
0.4	0.28
0.2	0.38
0.05	0.45

*A_1 = upstream flow area; A_2 = downstream flow area.

TABLE 4.9 Values of Bend Loss Coefficient for Various Geometric Configurations*

Smooth bends		Mitred bends	
r/D	K_B	Θ	K_B
1.0	0.4	20°	0.06
2.0	0.27	40°	0.21
4.0	0.20	60°	0.50
		90°	1.10

*r = bend radius; D = inside conduit diameter; Θ = angle between branches of the bend.

where v is the velocity at the bend, and values of the bend loss coefficient K_B are given in Table 4.9 for various geometric configurations.

Gates and Valves:

$$h_v = K_v \frac{v^2}{2g} \tag{4.24}$$

where v is the velocity at the gate or valve, and values of the loss coefficient K_v are shown in Table 4.10 for various gate or valve regimes.

Gradual Expansions:

$$h_{ge} = K_{ge} \frac{v_1^2}{2g} \tag{4.25}$$

where v_1 is the upstream velocity, and values of the loss coefficient for gradual expansions are given in Table 4.11 for various geometric configurations.

Penstock Design

Although almost all modern penstocks are built from steel, many have been built in the past from wooden staves reinforced with steel bands. The penstock,

TABLE 4.10 Values of Valve Loss Coefficient for Various Value Regimes

Gate valve		Butterfly valve	
Position	K_v	t/D*	K_v
Fully open	2.3	0.1	0.1
Half open	4.3	0.2	0.3
Quarter open	10.0	0.3	0.75

*t = thickness of the butterfly; D = pipe diameter.

TABLE 4.11 Values of Gradual Expansion Loss Coefficient for Various Geometric Configurations

	K_{ge}		
		Θ	
A_2/A_1	20°	15°	10°
3.0	0.4	0.30	0.2
2.5	0.3	0.25	0.15
2.0	0.2	0.15	0.12
1.5	0.15	0.1	0.08

*A_1 and A_2 = upstream and downstream flow areas respectively; Θ = expansion angle.

as pointed out earlier, must be designed to provide the required flow rate for the hydroelectric plant. Energy losses in the penstock are calculated using the methods and equations set forth in the section on closed-conduit flow.[4]

The hydraulic characteristics of the penstock and the turbine are closely connected. Sudden closure of the wicket gates at the turbine will cause a positive pressure wave to move upstream in the penstock at the speed of sound. The speed of sound in a steel penstock filled with water will be approximately 4,000 ft/s (1,220 m/s). The pressure rise occurring behind this wave will be given approximately by

$$\Delta P = \rho v c \tag{4.26}$$

where

ΔP = pressure rise, N/m^2 (lb/ft^2)
ρ = density of water, 1,000 kg/m^3 (1.94 slugs/ft^3)
c = velocity of sound in penstock, 1,220 m/s (4,000 ft/s)
v = velocity of flow, m/s (ft/s)

The penstock must be designed with walls thick enough to withstand the normal static pressure during full flow plus this pressure rise unless some means of relieving this transient pressure is provided. Since the elasticity of the pipe will allow some expansion, the velocity of sound in the penstock is somewhat lower than is documented for pure water. When a positive pressure rise is generated it will move up the penstock and be reflected negatively at the penstock entrance. This reduction in pressure will then move down the penstock, again with the speed of sound. When the wave again, reaches the turbine control valve, or wicket gates, it will be reflected negatively again, resulting in a reduction in pressure below the original steady-state value.

In cases where the penstock is short, it will probably be desirable to design the penstock for full pressure rise. However, for long penstocks it may be more economical to provide a surge tank near the power house. The surge tank allows water moving down the penstock to flow upward out of the penstock

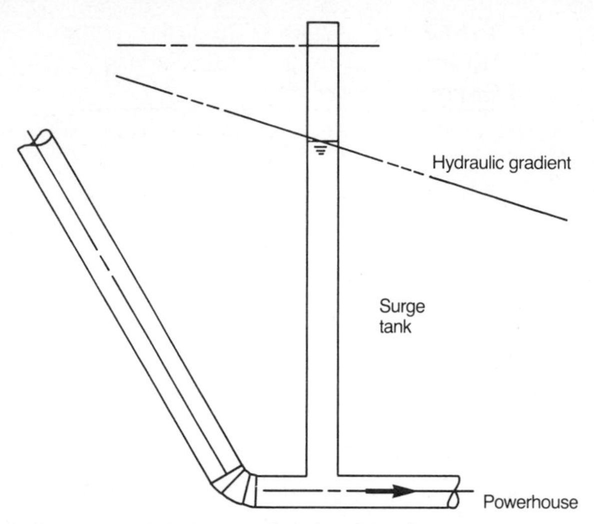

FIG. 4.13 Typical penstock and surge tank.

with gravity plus flow resistance, decreasing the velocity. In order to be most effective the surge tank should be located as near the turbine as possible. Figure 4.13 shows a typical penstock and surge tank arrangement.

The surge tank also functions when the turbine is started. Water standing in the surge tank will drop quickly when the turbine wicket gates are opened, relieving the quickly occurring low pressure which will arise without the surge tank. The transient pressures arising in the penstock are complex and a detailed analysis is likewise complex. Detailed final design should usually include a computerized analysis.[5] Graphical solutions have been presented by Allievi.[6]

At each change in direction, the penstock and its supporting structure must be designed to resist the forces resulting from changes in direction. Figure 4.14 shows a vertical bend in a penstock and free-body diagrams showing the resultant forces produced by weight, pressure, and change in momentum. The bend, the anchorage, and the footing supporting the penstock must be designed to resist these forces. Anchorage provided will vary according to need. If the vertical change in direction is from a steep to a flatter angle (Fig. 4.14, *A*) the net momentum and pressure force on the footing will act downward. Where the angle change is from flatter to steeper (Fig. 4.16) the net force may act upward. Weight of water and pipe must be assessed separately.

Net forces due to pressure and momentum change are computed as follows:

$$F_{Px} = P \cdot A\,(\cos \Theta_1 - \cos \Theta_2) \tag{4.27}$$

$$F_{Py} = P \cdot A\,(\sin \Theta_1 - \sin \Theta_2) \tag{4.28}$$

where

F_{Px} = horizontal component of pressure force
F_{Py} = vertical component of pressure force

P = pressure in penstock
A = cross-sectional area of penstock
Θ_1 = angle as shown in Fig. 4.14
Θ_2 = angle as shown in Fig. 4.14

$$F_{mx} = Q^2\rho\,(\cos\Theta_2 - \cos\Theta_1)/A \tag{4.29}$$
$$F_{my} = Q^2\rho\,(\sin\Theta_2 - \sin\Theta_1)/A \tag{4.30}$$

where

F_{mx} = horizontal component of momentum force
F_{my} = vertical component of momentum force
Q = rate of flow in penstock
A = cross-sectional area of penstock
Θ_1, Θ_2 = angles as shown in Fig. 4.14
ρ = density of water

The penstock must also be designed to allow for stresses produced by temperature changes, vibrations, buckling against external load, and differential settlement. In actuality, temperature variations are usually small when the plant is in operation and, unless the penstock is very long, expansion joints may not be required. Usually the stresses in the pipe produced by temperature variations are unimportant compared to movement of the anchorages. Where movement is expected it may be desirable to use expansion joints near major anchorages.

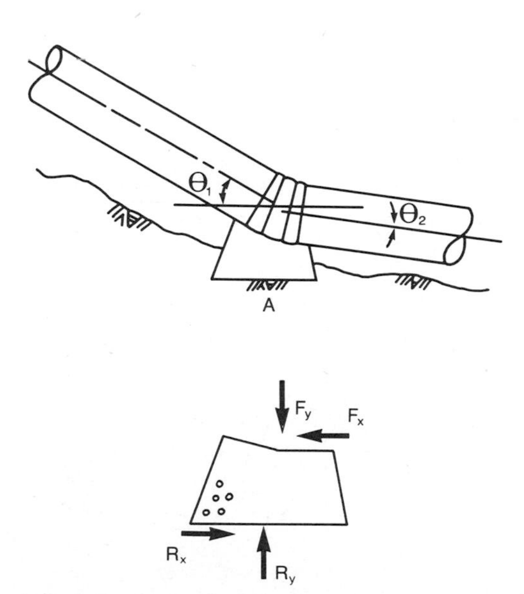

FIG. 4.14 Vertical bend in penstock.

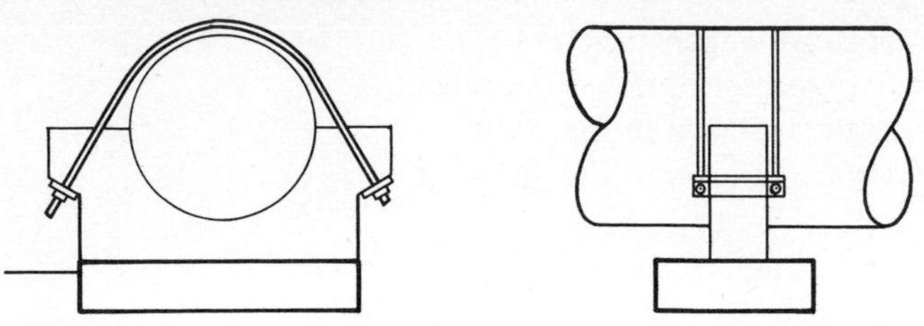

FIG. 4.15 Precast concrete penstock support.

Occasionally steel penstocks have been laid on the ground using continuous gravel or concrete bedding to provide for drainage of storm water. If a steel penstock is placed on the ground, the buried portion should be protected with a coat of primer paint, at least two coats of coal-tar epoxy resin, and a layer of bonded asbestos felt wrapping. Exposed portions of the penstock should always be protected with a coat of primer, a coat of coal-tar epoxy resin, and a coat of reflective aluminum paint.

Figure 4.15 shows one type of precast penstock supports which can be used. The cradles can be precast, transported to the site, and suspended from the penstock. The footing is then cast in place beneath the support cradle to complete the support.

The penstock must not be allowed to creep downhill. Properly designed anchorages, considering penstock weight, will provide the necessary resistance. In addition, the downstream end of the penstock should be embedded in a rock anchorage if at all possible. A "wye" joint should be used to divert water to the turbine as shown in Fig. 4.16.

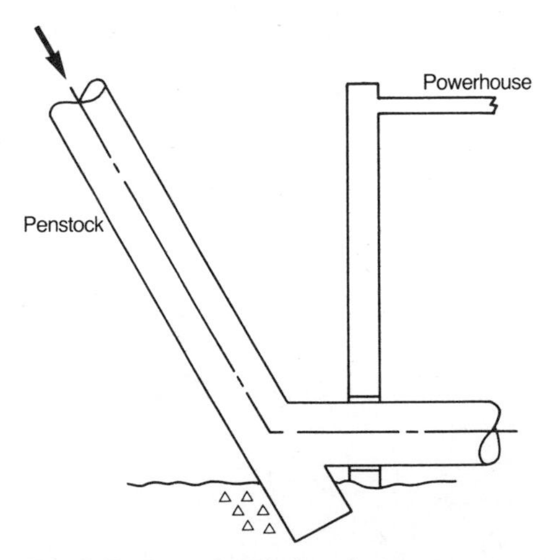

FIG. 4.16 Lower penstock anchorage.

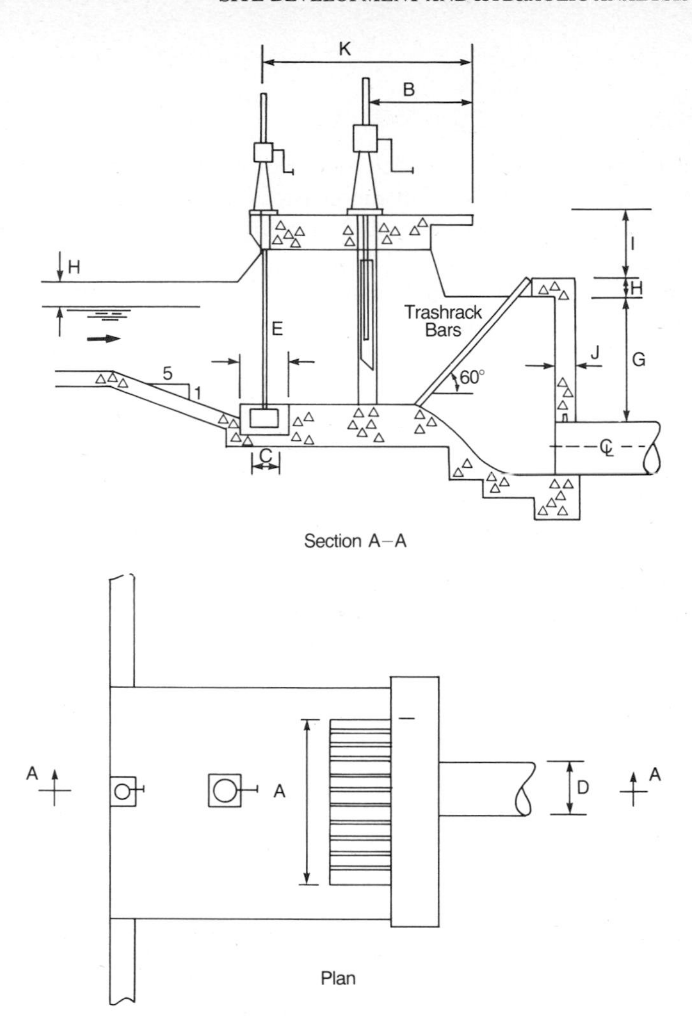

FIG. 4.17 Typical penstock intake structure.

Intake Structures for Penstocks

Figure 4.17 shows a typical intake structure for a steel penstock. A sluiceway channel should be provided to sluice out gravel or other deposition which may reach the structure. A slide gate can be used to cut off the flow to the penstock.

Proper submergence of the penstock entrance must be provided to prevent the formation of an air-entraining vortex at the penstock entrance. The required submergence in meters (shown as "G" in Fig. 4.17) can be computed from

$$G = 0.72vD \tag{4.31}$$

where

v = velocity in penstock, m/s
D = inside diameter of penstock, m

Sufficient room must be provided to allow for regular cleaning of the trash-racks. Placing the gate in front of the penstock will allow air to enter the penstock freely when the gate is closed and the penstock is drained, thus preventing the occurrence of possible penstock collapse due to negative pressures. Air valves can also be provided for this purpose, but their operation is less certain.

The penstock intake structure cannot be completely generalized because it must be designed to fit the particular site. However, Table 4.12 gives values of several typical dimensions which will guide the overall design of the structure. The values are given in meters in terms of the flow rate which must be passed.

4.5 TURBINE SETTING

Cavitation occurs to some extent in all hydraulic turbines and pumps when local pressure is reduced to the vapor pressure as a result of increases in velocity or an ambient drop in pressure. In this low-pressure area, water vaporizes forming bubbles which, when carried downstream into areas of higher pressure, collapse violently. The rapid collapse produces very high instantaneous pressures which fatigue and erode any nearby solid surfaces. If the cavitation is severe enough and occurs over a long enough period of time, serious damage to the turbine runner may occur.

In general, minor cavitation erosion will occur on most hydraulic turbines.

TABLE 4.12 Typical Dimensions for a Penstock Intake Structure*

Flow rate, m^3/s	1.0	2.0	3.0	4.0	6.0	8.0	10.0
A	1.9	2.3	2.5	2.9	3.5	4.2	4.6
B	0.80	0.90	0.95	1.00	1.10	1.15	1.20
C	0.90	0.90	0.90	0.90	1.00	1.10	1.30
D	1.0	1.2	1.3	1.5	1.9	2.3	2.4
E	0.50	0.60	0.60	0.70	0.80	0.90	1.0
G	1.5	1.7	1.9	2.1	2.4	2.8	3.3
H	0.4	0.5	0.6	0.65	0.7	0.8	0.9
I	0.9	1.1	1.3	1.5	1.7	1.9	2.0
J	0.35	0.40	0.45	0.50	0.53	0.57	0.6
K	1.6	1.8	1.9	2.0	2.2	2.3	2.4
Bar thickness, mm	45	50	50	50	50	60	70
Bar spacing, mm	25	25	30	30	32	34	35
Bar width, mm	5	5	6	6	6	6	6

*See Fig. 4.17 for the meaning of the symbols.

The eroded metal is replaced by welding during regular maintenance sessions for minor cavitation damages. To avoid major damage it is necessary to consider the unit speed properly, design all hydraulic passages properly, specify proper metals, and carefully establish the distance between the turbine centerline and the tailwater elevation. This distance is called the *turbine setting.* For a given turbine setting the tendency for cavitation increases with increasing rotational speed. For economy in equipment it is desirable to use a deep setting which allows the use of a smaller, less costly, higher-speed turbine for a given power.

Cavitation can occur on the buckets of impulse or Pelton turbines. However, on these units the tendency toward cavitation results from improper bucket design or construction since the turbine runners are normally exposed to atmospheric pressure. Cavitation is discussed further in Chap. 6.

4.6 POWERHOUSES

The purpose of the powerhouse is to support the hydraulic and electrical equipment and to provide protection from the adverse impacts of weather. In addition the powerhouse may provide a shop for maintenance work, a locker room for the use of operators and maintenance workers, and miscellaneous equipment including exciters, switchboards, oil switches, and sometimes high-voltage transformers. The layout of all this equipment will determine the overall dimensions of the powerhouse.

For larger turbines and generators, an overhead crane may be desirable. The crane must have sufficient capacity to allow handling of the turbine runner and generator rotor for initial construction and for future maintenance. Since the crane usually travels on rails mounted on the powerhouse superstructure, structural design of the powerhouse walls will frequently be governed by the crane loads. For smaller power plants it may be simpler and more economical to lift the runner or rotor using a mobile crane which can work from outside the powerhouse. For that alternative a properly sized opening in the powerhouse roof must be designed, or the roof itself can be designed for removal.

General Arrangement

For powerhouses with multiple turbines and generators, the generators are usually placed in a straight line so that flow passages are as simple as possible. Clearances between different pieces of equipment is important. Where multiple units are involved, an unobstructed aisle, approximately 8 ft (2.4 m) should be provided along one side of the powerhouse.

In a small powerhouse the operating switchboard should be located as near the generators as possible. Other equipment items such as governors, pumps, exciters, and ventilating fans should be located so that they are easily inspected by the operator. The overall layout should provide for ease of inspection and maintenance and, thus, requires careful thought during planning.

Careful consideration must be given to an operations plan for major equipment repair and maintenance. For tasks involving removal of the generator or the turbine runner, sheltered space should be provided in which these pieces can be laid down for temporary storage or repair. Although this space can be provided in another building, or outside temporary shelter in mild climates, it is usually desirable to provide a working bay which may be 1 to 1½ times as wide as the operating bay. Providing this space within the powerhouse simplifies repair and maintenance making it possible to do emergency repairs in minimum time.

Where spillway gates or pumps may need to be operated, a source of power should be provided which is independent of the main generators or the transmission lines. For remote sites a gasoline- or diesel-operated emergency generator should be available.

The overall configuration of the powerhouse is very much dependent upon the type of turbine used and the turbine setting. The size of the turbine, and the size of the powerhouse as well, will be inversely proportional to the head available for a given power output. Figures 4.18 and 4.19 show typical layouts for small high-head and low-head units respectively. Final layout of the powerhouse must be made after the turbine and generator have been chosen. However, rough estimates of floor area for the powerhouse and volume of reinforced concrete required can be estimated from Fig. 4.20.

For powerplants with less than 600 kW capacity, transformers will be small enough that they may be located inside the powerhouse itself. In such cases they should be located outside the work area, preferably in a separate room. This separation is required for both comfort and safety. The high voltages involved make proper clearances absolutely necessary.

The generators themselves will normally produce enough heat to keep the powerhouse comfortable in cold weather. However, during warm weather good ventilation of the powerhouse will be required both for comfort and stable operation of instrumentation.

For plants with capacities greater than 600 kW the transformers and switching gear should be located in a yard outside. This switchyard should be

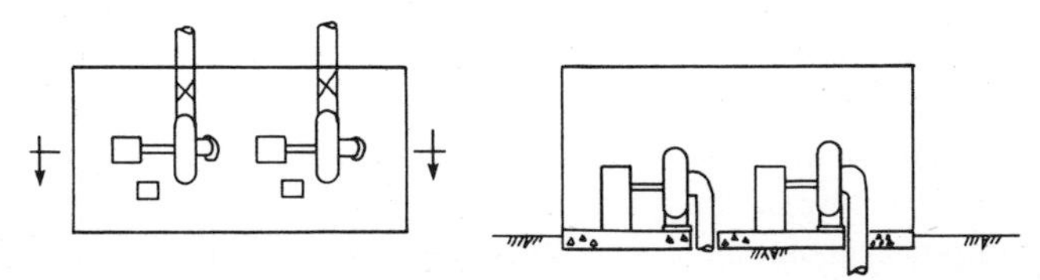

FIG. 4.18 Powerhouse layout for a high-head unit. Plan (left) and section of plan between arrows (right).

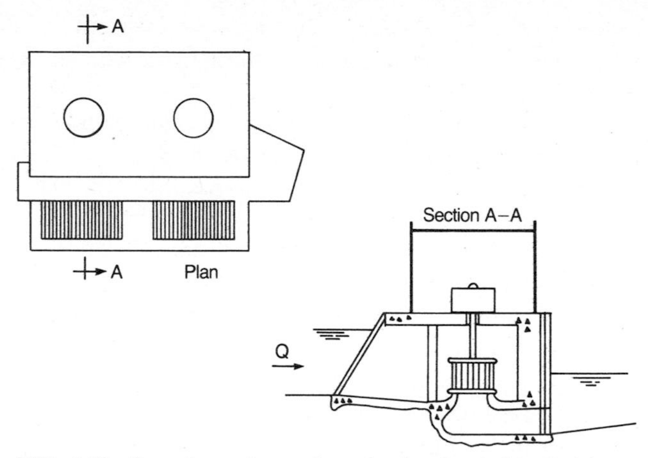

FIG. 4.19 Powerhouse layout for a low-head unit installed in a flume.

as close to the powerhouse as possible and should be surrounded by a sturdy, durable fence for safety. The yard should be located and graded to provide for good drainage without ponding. A gravel base should be provided around all equipment.

4.7 CAPITAL COSTS OF SITE DEVELOPMENT

Site development costs are very much site specific. Some sites will lend themselves to minimum excavation, short power canals, low dams, and short pen-

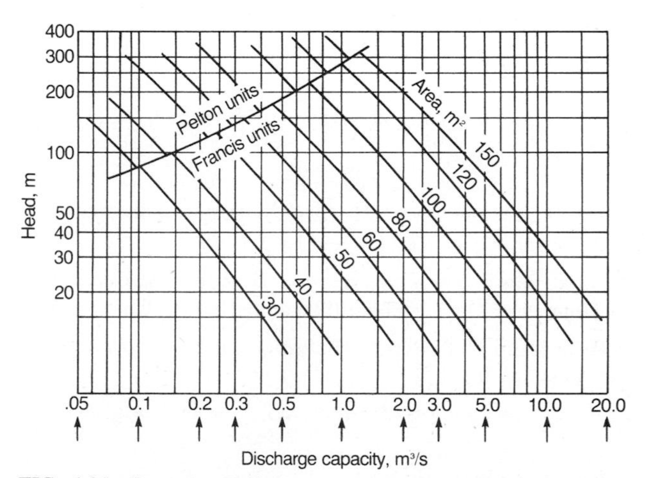

FIG. 4.20 Powerhouse floor area required for Pelton and high-head Francis turbine installations.

stocks; on others, rock may be deep, requiring large quantities of excavation, or soils may be unstable, requiring special designs. Costs will vary greatly with these varying conditions.

However, historical records can be used to develop capital costs for site development. This has been done by the U.S. Army Corps of Engineers[7] and by the United States Water and Power Resources Service.[8] The cost information given in these sources is intended to be for estimating purposes only, and particular installations may vary from this data. No consideration was given to the development of roads for access because those costs can vary by orders of magnitude depending upon topography and design standards.

Site development will include grading, foundation excavation, drainage, and parking facilities. It will also include construction of the hydraulic conveyance facilities and the powerhouse and switchyard. In general the site development will equal 15 to 45 percent of the total project cost. Other project costs will include the turbine-generator, accessory electrical equipment, and miscellaneous power plant equipment.

4.8 REFERENCES

1. Andrew Eberhardt, "An Assessment of Penstock Designs," *Water Power and Dam Construction,* June/July 1975, p. 249.
2. J. W. Ball, "Cavitation from Surface Irregularities in High Velocity Flow," *Journal of the Hydraulics Division,* ASCE, vol. 112, no. HY9, September 1967.
3. Donald Miller, *Fluid Flow,* British Hydromechanics Research Association, London, 1977.
4. H. G. Arthur and J. J. Walker, "New Design Criteria for USBR Penstocks," *Journal of the Power Division,* ASCE, January 1970, p. 129.
5. G. Watters, *Hydraulic Transients,* Ann Arbor Science Publishers, Ann Arbor, Mich., 1979.
6. W. P. Creager and J. D. Justin, *Hydroelectric Handbook,* Chap. 34 "Water Hammer," Wiley, New York, 1963.
7. U.S. Army Corps of Engineers, *Feasibility Studies for Small Scale Hydropower Additions, A Guide Manual,* vol. VI, *Civil Features,* Hydrologic Engineering Center and Institute for Water Resources, Davis, Calif., July 1979.
8. U.S. Water and Power Resources Service, *Development Costs for Small Hydroelectric Developments,* Denver Federal Center, Denver, Colo., 1982.
9. S. M. Woodward and C. J. Posey, *Hydraulics of Steady Flow in Open Channels,* Wiley, New York, 1958.
10. H. Rouse (ed.), *Engineering Hydraulics*, Wiley, New York, 1950.

5

Dams and Reservoirs

John J. Cassidy

A dam will be an important increment of nearly all small hydroelectric developments. For some installations the dam will be small and serves only to divert flow from a river or stream through an intake structure into a power canal or tunnel. Higher dams serve to develop head for low-head developments which may utilize bulb-type or flume-type turbine installations. Because the capital development cost of large dams is a major part of the total hydroelectric project cost, small hydroelectric developments utilizing high dams are generally not economically feasible. This chapter considers the various types of dams which are most often required as part of a small hydroelectric development.

5.1 TYPES OF DAMS

Dams are classified by type according to the materials used in construction and the physical nature of design. The major types are gravity, arch, buttress, earthfill, rockfill, and masonry. (See Fig. 5.1.)

Gravity dams utilize their own weight to overcome all forces applied to the dam and maintain stability. They are constructed from both concrete and masonry.

Arch dams normally are designed as structural members which transmit all loading to the abutments. Thus, abutments for arch dams must be sound rock,

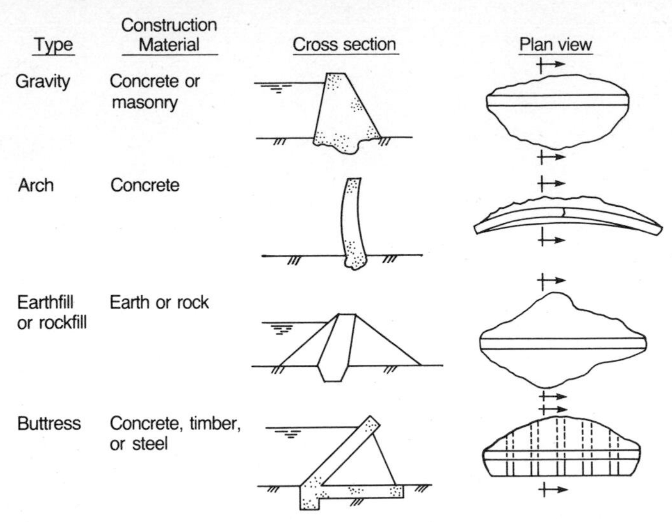

FIG. 5.1 Basic types of dams.

capable of supporting the rather large thrust loading with negligible displacement. Because of their required structural integrity, arch dams are normally constructed from reinforced concrete. The upstream curvature is designed to keep the concrete in compression throughout the dam.

Buttress dams of many types have been designed throughout history.[1] The simplest and most common consists of a sloping slab supported at intervals by buttresses. The buttress dam combines structural members whose weight and configuration combine to develop a gravity dam.

Earthfill dams are normally constructed entirely as earth embankments. Seepage through the dam is controlled by the designed use of upstream blankets or internal cores constructed using compacted soil of very low permeability.

Rockfill dams utilize dumped rock to form that part of the dam which provides for structural integrity. An impermeable section, normally built from concrete, provides seepage control.

Masonry dams differ from rockfill dams in that the stone is hand placed with mortar with the result that the entire dam is impermeable.

The type of dam to be used is chosen after considering the height of the proposed dam, the type of foundation material (earth or rock), the length of the dam, the availability of skilled and unskilled labor, the availability of cement, aggregate, and reinforcing steel, and the availability and cost of both permeable and impermeable soils and earthmoving and compacting equipment. A concrete dam generally requires a sound rock foundation. Some grouting of the foundation should be planned since rock will almost always be fractured to some degree.

One of the first considerations is the height of the dam and the requirements for a spillway. If little or no reservoir storage is required, a low overflow dam should be considered first. If the streamflow is highly variable, control gates may be required on the crest of the dam to force water into the diversion canal during low-flow periods. The dam must be designed so that overflow during the design flood can occur without serious damage.

If the dam is to be used to develop head for a low-head development, or if significant reservoir storage is required, a separate spillway section may be required. In this case the spillway must be able to pass the design flood without damage to the spillway, the dam, or the powerhouse. Whether the spillway is to be incorporated in the dam will have significant effect on the choice of type of dam. A long earthfill dam may incorporate a concrete-gravity section for the spillway. If a suitable off-channel site is possible the spillway may be constructed away from the dam itself. Figure 5.2 shows three typical dam and spillway configurations.

An earth dam of given height will almost always be cheaper than a concrete dam if earthmoving equipment is available and if sufficient quantities of the proper earthfill are available. However, a spillway cannot be constructed on top of an earthfill dam, whereas a concrete or masonry dam can be designed as an overflow section. Thus, consideration must be given to the relative cost of constructing a separate spillway as opposed to constructing a concrete or masonry dam. If the dam is relatively short, a concrete dam may be more economical. An earthfill dam must be provided with an impermeable zone to provide for control of seepage. Ideally, the earthfill dam site will have an underlying impermeable earth layer to which the impermeable core can be joined. However, an upstream blanket of impermeable earth can be used provided there are no horizontally stratified layers of gravel or sand under the dam.

The higher the dam the more critical will be the foundation requirements. Strength of the foundation for a high dam, 35 m (115 ft) or more, will be stringent for either earthfill or gravity dams. In addition, the higher the dam the more difficult it is to control seepage. In general a dam can be designed to be safe at any site. However, the resulting design may be prohibitively expensive. In actuality the number of feasible dam sites may be quite limited by topographic or geologic conditions. In general, a narrow stream flowing between steep rock walls will lend itself most economically to a concrete dam with an overflow section for a spillway, whereas a broad stream in rolling hills will lend itself most economically to an earthfill dam with a separate spillway.

A solid-rock foundation will offer good resistance to seepage and erosion and will provide good bearing capacity. Thus, it can support nearly any type of dam. The economy of available materials and labor, or the overall cost of the dam, will govern the choice of type of dam ultimately to be constructed.

Gravel foundations, if they are well compacted and free from soils subject to substantial settlement, can be satisfactory for earthfill or rockfill dams and

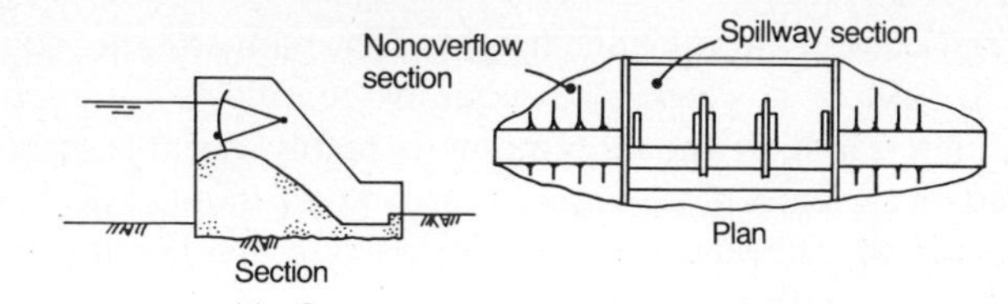

(a) Concrete or masonry with overflow section.

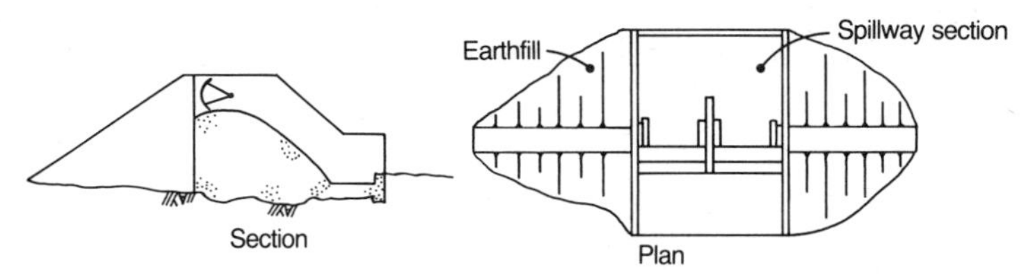

(b) Composite earthfill and concrete gravity.

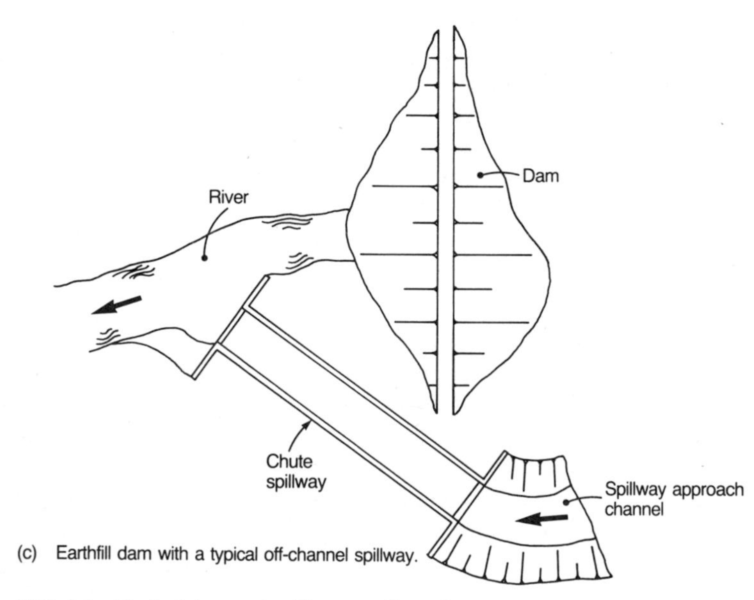

(c) Earthfill dam with a typical off-channel spillway.

FIG. 5.2 Typical dam and spillway configurations.

for low concrete gravity dams. However, because high percolation rates occur in gravels, special design precautions must be taken to cut off flow under the dam and through the abutments.

Foundations composed of fine sand or silt can sometimes support low concrete dams or earthfill dams. However, problems involving excessive settlement can arise and very special design precautions must be taken to prevent failure through piping. Because sand and silt are readily eroded, it is also necessary to make certain that erosion at the toe of the dam does not occur as a result of spillway or outlet works operation.

Clay foundations can support earthfill dams. However, if the clay is unconsolidated, and if its moisture content is high, objectionable settlement can occur. To prevent such settlement it is necessary to provide for special drainage and for preconsolidation of the foundation prior to construction of the dam. Tests of clay materials in their natural state are usually required to determine its consolidation characteristics and its ability to support the loads which will be superimposed during and after construction of the dam.

It is not at all uncommon to encounter foundations where more than one type of material and/or condition exists. Since differential settlements will almost always arise when a dam is constructed on such a site, special design considerations will be needed to choose the type of dam to be used and its final configuration.

The type of material available at the site will have strong bearing on the choice of type of dam through economics. If fractured rock can be quarried near the site, a rockfill or masonry dam can have an economic advantage. Similarly if suitable earth, free of organic materials, is readily available near the site, an earthfill dam will be favored. For economics to favor a concrete dam, clean gravel and sand and quality water must be available at the site. For all concrete, it is very important that both aggregate and water be free from alkali which reacts chemically with cement to produce serious deterioration of the concrete over a period of years.

5.2 EARTH DAM DESIGN

Failure of earth dams can occur as a result of overtopping during a flood, excessive erosion due to wave action, piping caused by flow through settlement produced cracks, or improperly designed or constructed filters. International studies of dam safety show that earthfill dams can be designed and constructed to stand safely at heights exceeding 1,000 ft (305 m).[2] Failures among small dams are, however, much more common, presumably because less importance is attached to investigations prior to design and less stringent design and construction requirements are adopted. This discussion is directed toward design procedures applicable for earthfill dams of the rolled-filled type which in general do not exceed 20 m (65 ft) in height.

There is no typical earthfill dam. However, some general similarities exist. Figure 5.3 shows schematic sections of several earthfill dams for particular foundation conditions. Seepage through the dam is controlled by a core of impervious soil located in the central or upstream portion. Upstream and downstream shells provide structural support for the impervious zone. Thickness of the impervious zone depends largely upon the availability of the impervious fill

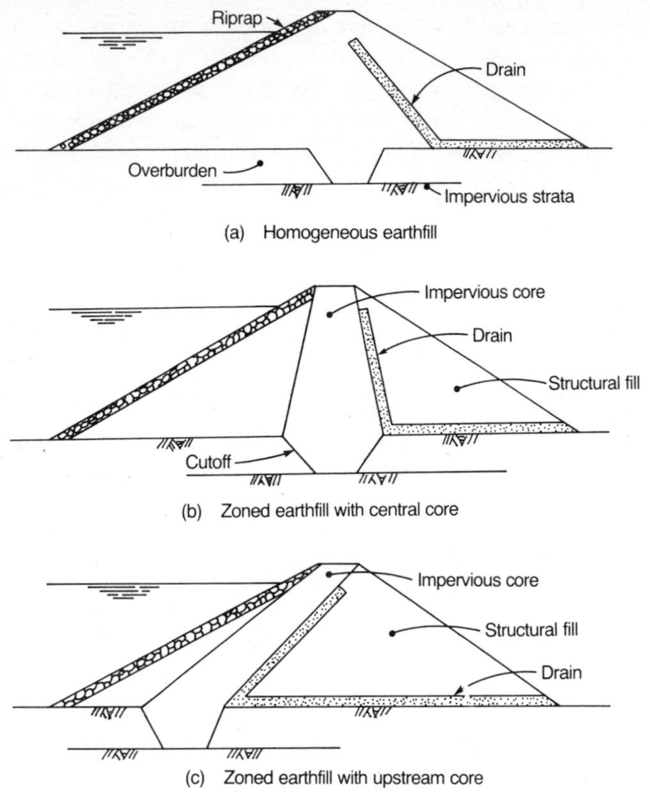

FIG. 5.3 Typical earthfill dam sections.

material. When borrow areas are such that selection of particular types of soils is impossible, a homogeneous fill can be constructed. However, even then care should always be taken to place the more pervious material toward the downstream or upstream limits. The more impervious fill should be placed toward the center of the fill.

Seepage through an earthfill dam always occurs. The embankment design must ensure that the seepage is a minimum and that the fill material is not carried away by the seepage (piping). To protect against piping it will always be necessary to provide drains with properly designed filters on the downstream side of coves or near the downstream limit of a homogeneous fill. A filter must be constructed from properly graded sands and gravels. Gradation requirements were originally set by Terzagi.[3] The following requirements have been established by the U.S. Army Corps of Engineers and are a modification of Terzagi's original standards:[4]

$$D_{15} < 5d_{85} \tag{5.1}$$

$$4d_{15} < D_{15} < 20d_{15} \tag{5.2}$$

$$D_{50} < 25d_{50} \tag{5.3}$$

where D and d refer respectively to the grain size of the filter material and the material to be protected. The subscripts indicate the percent of material fines.

To use these filter requirements it will be necessary to sample both the fill material and the filter material and to determine the grain-size distribution for both.

Where an earthfill dam is to be built on an impervious foundation, the foundation can be thoroughly cleaned and the earthfill placed directly on the foundation. In such cases, the contact zone may be the most critical portion of the dam, since irregularities there may provide a weak zone where piping can begin. Thus, it is vital to provide a properly designed filter along the downstream side of the impervious zone.

If the impervious foundation is rough rock it will be necessary to trim the rock and place a lean concrete layer over the foundation beneath the impervious zone. The concrete layer will ensure a uniform contact between the foundation and the impervious zone minimizing the chances for piping. Any badly fractured rock should be removed back to the solid foundation. Open joints and seams should be cleaned as deep as possible and backfilled with lean concrete.

Earth foundations must be stripped to a depth that all roots and other organic material are removed. Any pockets of soft compressible soils should be removed and any holes left by excavation of stumps or boulders should be backfilled and hand tamped. Once clearing and stripping of the foundation has been completed, the entire foundation should be plowed and then rolled to provide a uniform compaction. As heavy a roller as possible should be used for this operation in order that any previously undetected soft-soil deposits will be located.

If a cutoff trench is not required, an inspection trench, at least 6 ft deep should be excavated from one abutment to the other. This excavation will expose any abandoned pipes or pervious gravel deposits not detected during the foundation exploration. When inspection is complete the trench should be backfilled and compacted.

Where a pervious layer overlies an impervious zone, a cutoff trench will be desirable as shown in Fig. 5.3. The required bottom width w of the cutoff trench can be estimated with the equation

$$w = h - b \tag{5.4}$$

where h is the maximum depth of water in the reservoir and b is depth of the trench below ground surface. The minimum width of the trench should be wide enough to accommodate equipment for excavation and compaction.

A central impervious core, a cutoff, is very common in earthfill-dam design. It is preferable where proper soils may be readily obtained from the borrow areas. The impervious core must be wide enough to ensure that the impervious material will not arch between the upstream and downstream shells as settlement occurs. An inclined core tends to minimize this arching potential. Its width should in general not be less than is given by Eq. (5.4). The top width of any zone should be great enough to provide for ease of placement and compaction by the available equipment.

Slopes of the outer shells of the embankment are controlled by the shear strength of the fill material itself and that of the foundation. Stability of the slopes should be analyzed using the shear strength which will exist both during construction and after the reservoir fills. Conditions arising as a result of rapid drawdown of the reservoir, when much of the fill will be saturated, should also be analyzed. Earthquake loadings should also be considered in zones where seismic activity can be expected.

Analysis of slope stability can be performed in several ways.[5] The "slide-circle" approach is perhaps the simplest and most direct method to apply. Application of any of these methods requires that soil properties of both embankment and foundation material be determined through field and laboratory investigations. For a small dam, all this information may not be available, and it may be difficult or prohibitive to obtain. The U.S. Water and Power Resources Service (formerly the U.S. Bureau of Reclamation) has developed some guidelines for maximum upstream and downstream slopes to be used for small earthfill dams.[6] Table 5.1 lists those recommended slopes in terms of the materials used for the shell and core material. Table 5.2 defines the soil classification used in accordance with general properties. Table 5.1 should be used with care unless the user is experienced in soils engineering. Every effort should be made to classify the soils carefully in agreement with Table 5.2.

Figure 5.4 defines the slopes and the sizes of "maximum" and "minimum" cores as used in Table 5.1. The desired top width of an earth dam W can be approximated by Eqs. (5.5) and (5.6)

$$W = \frac{H}{5} + 10 \qquad \text{(USCS units)} \tag{5.5}$$

$$W = \frac{H}{5} + 3 \qquad \text{(SI units)} \tag{5.6}$$

where H is the height of dam; H and W are in feet for Eq. (5.5) and meters for Eq. (5.6).

Abutments

Abutments should be inspected as thoroughly as the foundation. Rock abutments require the same treatment as rock foundations. Where jagged or fractured rock exists on the abutment, particular care must be taken to trim the rock to a smooth line. If this is not done, differential settlement will occur and

TABLE 5.1 Recommended Slopes for Small Earth Dams on Stable Foundations

A. Zoned earthfill dams

Type	Purpose	Subject to rapid drawdown*	Shell material classification	Core material classification†	Upstream slope	Downstream slope
Zoned with "minimum core"‡	Any	Not critical	Rockfill, GW, GP, SW (gravelly), or SP (gravelly).	GC, GM, SC, SM, CL, ML, CH, or MH	2:1	2:1
Zoned with "maximum core"‡	Detention or storage	No	Rockfill, GW, GP, SW (gravelly), or SP (gravelly)	GC, GM	2:1	2:1
				SC, SM	2¼:1	2¼:1
				CL, ML	2½:1	2½:1
				CH, MH	3:1	3:1
Zoned with "maximum core"‡	Storage	Yes	Rockfill, GW, GP, SW (gravelly), or SP (gravelly)	GC, GM	2½:1	2:1
				SC, SM	2½:1	2¼:1
				CL, ML	3:1	2½:1
				CH, MH	3½:1	3:1

B. Homogeneous earthfill dams

Type	Purpose	Subject to rapid drawdown*	Soil classification†	Upstream slope	Downstream slope
Homogeneous or modified-homogeneous	Detention or storage	No	GW, GP, SW, SP	‡	‡
			GC, GM, SC, SM	2½:1	2:1
			CL, ML	3:1	2½:1
			CH, MH	2½:1	2½:1
Modified-homogeneous	Storage	Yes	GW, GP, SW, SP	‡	‡
			GC, GM, SC, SM	3:1	2:1
			CL, ML	3½:1	2½:1
			CH, MH	4:1	2½:1

SOURCE: Adapted from Ref. 6.

*Drawdown rates of 6 in (150 mm) or more per day following prolonged storage at high reservoir levels.

†See Table 5.2 for description of soil types. OL and OH soils are not recommended from major portions of homogeneous earthfill dams. Pt soils are unsuitable.

‡Pervious, not suitable.

a crack can open up as shown in Fig. 5.5. The crack can induce piping and resulting failure of the dam. Where abutments are steep, dam slopes should be flattened to avoid potential arching during settlement.

Embankment Soils

Most soils can be used for embankment construction unless they have objectionable physical or chemical properties. For example, soils with high salt contents or soils with significant organic content should not be used. Organic material increases a soil's compressibility and tends to lower its shear strength. Some soils have properties which make them difficult to use. Fat clays, which

TABLE 5.2 Description of Soils as Used in Table 5.1*

	Field inspection procedures	Group symbol	Typical names
	Coarse grained soils		
Gravels: Clean	Wide range in grain size and substantial amounts of all intermediate particle sizes	GW	Well-graded gravels, gravel-sand mixtures, little or no fines
	Predominantly one size or a range of sizes with some intermediate sizes missing	GP	Poorly graded gravels, gravel-sand mixtures, little or no fines
With fines	Nonplastic fines (for identification procedures see ML below)	GM	Silty gravels, poorly graded gravel-sand-silt mixtures
	Plastic fines (for identification procedures see CL below)	GC	Clayey gravels, poorly graded gravel-sand-clay mixtures
Sands: Clean	Wide range in grain sizes and substantial amounts of all intermediate particle sizes	SW	Well-graded sands, gravelly sands, little or no fines
	Predominantly one size or a range of sizes with some intermediate sizes missing	SP	Poorly graded sands, gravelly sands, little or no fines
With fines	Nonplastic fines (for identification procedures see ML below)	SM	Silty sands, poorly graded sand-silt mixtures
	Plastic fines (for identification procedures see CL below	SC	Clayey sands, poorly graded sand-clay mixtures

Fine grained soils

Soil type	Dry strength	Dilatancy	Toughness	Group symbol	Typical names
Silts and clays	None to slight	Quick to slow	None	ML	Inorganic silts and very fine sands, rock flour, silty or clayey fine sands with slight plasticity.
	Medium to high	None to very slow	Medium	CL	Inorganic clays of low to medium plasticity, gravelly clays, sandy clays, silty clays, clays
	Slight to medium	Slow	Slight	OL	Organic silts and organic silt clays of low plasticity
	Slight to medium	Slow to none	Slight to medium	MH	Inorganic silts, micaceous or diatomaceous fine sandy or silty soils, elastic silts
	High to very high	None	High	CH	Inorganic clays of high plasticity, fat clays
	Medium to high	None to very Slow	Slight to medium	OH	Organic clays of medium to high plasticity
Highly organic soils	Readily identified by color, odor, spongy feel and frequently by fibrous texture			Pt	Peat and other highly organic soils

*After Ref. 6.

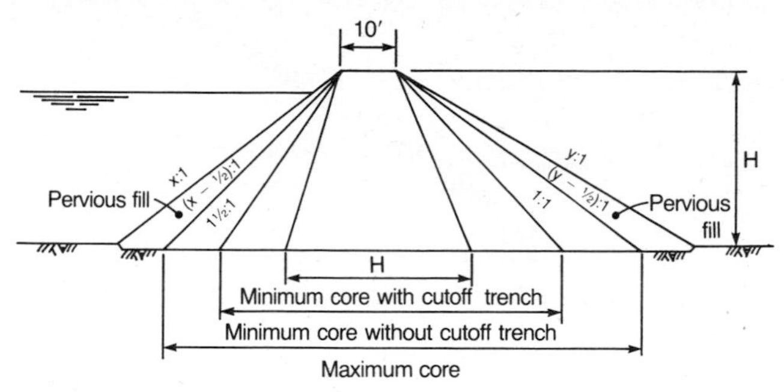

FIG. 5.4 Sizes of impervious cores as used in Table 5.1. *(After Ref. 6.)*

may have high liquid limits, will generally be very difficult to work. Fine silts with uniform gradation and rock flours are very sensitive to water content and are usually difficult to compact. Moisture content of silts and silty soils is low for optimum compaction, requiring special care to control moisture both on the embankment and in borrow areas. They are particularly difficult to use in rainy climates. In dry climates it will usually be necessary to add water to obtain proper compaction. In-site moisture content should be studied when a borrow area is being selected.

Wave Action and Freeboard

An earthfill dam can stand little or no overtopping. Overtopping during a flood, even for a very short time, can result in erosion and breaching of the dam.

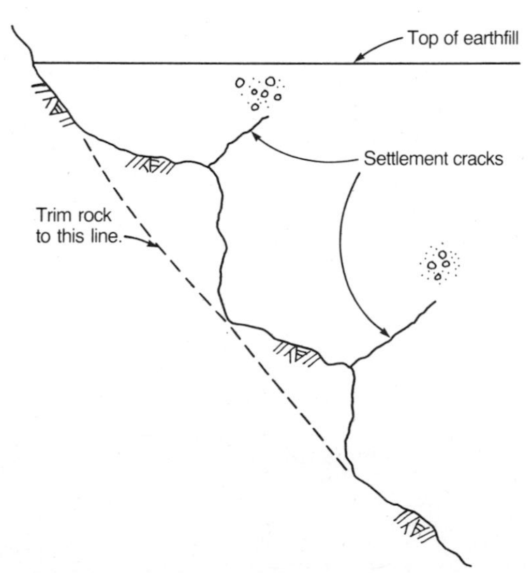

FIG. 5.5 Trimming required at a jagged rock abutment.

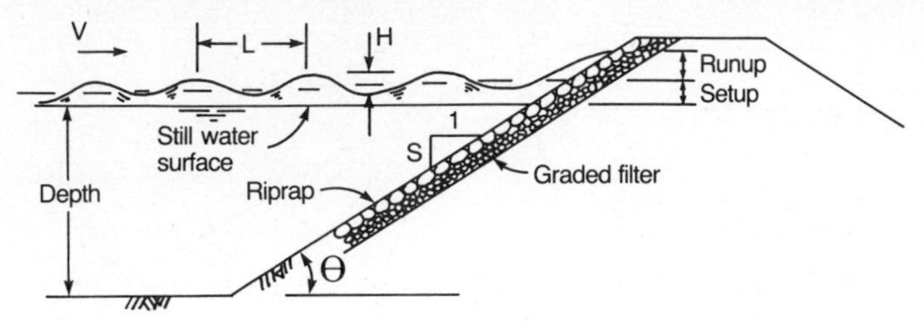

FIG. 5.6 Schematic illustration of wave height H and runup R.

Wave action can also be a threat to the integrity of the embankment. A layer of riprap must be placed on the upstream side of the dam to prevent the waves from eroding the fill. The size, gradation, and thickness of the riprap layer depends upon the height of wave which can be expected to develop on the reservoir. Figure 5.6 illustrates wave runup on a riprapped dam surface.

Waves are generated in inland reservoirs by sustained winds. The distance which the wind passes over water is called the *fetch*. Design fetch for a reservoir is determined by the reservoir configuration as well as the maximum over-water distance. Figure 5.7 shows a reservoir configuration as well as the max-

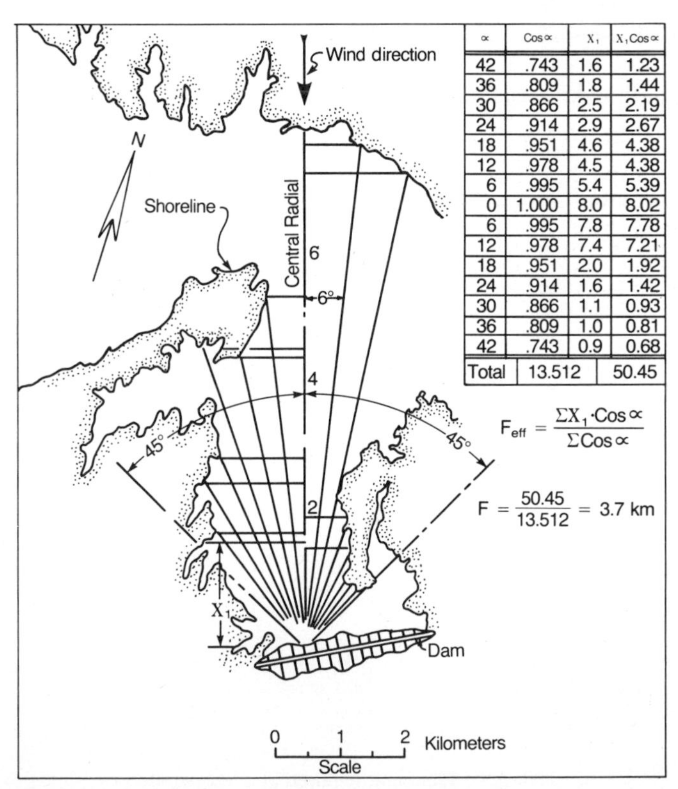

∝	Cos∝	X_1	X_1Cos∝
42	.743	1.6	1.23
36	.809	1.8	1.44
30	.866	2.5	2.19
24	.914	2.9	2.67
18	.951	4.6	4.38
12	.978	4.5	4.38
6	.995	5.4	5.39
0	1.000	8.0	8.02
6	.995	7.8	7.78
12	.978	7.4	7.21
18	.951	2.0	1.92
24	.914	1.6	1.42
30	.866	1.1	0.93
36	.809	1.0	0.81
42	.743	0.9	0.68
Total	13.512		50.45

$$F_{eff} = \frac{\Sigma X_1 \cdot \text{Cos} \propto}{\Sigma \text{Cos} \propto}$$

$$F = \frac{50.45}{13.512} = 3.7 \text{ km}$$

FIG. 5.7 Calculation of effective fetch for a reservoir.

imum over-water distance. It also shows the procedure by which effective fetch is calculated. The central ray in Fig. 5.7 is the maximum over-water distance between the dam and the far end of the reservoir. Supplementary rays are drawn as shown and the effective fetch is then calculated in accordance with the tabulated values.

Wind speed to be used in wave-height determinations must be chosen judiciously. Wind speed in the direction of maximum fetch is required. Data on wind speed is seldom available in sufficient quantity or quality to enable the development of a wind rose which would yield wind speed, direction, and duration. If possible 100-year and maximum wind speeds should be estimated on the basis of the nearest available records.

Wind-generated waves on an inland reservoir are not of uniform height. Instead, individual heights are part of a spectrum. If the heights of 33⅓ percent of the spectrum of waves are averaged, that average height is called the *significant wave height* a_s. Thirteen percent of the wind-generated waves will have heights greater than a_s.[7] Only 0.4 percent of all waves will have a height exceeding 1.67 times the significant wave height. Such a wave is referred to as the maximum wave.

The *maximum wave height* (1.67 a_s) is sometimes used as the design wave for freeboard and riprap design on large dams. However, for small dams the significant wave should be satisfactory.

Figure 5.8 can be used to estimate the height of the significant wave, once the design wind speed has been selected and the effective fetch has been calculated. Figure 5.8 also shows the minimum duration of the design wind speed required to produce the corresponding significant wave height.

The height to which a wave will run up the face of the dam depends upon the steepness of the wave (the wave height a divided by the wave length L), the roughness of the riprap surface, and the slope of the embankment. Rougher

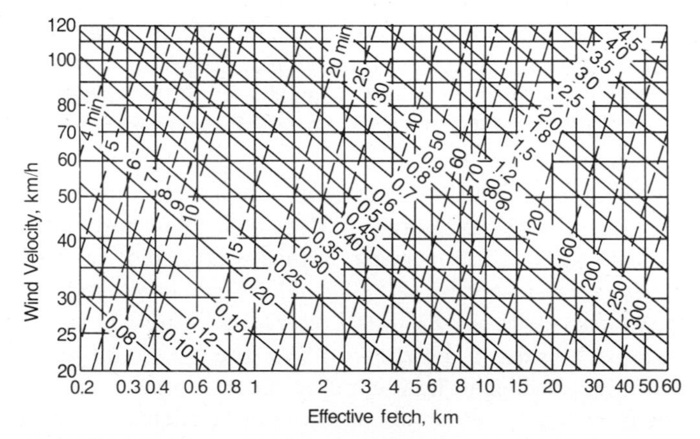

FIG. 5.8 Significant wave height as a function of fetch-sustained wind velocity (dashed lines indicate required wind duration in minutes).

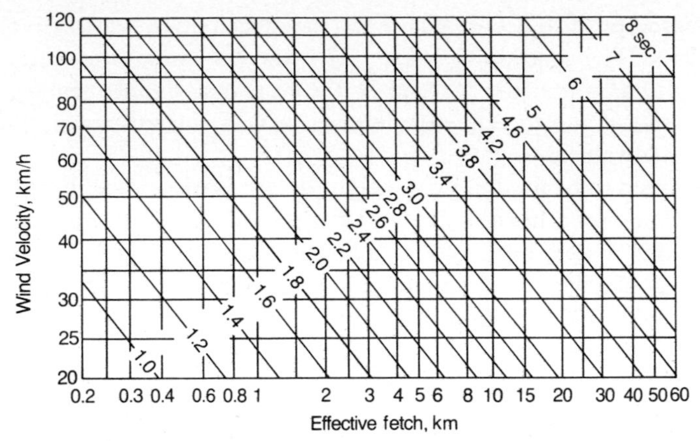

FIG. 5.9 Period of wind-generated waves as a function of sustained wind velocity and fetch.

riprap achieves greater energy dissipation and a lesser runup. Wave lengths for wind-generated waves can be computed from Eq. (5.7) or Eq. (5.8).

$$L = 5.12\,T^2 \qquad \text{(USCS units)} \tag{5.7}$$

$$L = 1.56\,T^2 \qquad \text{(SI units)} \tag{5.8}$$

where T is the wave period in seconds and L is the wave length in feet for Eq. (5.7) and meters for Eq. (5.8). Figure 5.9 can be used to obtain the wave period using the design wind speed and the calculated effective fetch.

When a deep-water wave approaches the face of the dam, its height increases because of the decreasing depth. When the depth becomes less than 0.78 a, the wave will break. For an earthfill dam, the face is steep enough that breaking of the wave effectively occurs on the face of the dam, and new smaller waves do not have time to form downwind from the breaking wave. Only runup of the fully developed wave needs to be considered. Relative runup R/a can be determined through use of Fig. 5.10 for both smooth and rough dam forces as a function of wave steepness a/L and the slope of the dam. Figure 5.10 shows that runup for a wave of given height is smaller for flatter slopes. Because the wave travels farther on a flatter slope, energy dissipation reduces the wave's potential to runup.

Winds blowing over an inland reservoir produce a shear over the water surface which in turn causes the reservoir surface to slope upward in the direction the wind is blowing. The rise in water surface produced at the dam is called *setup*. For very large reservoirs setup should be calculated and used in freeboard calculations. However, for small reservoirs, setup will be insignificant and need not be considered. Freeboard above the maximum reservoir water surface should be provided at least equal to the runup distance R. Allowance must also be provided for settlement of the dam. For dams constructed on relatively incompressible foundations a camber should be provided at least equal

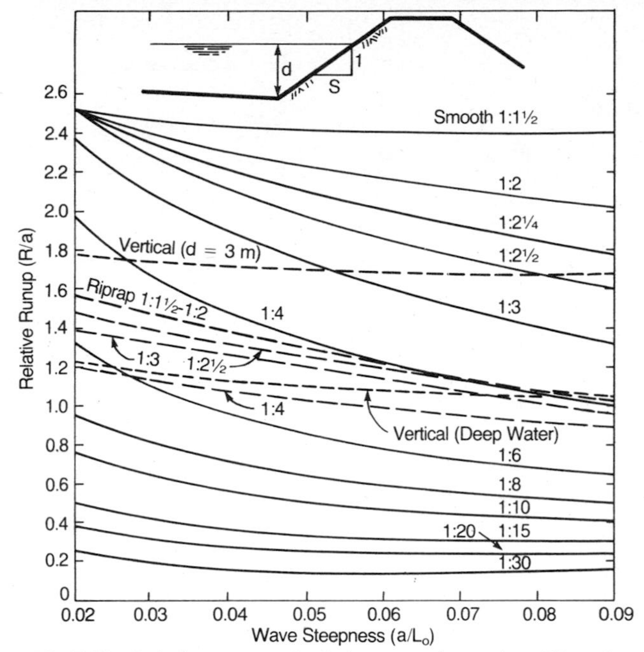

FIG. 5.10 Relative runup of wind-generated waves striking the face of a dam.

to 1 percent of the dam height. For dams on compressible foundations, 1 m (3.28 ft) or more may be required.

Riprap Design

Durable rock is desirable for use as riprap to protect the face of the dam from erosion by waves. Historical studies show that failure of riprap is very common on earthfill dams. Riprap normally fails by being displaced by wave action, in turn allowing erosion of the embankment. Riprap large enough and heavy enough to resist movement by wave action is necessary. A distribution of sizes is also required to prevent piping of the embankment through the interstices of the rocks. If soft stone, such as sandstone is used, waves and weather will readily break and erode the riprap to the point where it is no longer effective in damping wave action. Granite, basalt, and sound limestone are examples of rock which will be effective as riprap. Some maintenance should be expected for riprap, and it should be inspected at least annually and after any severe storm.

Weight of the median riprap rock is calculated in terms of the stability number N_s with Eq. (5.9).

$$W_{50} = W_R \left(\frac{a}{N_s(S - 1.0)} \right)^3 \tag{5.9}$$

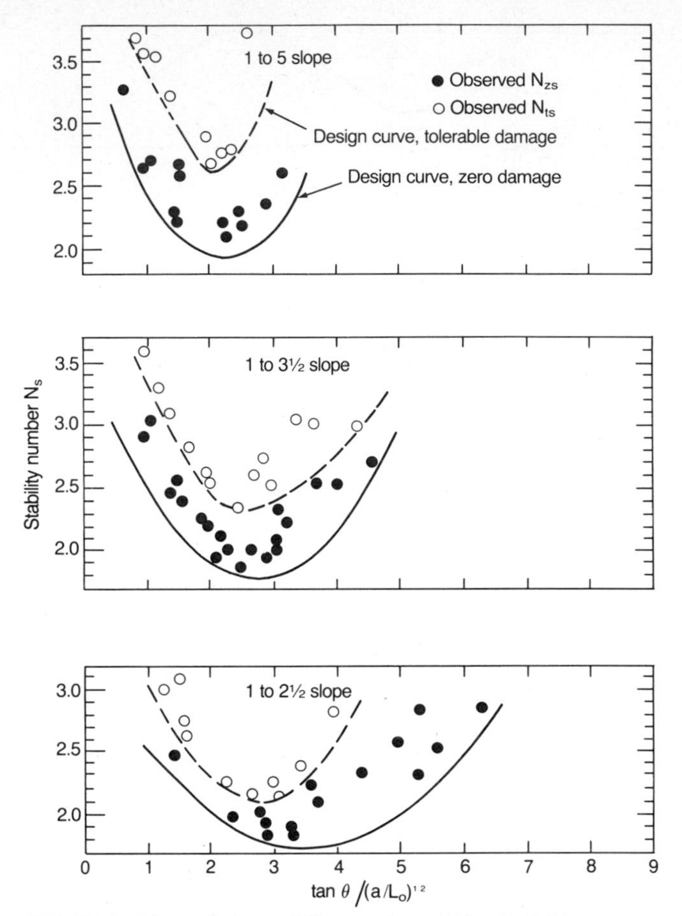

FIG. 5.11 Riprap design stability numbers. *(After Ref. 8.)*

where S is the specific gravity of the stone, W_R is the unit weight of the stone being used, and a is the design wave height.

Figure 5.11 can be used to select a proper stability number in terms of the wave steepness a/L and the tangent of the embankment slope angle (see Fig. 5.6). A range of values of N_s can be chosen from Fig. 5.11. The upper curve in each case will result in a stone size which will suffer a "tolerable" damage due to wave action.[8] Tolerable damage may involve some displacement of rock during a severe storm but no embankment erosion.

The stone diameter can be calculated from the stone weight as

$$D = 13.8 \left(\frac{W}{W_R} \right)^{1/3} \qquad \text{(USCS units)} \qquad (5.10)$$

where D is in inches, W is in pounds, and W_R is in pounds per cubic foot.

$$D = 114.9\left(\frac{W}{W_R}\right)^{1/3} \qquad \text{(SI units)} \qquad (5.11)$$

where D is in cm, W is in metric tons, and W_R is in metric tons per cubic meter.

Thickness of the riprap layer should be 1.5 to 2.0 times the size of the median stone. For proper riprap gradation the weights of the maximum stone and minimum stone should be 3.6 and 0.2 times the weight of the median stone respectively. Equation (5.10) or (5.11) can be used to calculate the stone size using stone weight.

Normally riprap will be obtained from a quarry where the maximum stone size is available, probably by blasting. The quarried material is run over a bar screen to segregate the large sizes. The finer material, which falls through the bar screen, is first placed on the face of the dam by dumping. Coarser stone is then dumped on top of the finer material. This procedure forms a natural filter between the riprap and the embankment. Ideally this filter layer should satisfy the gradation requirements of Eq. (5.1). Figure 5.6 illustrates the positioning of riprap and filter.

Appurtenant Structures

Under no condition should any type of conduit be placed in the embankment. As the embankment settles during and after construction, cracks will form in the fill around the conduit resulting in piping and, ultimately, in failure of the embankment. Many small dams have been constructed with corrugated metal pipes installed as culvert or conduit spillways within the embankment. Even with collars installed to control seepage, dam failures are very common among these structures.

If a conduit must be placed under an earthfill dam special care must be exercised. The conduit should be placed in a trench excavated in the undisturbed foundation. Backfill must be carefully placed around the conduit and hand tamped prior to placing any embankment. Joints must be provided which allow for extension when the conduit stretches as the foundation deforms under the load of the dam. The shape of the conduit should be such that fill will settle onto the conduit not away from it. The horizontal drainage blanket at the downstream side of the dam should be carried around the conduit to collect seepage and drain it safely. If the foundation is rock, the conduit should be backfilled with concrete.

If the conduit is to operate under pressure beneath the impervious core of the dam, it must be steel lined to prevent water under pressure from flowing out of possible leaks in the conduit and initiating a piping failure. Figure 5.12 illustrates a simple discharge conduit connected to a powerhouse. The slide gate shown on the upstream end of the conduit is used only for emergencies or for closure during maintenance of the turbine or other downstream structures.

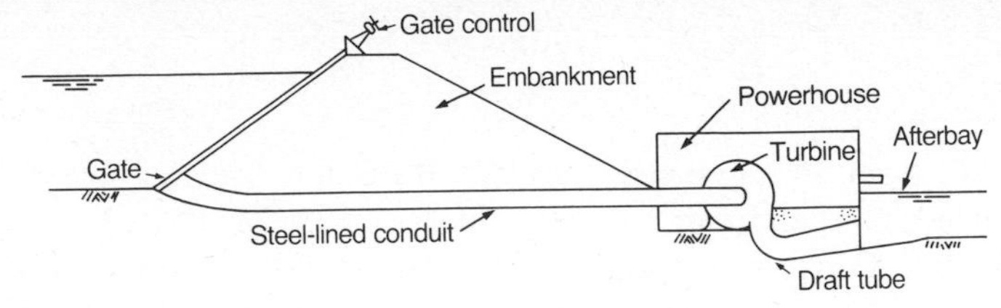

FIG. 5.12 Discharge conduit connected to a turbine.

Often a guard gate, such as a butterfly valve, is provided immediately ahead of the turbine for flow control.

Figure 5.2*c* shows a possible location for a spillway for an earthfill dam. Care should be taken to ensure that flow from the spillway does not impinge upon and erode the base of the dam. The crest for a controlled spillway should be located as far upstream as possible to minimize any possibility for seepage into the embankment or abutment while the reservoir water surface is high.

Figure 5.2*b* shows an earthfill dam and an incorporated concrete overflow section used as a spillway. The earth contact between the dam and the walls of the spillway must be compacted carefully to ensure continuous contact.

Maintenance

Earthfill dams do require maintenance. Because it is difficult to completely identify all geotechnical problems which can arise in the foundation and the abutments, it is necessary to inspect the completed dams on a regular basis. During the first year of operation the abutments, the downstream slope of the embankment, and the valley floor should be inspected at least weekly (daily during filling if the reservoir fills quickly). All seepage should be noted carefully and the volume rate of seepage flow should be measured using a small weir. Comparison of each measurement with preceding values will call attention to any increase in seepage rate. If seepage is normal and all drains and filters are working properly, seepage water will be clear. A qualitative assessment of seepage clarity should be recorded during each inspection. If the seepage becomes muddy, there will be cause for concern, since muddy water indicates that piping is occurring within the abutment, the embankment, or the foundation. If muddy discharge continues, the reservoir should be drawn down as quickly as possible and every effort should be made to grout or otherwise control the piping.

A postconstruction survey of the embankment configuration should be made and recorded. Concrete benchmarks should be established on the dam crest, near the downstream toe, and on the valley floor near the dam's toe. Annual measurements should be made of the elevation and location of these benchmarks to detect any bulging which might herald a slide of the embankment.

Riprap should also be inspected annually to document any movement or deterioration. Replacement of riprap in damaged locations should be made as soon as possible to prevent possible erosion of the embankment.

Erosion often occurs along the *groin,* the line of contact between the embankment and the abutment. A gravel layer can be placed along the groin to reduce potential for erosion. Monthly inspection should also include observation of the groins.

Any emergency equipment, such as guard gates at the penstock entrance, or guard valves within the penstock should be operated monthly to make certain that excessive rust or other damage does not prevent operation during emergencies.

5.3 ROCKFILL DAMS

Rockfill dams are defined as embankment dams for which at least half the material in the maximum cross section is composed of rock. Rockfill dams may have impervious-face membranes, sloping earth cores, thin central cores, or thick central cores. Rock used for the embankment may be composed of angular fragments produced by quarrying rocks in naturally occurring talus slopes, or stone obtained from natural gravel or boulder deposits. Rockfill may be loosely dumped in lifts up to 10 ft (3.3 m) in thickness, or it may be placed in 3.28-ft (1.0-m) layers and compacted mechanically. In dumped-rock the fill is best developed from unsegregated rock free of small gravel and fines since fine material will settle through the coarse rock resulting in settlement of larger rock above it and possibly excessive settlement of the dam. Ideally the rock should be of uniform-size stones which develop rock-to-rock contact. In compacted rock fills it is not necessary to use such well-graded rock, since vibrator compaction of thin lifts produces little or no tendency to segregate sizes.

Foundations for nearly all rockfill dams will be rock. To control seepage, the bottom of the impervious section must be carefully constructed.

Impervious Facings

Figure 5.13 shows a typical section for a rockfill dam with an upstream impervious facing. The impervious facing is most frequently made from concrete. However, facings have been successfully constructed from both timber and steel. A layer of hand-placed rock or a layer of compacted small rock is constructed on the upstream side of the fill as a backing for the concrete impervious facing. This backing serves to transmit force exerted by water in the reservoir uniformly to the rockfill. Dumped rockfill should be sluiced with water as it is placed in order to accelerate settlement of the rock. High-pressure jets should be directed into the fill at velocities up to 100 ft/s (30.5 m/s). Water volumes of 2 to 4 times the rockfill volume should be used. Settlement of the rockfill during construction and sluicing can be as much as 4 to 6 percent of

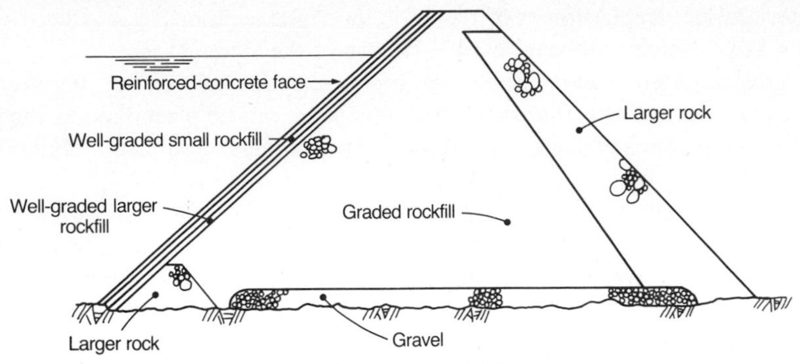

FIG. 5.13 Typical maximum section for a rockfill dam.

the height of the fill. Most settlement will occur during construction, but settlement of as much as 1 percent of the dam's height can be expected to occur within several years after completion. The impervious facing must be designed to allow for this settlement. Lateral movement of the fill will occur at abutments during settlement. The top of the fill should be constructed higher than the designed height by 1 to 2 percent of the dam's height to allow for settlement.

The upstream and downstream slopes of rockfill dams are quite steep when compared to earthfill dams of equal height. The slopes are constructed at approximately the angle of repose of the dumped rock which may vary from 1.4 to 1.2 horizontal to 1.0 vertical. Careful placing of a layer of rock on the upstream side may make it possible to construct slopes which are steeper than the natural angle of repose of the rockfill.

For the small dams discussed herein, the impervious concrete facing should not be less than 1.0 ft (305. mm) thick throughout. Reinforcing-steel area in the slab should be 0.5 percent of the concrete area in both the horizontal and vertical directions. Vertical expansion joints should be provided at approximately 30-ft (9.1-m) intervals in the concrete facing. Plastic or rubber water stops should be provided in the joints. Figure 5.14 illustrates expansion joint details. A concrete beam should be cast behind each joint so that the facing slab can move without damage during settlement of the dam or larger temperature changes.

Asphaltic concrete facings have been used successfully on several high dams

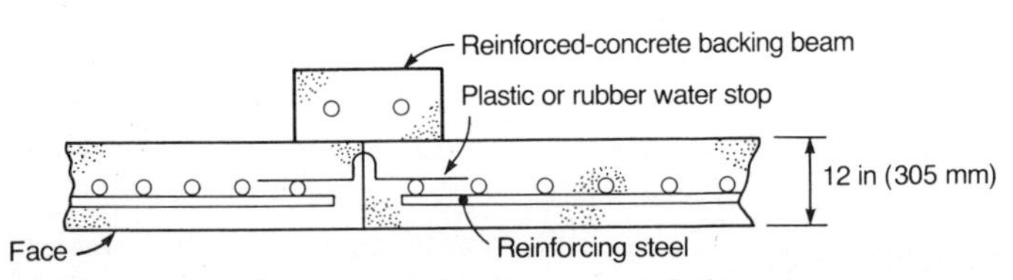

FIG. 5.14 Typical expansion joints for concrete facings.

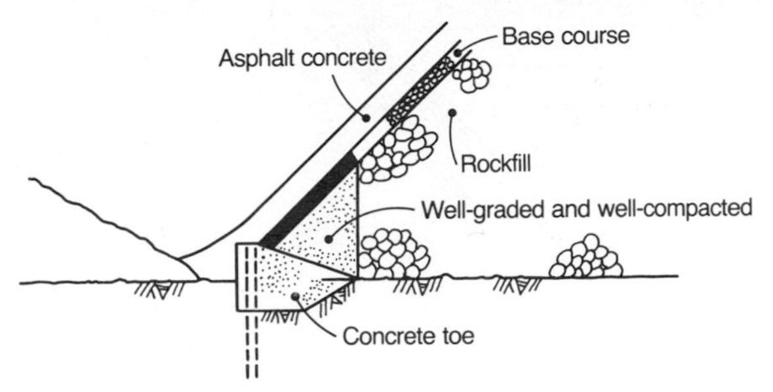

FIG. 5.15 Typical asphaltic concrete facing.

in the United States and in Algeria.[11] Figure 5.15 shows a typical section for an asphaltic concrete facing. Since the asphaltic membrane is unreinforced, special care must be taken to ensure its flexibility to prevent the formation of cracking during settlement. It must also be more impervious than is normally developed on highway surfaces. The aggregate used must have a high percentage of fines. A gradation used for the asphalt facing of Homestake Dam was[10]

Size	Grading % passing
1½ in (38.1 mm)	100
No. 4 (4.76 mm)	48–75
No. 200 (0.074 mm)	7–15

Using dry aggregate weight, 8 percent of 50–60 penetration asphalt was used in the mix. Careful tests should be made to determine a satisfactory mix. The asphalt should be rolled in layers and total thickness should not be less than about 7.0 in (178 mm). Proper backing of the asphalt facing is extremely important.

Timber facings on rockfill dams were very common in the early twentieth century. If a lasting wood, such as redwood, is available, the facing can be quite durable.

Steel facings have also been used on a number of older dams.[11] Its desirable characteristics are strength and ease of repair. However, current costs (1983) tend to make steel much more expensive than either reinforced concrete or asphaltic concrete. In addition leaks can develop rapidly as a result of corrosion.

Impervious Cores

A relatively thin zone of compacted earth is frequently used as the impervious zone on a rockfill dam. Rockfill sections on the upstream and downstream sides of the impervious core provide the necessary structural support. Figure 5.16

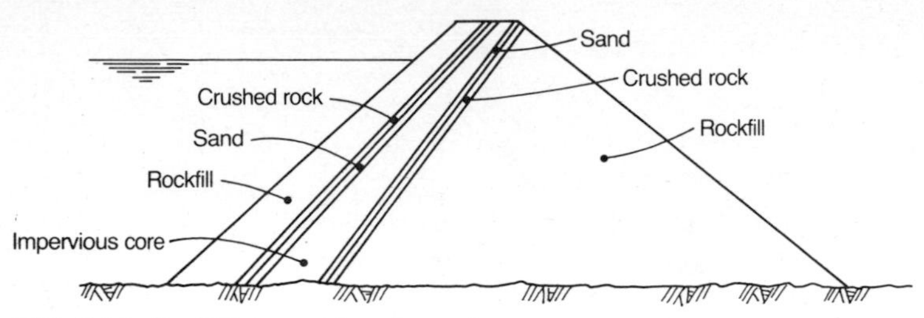

FIG. 5.16 Rockfill dam with an impervious core.

illustrates this combination. The impervious core is designed and constructed as in an earthfill dam. However, a filter section is required on both the upstream and the downstream side of the core. These filters require special considerations, since the rockfill zones will have much larger voids than an earthfill section. The upstream filter is required to prevent wave action and rainfall from eroding the upstream face of the core. The downstream filter protects against piping as well as erosion from rainfall percolation. To provide adequate piping protection, the filter section should be graded outward from sand, through gravel, to rock as shown in Fig. 5.16.[12]

Although vertical impervious cores have been used on many dams, sloping cores provide protection against piping cracks in the core created by settlement. The sloping core also makes construction of the various filter layers easier since the entire dam does not need to be raised simultaneously. Each layer can be placed and compacted individually, progressing in the upstream direction. It is very important to develop good compaction in all layers and particularly in the downstream rockfill. Optimum slope of the face of the sloping core is from 0.5 to 0.6 horizontal to 1.0 vertical. However, the contact between the foundation and the impervious core occurs near the upstream side of the dam, which tends to reduce the normal force of contact. Thus, particular care must be taken there in cleaning, grouting, and preparing the foundation.

Concrete core walls have been used in the past on both earthfill and rockfill dams. Although most of those dams still exist many problems have arisen including cracked and leaking core walls. The cracking is the result of stresses induced by settlement of the dam. In general, brittle materials, such as concrete, should be avoided in core walls.

Inspection

Rockfill dams, because of possibly large settlement need to be observed carefully during first filling. Concrete surveying monuments should be placed along the crest of the dam and along the downstream toe so that settlement and movement can be detected and measured. As pointed out earlier, cracks in the impervious core can occur as a result of settlement of the dam. Seepage flows should also be measured to detect any significant changes in seepage rate or muddiness of the flow.

Spillways and Outlet Works

Design, construction, and configuration of outlet works and spillways are the same as for earthfill dams. However, rockfill dams can be designed for overflow if necessary. Thus, there is a tendency to construct concrete-slab spillways on the completed dam embankment. If this is to be done, the embankment should be allowed to settle before spillway construction begins.

5.4 CONCRETE DAMS

Concrete dams are most often one of three types: gravity, arch, and buttress (Fig. 5.1). Each type of dam lends itself to particular situations. The *arch dam* is most likely used in a narrow site with steep walls of sound rock. *Gravity dams* depend upon their own weight to be stable and are generally used where the foundation is rock and earthfill in proper quality and quantity is not available. Low concrete dams of height less than 10 m (32.8 ft) can also be built on earth foundations, but particular precautions must be taken to ensure that allowable foundation stresses are not exceeded and that the possibility of piping due to seepage under the dam is minimized. When a concrete dam is constructed on an earth foundation, uplift forces created by reservoir water pressure must be controlled through the use of cutoffs or aprons.

Prior to final design of a concrete dam, it is necessary to determine the characteristics of the material which will be used in the construction. Compressive, tensile, and shear strengths must be determined for the design mix. In addition the instantaneous and sustained moduli of elasticity, Poisson's ratio, thermal conductivity, specific heat, and diffusivity must be determined for the mix. For the foundation material, it will be necessary to know the shear strength, permeability, compressive strength, and Poisson's ratio. When the foundation is not homogeneous, it will be necessary to determine the preceding quantities for each material. In-depth structural analysis of arch and buttress dams has not been presented because of the detail required and because such dams will very seldom be economical. (Design details for arch and buttress dams are given in Ref. 1.)

Design Loads

Each concrete dam must be analyzed considering the following structural loads (See Fig. 5.17):

1. The weight of the dam itself (dead load). All weight is assumed to be transferred directly to the foundation without shear stress between adjoining blocks. The force due to weight acts at the centroid of the section. Concrete is assumed to weigh 150 lb/ft^3 (2,402 kg/m^3).
2. Hydrostatic forces produced by water in the reservoir. These may be sepa-

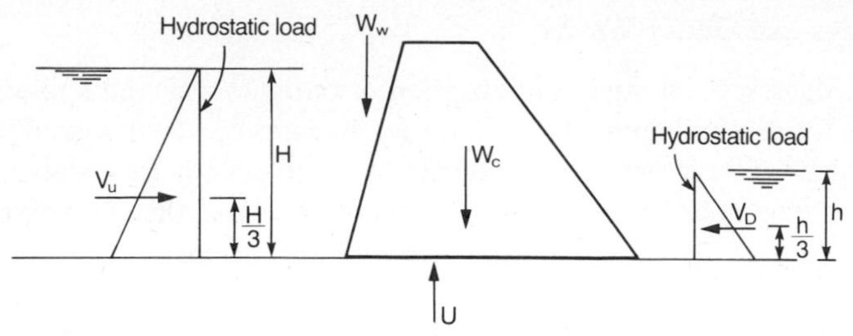

FIG. 5.17 Forces acting on a dam.

rated into horizontal components and vertical components as shown in Fig. 5.17.

3. Hydrostatic uplift acting under the base of the dam. The uplift pressure is assumed to vary linearly from that created by the full reservoir at the upstream face to the pressure created by tailwater depth at the downstream toe. The integrated pressure over the base should reflect the presence of any drains.
4. Forces due to silt deposited in the reservoir, but in contact with the dam. Normally 85 lb/ft^3 (1,362 kg/m^3) is used as specific weight of silt.
5. Ice loads should be considered in cold climates where thick ice will form and can be expected to remain frozen for weeks or months. A horizontal load of 10,000 lb/ft (14,882 kg/m) is assumed to act on the dam, at the level where ice will form if ice depths of 2 ft (0.6 m) or greater can be expected.
6. Earthquake forces must be assumed to act both horizontally and vertically through the center of gravity of the dam if the site is in a seismically active area. The forces are applied statically unless horizontal accelerations exceed 0.10 *g*. Dynamic analysis of the structure must be done for larger seismically produced accelerations. Local authorities must be contacted to obtain maximum expected horizontal and vertical accelerations for earthquakes. The forces applied are actually inertia forces equal to the product of the earthquake acceleration and the mass of the dam. Their direction should be assumed so as to produce critical stress.
7. Earthquake forces are also induced by the relative movement of the dam and reservoir. The horizontal pressure force for a unit width of dam due to earthquake is given by:

$$F_e = 0.726\ C\alpha\gamma h^2 \tag{5.12}$$

where

F_e = the force per unit width of dam

α = coefficient which when multiplied by the acceleration due to gravity, 32.2 ft/s^2 (9.8 m/s^2) gives the horizontal acceleration in an earthquake. It is usually less than 0.1 except in areas of strong seismic activity.

γ = specific weight of water, 62.4 lb/ft^3 (1,000 kg/m^3)

h = depth of water in the reservoir

C = a constant depending on the upstream slope of the dam.

The constant C is given approximately by

$$C = 0.7(1 - \theta/90) \tag{5.13}$$

where θ = the angle between the upstream face of the dam and a vertical line (in degrees).

The moment about the base of the dam is given by

$$M_e = 0.299\ C\alpha\gamma h^3 \tag{5.14}$$

where M_e = the moment per unit width of dam.

A *normal* combination of forces to be considered during design includes hydrostatic forces produced by a reservoir at maximum normal elevation, temperature stresses produced by temperatures normal for that time of year, weight of the dam, and ice and silt load. An *unusual* load combination could include loads produced by the reservoir at maximum elevation coupled with stresses produced by minimum temperatures and appropriate dead load, hydrostatic forces, and silt loadings. A load combination including an earthquake would be classified as a *severe* loading. Factors of safety for unusual and severe loadings should be smaller than those acceptable for usual loads.

For a small concrete dam designed for overflow, full consideration should be given to hydrostatic forces upstream and downstream, but the weight of water flowing over the structure should not be considered.

Analysis of Stability

The dam must first of all be safe from overturning for all loading conditions. The contact stress between the foundation and the dam must be greater than zero at all points or the dam is unsafe against overturning. That stress is calculated considering all overturning moments and the weight of the dam, and assuming a straight-line distribution of stress on the base. To assure that the stress does not go to zero, the resultant of all horizontal and vertical forces must pass through the middle one-third of the base.

The maximum compressive forces in the concrete and on the foundation must be less than the specified concrete compressive strength divided by 1.0, 2.0, and 3.0 for the *severe, unusual,* and *normal* loading cases respectively.

Shear stresses on any plane within the dam must be less than the specified concrete shear strength divided by 1.0, 2.0 and 3.0 for the *severe, unusual,* and *normal* loading cases respectively.

The dam must be safe against sliding. Equation (5.15) can be used to determine the static coefficient of friction between the foundation and the dam.

$$f = \frac{\Sigma F_h}{\Sigma F_v} \tag{5.15}$$

where

f = static coefficient of friction
ΣF_h = sum of all horizontal forces acting on dam
ΣF_v = sum of all vertical forces acting on dam.

The coefficient of static friction may vary in the range from 0.6 to 0.75. However, a value should be determined by laboratory test for the actual foundation material.[11] The acceptable coefficient of friction should be not more than the specified value divided by 2.0, 1.5, and 1.25 for the *severe, unusual,* and *normal* loading combinations respectively.

Configuration

The concrete gravity dam will frequently have a vertical upstream face to concentrate the weight of the dam upstream. The downstream face usually has a constant slope. The dam is usually laid out and then an analysis of all loads and stresses is conducted. The section is then increased or decreased as required to meet the stability requirements discussed under the previous heading Analysis of Stability. Each successive layout should be an improvement on the previous section.

Width of the top of dam is usually determined by the width required to move any necessary equipment over the dam.

For an overflow dam, the layout is similar except that the downstream surface should correspond to that given for spillways in Chap. 4. Width of the base of the overflow section is increased or decreased as required for stability. If gates are incorporated, forces produced by water in the reservoir should be applied with the gates closed.

Freeboard for a concrete dam can be economically provided with a thin concrete or timber parapet wall constructed along the upstream side of the top of the dam.

Foundation

Foundation for a concrete dam must be very carefully prepared. All alluvial overburden must be excavated to expose the rock. The rock must be thoroughly cleaned and all loose or cracked rock removed. All cracks should be grouted. All weak or badly fractured rock must be removed and the hole backfilled with concrete. In general all seams should be excavated to a depth of at least 0.1 times the dam height. However, final judgment on depth of excavation should be made at the site. If the crack is large, or if the material filling it is quite soft, greater excavation may be required. Rock in the foundation should be excavated to eliminate any sharp changes in elevation where stress concentrations can occur.

If the foundation is shale, chalk, mudstone, or siltstone it will be desirable to cover the exposed rock with 4- to 6-in (100- to 150-mm) thickness of mortar to prevent rapid deterioration of the foundation by weather.

The foundation must be carefully examined to determine its permeability. If joints, crevices, or strata of permeable material exist, excessive uplift and/or excessive leakage can occur and it may be necessary to grout the foundation. In grouting, holes are drilled into the foundation and cement grout is pumped into the holes under pressure. The decision of whether to grout and how deep to grout can only be made by an experienced geologist after studying the detailed geology of the site. Because every foundation is unique it is always advisable to have an experienced geologist or engineer inspect the foundation prior to design but particularly prior to construction.

Spillways and Outlet Works

One basic advantage of a concrete dam is that it can be designed as an overflow structure. Figure 5.2*b* shows a typical concrete overflow structure flanked by an earthfill dam on each side. The overflow structure is constructed of reinforced concrete and must be designed in accordance with the criteria for any concrete dam. Spillway sections are usually located near the old streambed.

Bottom outlet works are usually incorporated in the concrete dam section. They are required to drain the reservoir for maintenance or to furnish required flow in the stream. For a small dam, such as considered herein, the outlet works is usually a steel conduit cast into the dam. A bulkhead or guard gate is frequently provided on the upstream face.

Power outlets may be cast into the dam if the powerhouse is immediately downstream from the dam. However, most small dams provide a means of diverting water into a power canal, a flume, or a power tunnel. The location of such structures is usually on one bank immediately upstream from the dam.

5.5 MASONRY DAMS

Masonry dams are extremely labor intensive and, thus, are economical only where labor is relatively inexpensive and fractured stone (rubble) of durable quality is readily available. Cement mortar is used to embed each stone. Many notable masonry dams have been built in the past century and have provided excellent service. The theory used in designing a masonry dam is the same as described for a concrete dam. However, allowable stresses for masonry are much more uncertain than for concrete. In 1910, Baker (cited in Ref. 13) rec-

TABLE 5.3 Allowable Stress Levels for Masonry Dam Materials

Material	Allowable stress	
	lb/ft²	kg/m²
Rubble	20,000–30,000	97,650–146,470
Squared stone	30,000–40,000	146,470–195,300
Limestone	40,000–50,000	195,300–244,120
Granite	50,000–60,000	244,120–292,950

ommended the data in Table 5.3 as allowable stresses for rock and mortar of excellent quality.

All forces considered in design of a concrete dam should be considered in the design of a masonry dam. Since allowable stresses for masonry are lower than for concrete, the final structure will usually have a wider base than a concrete dam. Unit weight of the rock to be used must be determined by test prior to design.

Foundation preparation for a masonry dam is equally important as it is in a concrete dam and the same precautions should be followed. In addition, a 6-in (150-mm) layer of mortar should be placed over the foundation prior to laying of the masonry. Particular care must be taken to obtain clean sand, free of earth and organic material, for use in mixing the mortar. Clean water is also vital. In addition, rocks should be well cleaned of mud or other adherents prior to being placed in the dam.

5.6 SPILLWAY DESIGN

The spillway is designed to pass flows larger than can be used for hydroelectric generation. Importance of the spillway cannot be overemphasized. Many dam failures have occurred because the spillway was incapable of passing a particular flood, with the result that the dam was overtopped and breached. Many spillways with adequate capacity have failed because severe erosion occurred at the base of the spillway, resulting in damage to the spillway, the dam, or both.

Overtopping of earthfill dams will usually result in their breaching, unless the overtopping is of very short duration due to waves. Concrete dams may, however, withstand moderate overtopping. Rockfill dams may withstand minimal overtopping, unless they have an impervious earthfill core.

The flood for which a spillway should be designed has not been uniformly accepted by the profession. Criteria specified by the U.S. National Dam Safety Act (1972) provides the guidelines shown in Table 5.4, where PMF indicates probable maximum flood. Thus a small dam located in a drainage area not far above an inhabited village might require a spillway designed for a PMF. However, the designer must also consider that, in the event of a PMF, the failure of the small dam may have only an insignificant effect upon the depths and velocity of flood-flow downstream. On the other hand the incremental cost required to increase a spillway's capacity is often much less than directly proportional to the increase in capacity. An economic analysis to minimize costs and optimize spillway size is usually not feasible because of difficulty in assigning a value to human life where at least significant hazard is involved.

Frequently two spillway structures will be provided. The first, a service spillway, may be a small overflow concrete structure as shown in Fig. 5.2*b*. This spillway will be designed to pass all small floods which produce no danger of overtopping the dam. A second spillway, an emergency spillway, may be located off the dam and will be designed to supplement the service spillway during the large design flood. Some damage should be expected if the emergency spillway operates.

The simplest form of spillway is an uncontrolled crest. Coordinates for shaping such a crest, and equations for determining the hydraulic capacity of the crest, are given in Chap. 4. If the design discharge for the spillway is large, the length of the uncontrolled crest may become very long. If the length cannot be developed at the site, it may be necessary to provide gates on the crest which

TABLE 5.4 Criteria Specified by the U.S. National Dam Safety Act

Size classification

Category	Reservoir storage, s		Dam height, H	
	Acre-feet	Cubic meters	Feet	Meters
Small	50–1,000	0.06×10^6 to 1.23×10^6	25–40	7.6–12.2
Intermediate	1,000–5,000	1.23×10^6 to 61.7×10^6	40–100	12.2–30.5

Hazard potential

Category	Loss of life	Economic loss
Low	None expected	Minimal
Significant	Few	Appreciable
High	More than a few	Excessive

Recommended design flood

Hazard	Size	Design flood
Low	Small	50–100 year
	Intermediate	100–½ PMF*
Significant	Small	100-year–½ PMF*
	Intermediate	½ PMF*–PMF*
Large	Small	½ PMF*–PMF*
	Intermediate	PMF*

*Probable maximum flood.

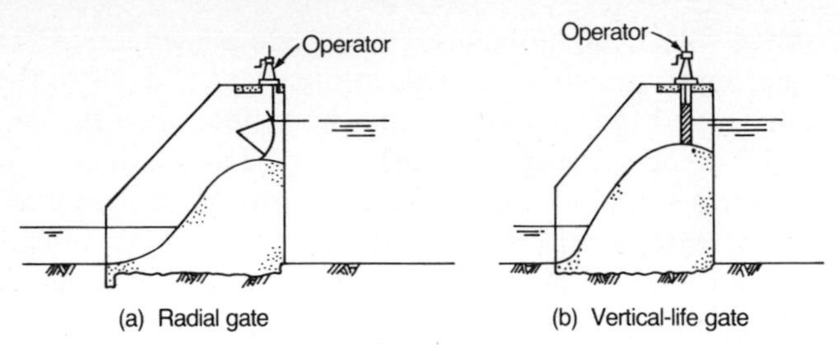

FIG. 5.18 Gate installations on a concrete spillway.

control the flow, except when large flood flows must be passed. Flashboards, stop logs, rectangular gates and radial gates are commonly used on small dams.

From the standpoint of operation, radial gates are easiest to operate. The resultant of pressure forces acting on the radial gate is normal to the circular surface, thus causing no moment. Only the weight of the gate itself must be lifted when the gate is opened. Figure 5.18 shows a typical radial gate installation.

Rectangular vertical gates can be constructed as roller gates which makes it possible to raise them under full hydrostatic pressure. However, slide gates without rollers can be very difficult to operate under pressure. Figure 5.18 shows a typical vertical gate installation.

Stop logs are narrow rectangular beams which can be placed in slots on the spillway crest to raise the reservoir surface. Because of friction in the slots, they are very difficult to install or remove under overflow conditions. They should not be used in situations where unexpected floods can occur since advance warning is required to remove the flashboards prior to a flood. Figure 5.19 shows a typical stop-log installation.

Flashboards consist of individual boards which are held on the spillway crest by vertical pipes or columns anchored to the crest. The flashboards can be designed to fail if the level in the reservoir reaches a given level, thus providing an automatic operation. If they are not designed to fail automatically, sufficient warning time must be possible to allow removal of the flashboards before a flood arises.

Riprap will be required in the channel downstream from the spillway to control erosion unless the channel is rock. To prevent erosion of the dam, the

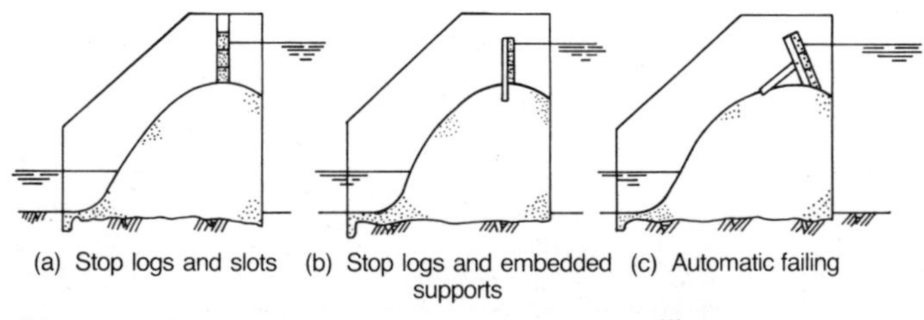

FIG. 5.19 Flashboard installations on a concrete spillway.

spillway should be extended downstream from the toe of the dam a sufficient distance to ensure that velocities near the embankment are below the magnitude which will produce scour.

5.7 DIVERSION DURING CONSTRUCTION

Except for unusual situations it will be necessary to divert the natural stream during construction. Diversion through a tunnel drilled in one abutment provides the most convenient diversion since the entire dam site can be excavated and the foundation cleaned in one operation. A cofferdam is constructed on the upstream and the downstream side of the site. Once construction on the dam has advanced far enough the tunnel can be plugged. A concrete bulkhead is

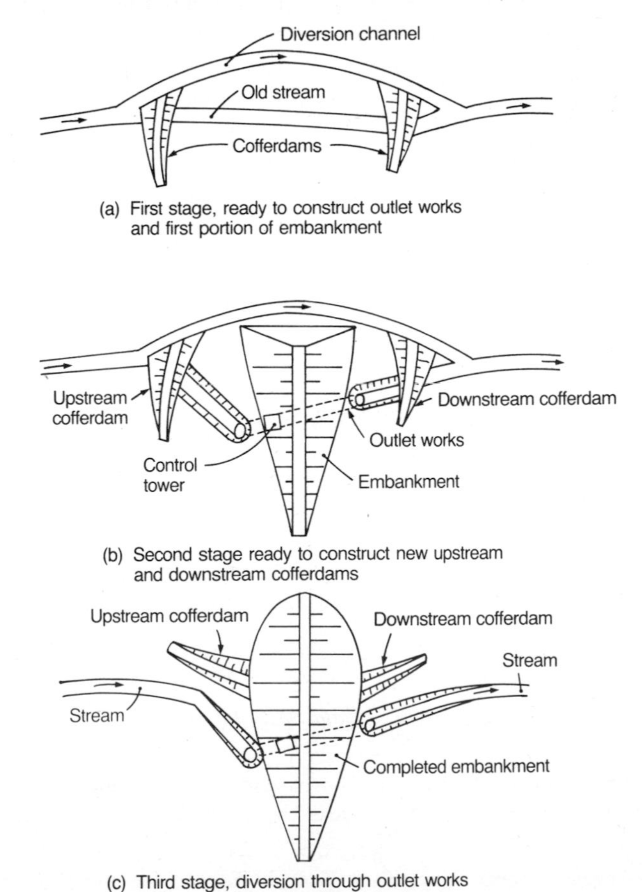

FIG. 5.20 Typical diversion plan.

usually constructed to make the initial closure on the tunnel. The tunnel can in some cases be used as an outlet works if a concrete plug is constructed in the tunnel with a conduit and valves installed in the plug.

More commonly on small dams, a diversion channel is excavated in one bank. A cofferdam is constructed upstream and downstream of the site. The upstream cofferdam diverts the natural flow of the stream into the diversion channel. An outlet works is constructed in the first portion of the dam to be completed. A new cofferdam is constructed to divert the natural flow through the new outlet works. A new cofferdam is constructed around the remaining site and the dam is completed. Figure 5.20 illustrates this type of diversion in stages. Normally diversion structures and cofferdams are designed for a 10-year flood. In some cases the cofferdams are designed to be incorporated into the final section of the dam. Pumping will usually be required to keep the dam site dry since seepage through and around the cofferdams will be inevitable. The cofferdams, although low, should be considered to be earthfill dams and designed accordingly.

Closure of the diversion plan is the most critical time and should be planned for that time of year when the lowest flow is assured. Closures are frequently accomplished by dumping rock and earth into the channel rapidly. However, it may be easier and more economical to construct a closure structure into which stop logs can be lowered to close the section.

5.8 REFERENCES

1. Alfred R. Golze (Ed.), *Handbook of Dam Engineering,* Van Nostrand Reinhold, San Francisco, 1977.
2. Robert B. Jansen, *Dams and Public Safety,* U.S. Department of the Interior, Water and Power Resources Service, Denver Federal Center, Denver, Colo., 1980.
3. K. Terzagi and R. B. Peck, *Soil Mechanics in Engineering Practice,* Wiley, New York, 1948.
4. C. J. Posey, "Terzaghi's Inverted Filter—An Important Tool for the Civil Engineer," *Civil Engineering,* April 1976.
5. J. L. Sherard, R. J. Woodward, S. F. Griziensky, and W. A. Clevenger, *Earth and Earth-Rock Dams,* Wiley, New York, 1963.
6. U.S. Bureau of Reclamation, *Design of Small Dams,* chap. VI, Denver, Colo., 1974.
7. U.S. Army Corps of Engineers, *Computation of Freeboard Allowance for Waves in Reservoirs,* ETL 1110-2-8, Office of the Chief of Engineers, Washington, D.C., August 1, 1966.
8. J. P. Ahrens, "Large Wave Tank Tests of Riprap Stability," *Technical Memorandum No. 51,* U.S. Army Corps of Engineers, Coastal Engineering Research Center, Fort Belvoir, Va., May 1975.

9. U.S. Army Corps of Engineers, *Hydraulic Design Criteria,* Waterways Experiment Station, Vicksburg, Miss.
10. H. H. Burke, M. M. Forest, and R. L. Perley, "Design and Construction of Homestake Dam," preprint 346, *American Society of Civil Engineers, Annual Water Resources Conference,* Denver, Colo., 1966.
11. K. V. Taylor, "Slope Protection on Earth and Rockfill Dams," *Transactions of the Eleventh International Congress on Large Dams,* vol III., Madrid, Spain, 1973.
12. M. Maksimovic, "Optimum Position of the Central Clay Core of a Rockfill Dam in Respect to Arching and Hydraulic Fracture," *Transactions of the Eleventh International Congress on Large Dams,* vol. III, Madrid, Spain, 1973.
13. W. P. Creager, *Masonry Dams,* Wiley, New York, 1929.

6

Hydraulic Turbines

Roger E. A. Arndt
Cesar Farell
Joseph M. Wetzel

6.1 INTRODUCTION

The development of hydraulic turbines as we know them is in a mature state. Much of the more recent developmental work has been directed toward the design of very large, highly efficient units. Unfortunately, smaller turbine technology has not benefited to any great degree from the research and development of larger units. Smaller turbines, essentially scaled-down units, have not reached the same degree of perfection as their larger counterparts for purely economic reasons.

Smaller turbines are essentially built-to-order as are larger units, but their smaller size makes them less profitable. Rapid escalation in the cost of energy made many smaller sites economically feasible and greatly expanded the market for small turbines. Added to this is the rapidly increasing need for turbines in the less developed countries (LCDs) where hydropower represents an attractive source of energy that can be easily developed. Because of differences in the design objectives for the smaller units, it is anticipated that new developments will occur relatively rapidly as the marketplace creates new competitive pressures for the manufacturers.

Since we are in transition from a relatively stagnant period of design and development to an anticipated renaissance of small-turbine development, this chapter is structured to provide the basics of turbine performance as well as a summary of turbine characteristics which affect the operation of a hydropower site. Cookbook procedures for turbine selection are avoided. Instead, a firm grounding in the basic principles of turbine hydrodynamics is given to provide the necessary tools not only to select a turbine from the current offerings but also to evaluate new turbine types which may appear.

6.2 HISTORICAL PERSPECTIVE

Evolution of the Modern Turbine from the Water Wheel

The water turbine has a rich and varied history and has been developed as a result of a natural evolutionary process from the waterwheel.[1,2] The water turbine was originally used for direct drive of machinery; its use for the generation of electricity is comparatively recent. Much of its development occurred in France. England had cheap and plentiful sources of coal which sparked the industrial revolution in the eighteenth century, but Nineteenth-century France had only water as its most abundant energy resource. To this day *houille blanche* (literally white coal) is the French term for waterpower.

In 1826 the Société d'Encouragement pour l'Industrie Nationale offered a prize of 6,000 francs to anyone who "would succeed in applying at large scale, in a satisfactory manner, in mills and factories, the hydraulic turbines or wheels with curved blades of Bélidor."[2] Bernard Forest de Bélidor, a hydraulic and military engineer, authored (1737–53) the monumental 4-volume *Architecture Hydraulique,* a descriptive compilation of hydraulic engineering information of every sort.

The watcrwheels described by Bélidor departed from convention by having a vertical axis of rotation and being enclosed in a long cylindrical chamber approximately 1 m (3.28 ft) in diameter. Large quantities of water, supplied from a tapered sluice at a tangent to the chamber, entered with considerable rotational velocity. This pre-swirl combined with the weight of water above the wheel was the driving force. The original tub wheel had an efficiency of only 15 percent to 20 percent.

Water turbine development proceeded on several fronts from 1750 to 1850. The classical horizontal-axis waterwheel was improved by such engineers as John Smeaton (1724–92) of England, who used the first avowed model experiments in this endeavor and also played an important role in windmill development, and Jean Victor Poncelet (1788–1867) of France. These improvements resulted in waterwheels having efficiencies in the range of 60 to 70 percent.

At the same time, reaction turbines (somewhat akin to the modern lawn sprinkler) were being considered by several workers. The great Swiss mathematician Leonhard Euler (1707–83) investigated the theory of operation of these devices. A practical application of the concept was introduced in France in 1807 by Mannoury de Ectot (1777–1822). His machines were, in effect, radial outward-flow machines. The theoretical analyses of Claude Burdin (1790–1893), a French professor of mining engineering who introduced the word *turbine* in engineering terminology, contributed much to our understanding of the principles of turbine operation and underscored the principal requirements of shock-free entry and exit with minimum velocity as the basic requirements for high efficiency.

A student of Burdin, Benoît Fourneyron (1802–67), put his teacher's theory to practical use, which led to the development of high-speed outward-flow turbines with efficiencies of the order of 80 percent. The early work of Fourneyron resulted in several practical applications and won the coveted 6,000 franc prize in 1833. After nearly a century of development, Bélidor's tub wheel had been officially improved.

Fourneyron developed some 100 turbines in France and elsewhere in Europe. Some even were sent to the United States, the first in about 1843. The Fourneyron centrifugal turbines were designed for a wide range of conditions, with heads as high as 114 m (374 ft) and speeds as high as 2,300 r/min. Very low-head turbines were also designed and built.

Fourneyron turbines, successful as they were, lacked flexibility and were only efficient over a narrow range of operating conditions. S. Howd and U. A. Boyden (1804–79) addressed this problem, and their work evolved into the concept of an inward-flow motor due to James B. Francis (1815–92). The modern Francis turbine is the result of this line of development. At the same time, European engineers addressed the idea of axial-flow machines, which today are represented by "propeller" turbines of both fixed-pitch and the Kaplan type.

Just as the vertical-axis tub wheels of Bélidor evolved into modern reaction turbines of the Francis and Kaplan type, development of the classical horizontal-axis waterwheel reached its peak with the introduction of the impulse turbine. The seeds of development were sown in 1826 when Poncelet described the criteria for an efficient waterwheel. These ideas were cultivated in the late nineteenth century by a group of California engineers that included Lester A. Pelton (1829–1908). His name was given to the Pelton wheel, which consists of a jet or jets of water impinging on an array of specially shaped buckets closely spaced around the periphery of a wheel. Thus, the relatively high-speed reaction turbines trace their roots to the vertical-axis tub wheels of Bélidor, whereas the Pelton wheel can be considered as a direct development of the more familiar horizontal-axis waterwheel.

Turbine configurations as we know them are generally in the form as originally developed. For example, the overwhelming majority of Pelton wheels have horizontal axes. Vertical-axis Pelton wheels are a relatively recent development. In more than 250 years of development many ideas were tried; some were rejected, and others were retained and incorporated in the design of the modern hydraulic turbine. This development resulted in devices with efficiencies as high as 95 percent in the larger sizes. In terms of design concept, these fall into roughly three categories, reaction turbines of the Francis type, reaction turbines of the propeller design, and impulse wheels of the Pelton type.

The rest of this chapter reviews the principles of operation, the classification and selection of turbines for given operating conditions, and performance characteristics and operational limitations. Most development efforts have concentrated on large turbines, with small-turbine technology consisting chiefly of scaling down larger turbines. The validity of this concept is reviewed, and areas where improvements can be made are addressed.

Overview of Existing Installations

The majority of small-scale hydropower installations in the United States may be classified as low head. On the whole, these are considered run-of-river plants for which little or no long-term storage is available. Therefore, the head and discharge can vary over a wide range, and this variability should be considered in selecting a turbine.

Recently, extensive interest developed in the restoration of power generation at sites that have been retired for some time. In such cases, the integrity of the dam may still be adequate, and the powerhouse and associated equipment may still be serviceable after some rehabilitation. Each site must be examined separately, and the condition of the components as well as the economics of rehabilitation or replacement with more modern turbomachinery must be evaluated. Analysis of some specific retired low-head small-scale installations has indicated that the potential power output is considerably greater than that previously developed. This may be associated with the lack of hydrologic data at the time of the original design, with economics, or with a limited selection of turbomachinery available.

The power output for small-scale hydropower is relatively low in comparison to large installations with high heads, large reservoirs, and large discharges. As a result, the small-scale installations have different operational requirements. Larger plants supply baseload power to a large population and therefore take every precaution to prevent power outages. If a small-scale plant is tied into the large grid system, its contribution to the total power capacity is relatively minor, and the consequence of power failure is less severe. The entire small-scale installation is therefore much simplified, and remote monitoring of the output may be adequate with an operator on call to restart the equipment.

Recent advancements in control and microprocessor technology may make it possible to apply automation even to small-scale hydropower installations. Variation of the parameters may be sensed and adjustments automatically made to ensure the most efficient operation of the turbine. Typical efficiency curves for various turbines are presented in Sec. 6.4, under Performance Characteristics, which should more clearly define the problem. Since flexibility and control are reflected in the initial cost of the installation, an economic analysis should be made to determine their practicality. As the number of installations and demand for units increases, the economics associated with such innovations should become more attractive.

6.3 BACKGROUND MATERIAL

Basic Principles of Fluid Mechanics and Exchange of Mechanical Energy at the Runner: Euler's Equation

A brief account of basic principles is presented here to have the equations of interest available for reference. Flow will be assumed to be steady and incompressible. The Euler equations of motion (a particular case of the Navier-Stokes equations for zero viscosity) projected along the streamline direction, yields

$$V\frac{\partial V}{\partial s} = -\frac{1}{\rho}\frac{\partial p}{\partial s} - \frac{\partial}{\partial s}(gh) \tag{6.1}$$

where V is the velocity magnitude, s denotes the streamline coordinate, p is the pressure, ρ is the mass density of the fluid, g the acceleration of gravity, and h the elevation.

Integration of this equation along the streamline yields the Bernoulli relationship

$$h + \frac{p}{\gamma} + \frac{V^2}{2g} = \text{constant} \tag{6.2}$$

where $\gamma = \rho g$.

This equation can be generalized to account for viscous effects by proceeding from the Navier-Stokes equations rather than from the inviscid flow Eq. (6.1). A one-dimensional form of Eq. (6.2), including viscous losses, applied between sections 1 and 2 of a fixed channel, is

$$\left(h + \frac{p}{\gamma} + \frac{V^2}{2g}\right)_1 = \left(h + \frac{p}{\gamma} + \frac{V^2}{2g}\right)_2 + H_L \tag{6.3}$$

where the pressure distributions at both transverse flow sections are assumed to be hydrostatic, the kinetic energy flux coefficient (which accounts for the nonuniformity of the velocity distributions) has been taken equal to one, and H_L represents the energy loss between sections 1 and 2 per unit weight of fluid.

For a rotating channel (or if work is done on the surrounding by any other means) Eq. (6.3) becomes

$$\left(h + \frac{p}{\gamma} + \frac{V^2}{2g}\right)_1 = \left(h + \frac{p}{\gamma} + \frac{V^2}{2g}\right)_2 + H_L + H_u \tag{6.4}$$

where H_u is the work done (positive if an output) on the surroundings per unit weight of fluid.

Equation (6.4) is a mechanical conservation of energy equation (which can be derived by equating the work done by the forces acting on the system to the corresponding change in kinetic energy and transforming all terms to their control volume counterparts). Such a derivation can also be applied if the reference system is moving.

In the particular case that the reference system is rotating at constant angular velocity Ω about some axis, and assuming again steady relative motion, two essential changes occur in the equation. First, in the rotating reference system, a centrifugal force appears as an added body force. Per unit weight, the centrifugal force is $\Omega^2 r/g$, directed away from the axis of rotation, where r denotes the distance from this axis. The corresponding potential energy for this force is $-\Omega^2 r^2/2g$, to be added to the gravitational potential energy per unit weight, $\rho gh/\gamma = h$. Furthermore, there is no relative work done on the channel walls (zero relative velocity at the walls). With the absolute velocities V replaced by the relative velocities V_R, Eq. (6.4) becomes

$$\left(h + \frac{p}{\gamma} + \frac{V_R^2 - u^2}{2g}\right)_1 = \left(h + \frac{p}{\gamma} + \frac{V_R^2 - u^2}{2g}\right)_2 + H \tag{6.5}$$

where $u = \Omega r$ is the velocity of rotation at distance r from the axis. This can be referred to as the Bernoulli or mechanical energy equation in a rotating frame of reference.

To fix ideas and simplify the kinematics while retaining all essential concepts, we consider (see Fig. 6.1) a radial flow hydraulic turbine. Equations

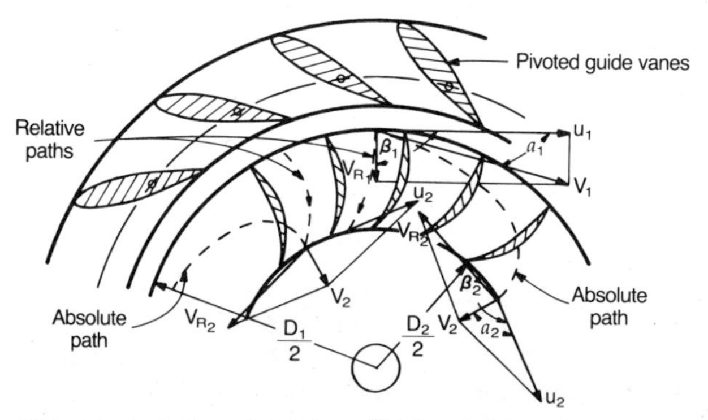

FIG. 6.1 Definition sketch for radial-flow turbine runner.

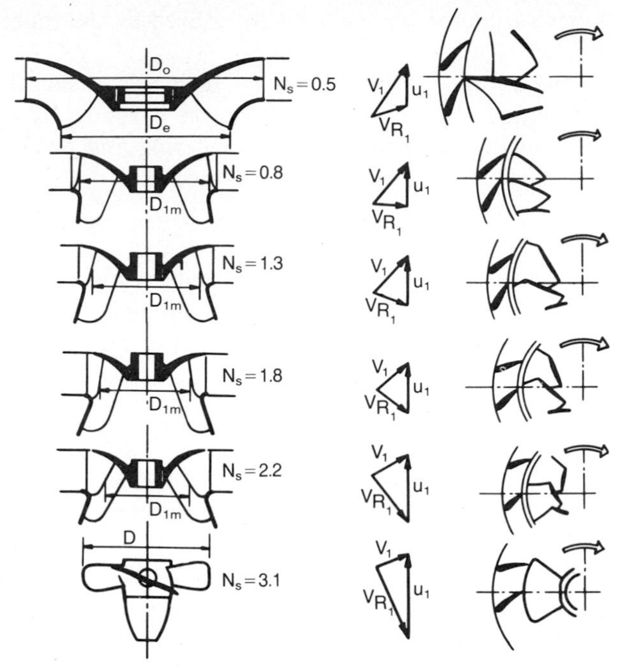

FIG. 6.2 Entrance velocity triangles as a function of specific speed. *(Adapted with permission from Mataix, C., "Turbomaquinas Hidraulicas," ICAI, Madrid, 1975.)*

(6.4) and (6.5) can be applied to the rotating channels between entrance and exit sections of the runner. The absolute velocity V is the vector sum of the relative velocity V_R and the moving frame velocity u

$$\mathbf{V} = \mathbf{u} + \mathbf{V}_R \tag{6.6}$$

Typical entrance and exit velocity triangles for a radial turbine are shown in Fig. 6.1, where the angles α and β are defined, respectively, as the angles that the absolute and relative velocities of the fluid make with the rotation velocity of the runner **u.** Figure 6.2 shows typical entrance velocity triangles as a function of specific speed N_s to be defined shortly. From the definition of the angles α and β, it follows that

$$\begin{aligned} V \sin \alpha &= V_R \sin \beta \\ V \cos \alpha &= u + V_R \cos \beta \end{aligned} \tag{6.7}$$

with $u = \Omega r$.

Combining Eqs. (6.4) and (6.5), one obtains

$$H_u = \frac{V_1^2 - V_2^2}{2g} + \frac{u_1^2 - u_2^2}{2g} - \frac{V_{R1}^2 - V_{R2}^2}{2g} \tag{6.8}$$

a form of Euler's equation for the head H_u utilized by the runner in the production of power.

An alternative form of Eq. (6.8) can be obtained by using the moment of momentum equation. In words, the torque on a runner is the difference between the moment of momentum fluxes entering and exiting the runner. Referring to Fig. 6.1, the torque T can be written as

$$T = \rho Q(r_1 V_1 \cos \alpha_1 - r_2 V_2 \cos \alpha_2)$$

assuming uniform velocity distributions over sufficiently small entrance and exit sections. Here Q is the volume rate of flow through the runner. The head H_u utilized by the runner in the production of power can be obtained from

$$T\Omega = \gamma Q H_u \tag{6.9}$$

We then have

$$H_u = \frac{u_1 V_1 \cos \alpha_1 - u_2 V_2 \cos \alpha_2}{g} \tag{6.10}$$

Equations (6.10) and (6.8) are different forms of Euler's equation for the head H_u. Their identity can be immediately demonstrated making use of the kinematic relationships

$$V^2 = V_R^2 + u^2 + 2V_R u \cos \beta \tag{6.11}$$

and

$$uV \cos \alpha = u(u + V_R \cos \beta)$$

which follow from Eqs. (6.6) and (6.7).

Turbine Efficiency and Losses

Definitions

The *hydraulic efficiency* of a turbine is defined by

$$n_h = \frac{H_u}{H} \tag{6.12}$$

where H_u is the head utilized by the runner and H is the net head on the turbine, defined as the difference between the total head at the entrance to the turbine proper (entrance to the spiral casing) and the total head at the tailrace.

The definition of *net head* is illustrated in Fig. 6.3. The hydraulic efficiency expresses the effectiveness of the transfer to the runner of the available power in the fluid that flows through it. Also illustrated in Fig. 6.3 is the definition of *gross head*, the difference between headwater and tailwater elevations.

The *volumetric efficiency* η_v of a turbine is defined as the ratio of the rate

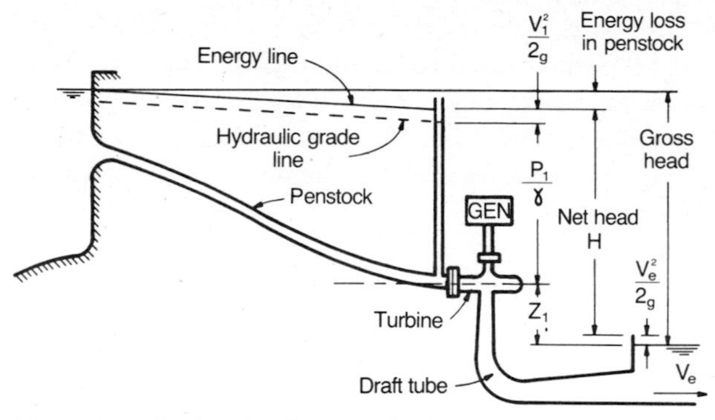

FIG. 6.3 Definition sketch for net head.

of flow that effectively acts on the rotor Q_e to the total rate of flow through the turbine. That is

$$\eta_v = \frac{Q_e}{Q} = \frac{Q - Q_L}{Q} \tag{6.13}$$

where Q_L represents leakage around the runner, which produces no work. The efficiency η_v is a measure of the effectiveness of the turbine seals.

The *mechancial efficiency,* finally, is defined as the ratio of the power available at the shaft of the machine to the power transmitted from the water to the runner. That is

$$\eta_m = \frac{bp}{bp + fp} \tag{6.14}$$

where bp is the brake power (available at the shaft) and fp represents the power consumed by mechanical friction and by viscous losses other than losses in the turbine passages. These losses include friction in the bearings and stuffing boxes and the so-called disk friction, the viscous losses in the fluid that fills the space between the runner and the adjacent casing.

The *overall efficiency* of the turbine is the product of the three efficiencies,

$$\eta = \eta_h \eta_v \eta_m \tag{6.15}$$

Energy Losses, Boundary Layers, and Separated Flows

The difference between the head H_u utilized by the runner [as given by Eqs. (6.8) and (6.10)] and the net head H on the turbine (defined in Fig. 6.3) is made up of viscous losses in various parts of the installation.

These include losses in the spiral casing, in the guide vanes and runner passages of the turbine, and in the draft tube; shock losses at the entrance to the runner if the relative velocity of the fluid leaving the guide vanes is changed abruptly in magnitude and/or direction as it enters the runner; and the losses at submerged discharge from the draft tube into the tailrace. Draft tubes are considered integral parts of the turbine and their performance is a contributing factor to the hydraulic efficiency η_h defined by Eq. (6.12).

Viscous energy losses occur in a flow field depending on the values of the rates of deformation (or equivalently of the deformation stresses) throughout the flow. For flow around bodies at sufficiently high Reynolds numbers, Re = V_0L/ν (V_0 = oncoming stream velocity, L = characteristic dimension of the body, ν = kinematic viscosity of the fluid), the zone of appreciable viscous action is limited to the vicinity of the body surface.

Recognition of this fact by Ludwig Prandtl in 1904 resulted in the concept of the boundary layer, where viscosity effects can be considered concentrated in the sense that they are comparable in magnitude with convection and other manifestations of acceleration terms or inertia forces, while the rest of the flow can be treated as inviscid. This hypothesis holds under broad conditions and applies to boundary layers and to certain other separated layers, with thicknesses approaching zero in all cases as Re tends to infinity. The thinness of these layers permits simplifications in the Navier-Stokes equations of motion that greatly increase the number of problems that can be treated analytically. Furthermore, the theory affords a physical explanation of the drag of bodies and of associated phenomena such as separation and wake formation and generally of energy losses in a flow.

Because of the thinness of the boundary layer, the pressure across it remains essentially constant, and the longitudinal pressure gradient along the boundary can be considered as imposed by the outer inviscid flow. This outer flow in which a boundary and its boundary layer may be considered immersed can be quite arbitrary in nature, and so can, therefore, the pressure gradients along the boundary, which have a decisive effect on the velocity distributions throughout the layer. In particular, if the pressure gradient is adverse (positive), corresponding to a deceleration of the outer flow, the low-velocity fluid particles close to the boundary may not have the kinetic energy necessary for the restoration of pressure imposed by the outer flow, which must then separate from the boundary to prevent physically impossible decelerations.

This is always the case for bluff-body flows, as opposed to slender (streamlined) bodies for which viscous effects and the possible flow turbulence are confined to the attached boundary layer and a thin wake behind the body. Separation in bluff-body flows leads to broad wakes, possibly exhibiting strong large-scale perodicities due to vortex shedding. The mean separation streamlines enclose a low-velocity region in which the (mean) pressure does not vary greatly from a value roughly equal to that in the irrotational flow just outside the separated boundary layer. This low pressure (negative relative to free stream) together with the positive pressure on the upstream side of the body near stagnation produces an asymmetry in the pressure distribution which results in an appreciable form (as opposed to boundary-shear) drag on the body and large energy dissipation rates.

A drag coefficient, defined as $C_D = F_D/(\frac{1}{2}\rho V_0^2 A)$, where A is the projected area of the body on a plane normal to the flow, can be represented as a function of Reynolds number (see similarity considerations below) and exhibits drastically different values for streamlined bodies (say, $C_D \sim 0.01$) and bluff bodies (say, $C_D \sim 1.0$) such as circular disks or cylinders. The aforementioned shock loss at the entrance to the runner, which arises if the relative fluid velocity is not properly aligned with the runner vanes, is related to this drastic increase in C_D values as separated flow zones appear and grow in size.

Flow in Closed Conduits and Diffusers

Separation occurs also in closed-conduit flows wherever conduit enlargements and/or alignment changes are present. For constant-diameter straight circular pipes with fully established flow, the energy dissipation per unit length of pipe is constant along the pipe. The head losses H_L in Eq (6.3) are customarily expressed in the form $KV^2/2g$. For fully developed pipe flow $K = fL/D$ and

$$H_L = \Delta\left(h + \frac{p}{\gamma}\right) = f\frac{L}{D}\frac{V^2}{2g} \tag{6.16}$$

where f is a function of Re and pipe roughness.

Energy dissipation occurs throughout the flow, but it is maximum close to the pipe walls and is directly related to the boundary shear. (For *uniform conduits* of noncircular cross section, it should be noted, the asymmetry of the conduit boundaries gives rise to secondary flows superimposed on the primary longitudinal flow, resulting in spiral flow patterns which increase the energy loss for a given cross-sectional area and flow rate.) For *nonuniform conduits*, the energy dissipation will depend in part upon similar mechanisms and in part upon the presence of separated regions in the flow.

While separation effects are analogous for both immersed body flows and pipe flow, there are significant differences in flow dynamics. First, immersed bodies are generally surrounded by great expanses of fluid and, second, while boundary layer growth begins at the leading edge of an immersed body, a boundary layer may already exist at the beginning of any change in conduit cross section or alignment. The presence of separated flow regions in conduit flow implies large local energy dissipation rates. These local rates can be expressed in terms of a Reynolds number and the parameters characterizing the boundary geometry, including also possibly a wall-roughness parameter.

The preceding concepts are of special interest in turbomachinery in regard to the various losses described in this section. An important element of a turbine installation is the draft tube, which is used to recover as much of the velocity head of the water as possible as it leaves the runner. The recovery of the velocity head is very important at high specific speeds because the velocity head at the exit from the runner may be 15 percent of the net head H of the installation. (For low specific speeds the velocity head is of the order of 3 to 4 percent of H.) Actually, an efficient recovery results in a decrease of the exit diameter D_2 of the turbine, and thus has a direct bearing on the cost of the turbine. A description of typical draft-tube geometries is presented later.

Diffuser performance is briefly discussed here. The efficiency of a diffuser η_D is defined in terms of the piezometric head difference between inlet (subscript 1) and exit (subscript 2). Using Eq. (6.3) and assuming $h_1 = h_2$ for simplicity, maximum pressure recovery for $H_L = 0$ would be $(p_2 - p_1)_{max} = (½)\rho(V_1^2 - V_2^2)$. The efficiency η_D can then be defined as

$$\eta_D = \frac{p_2 - p_1}{\frac{1}{2}\rho(V_1^2 - V_2^2)} \tag{6.17}$$

which implies

$$\gamma H_{L_{1-2}} = (1 - \eta_D)\frac{1}{2}\rho(V_1^2 - V_2^2) \tag{6.18}$$

For *straight diffusers* with circular cross sections there exists an optimum included angle of divergence, 2α, for given area ratio A_2/A_1 and Reynolds number at entry, that yields a maximum η_D. Obviously for large angles of divergence, separation of the boundary layer produces large losses; for very small included angles the length of the diffuser becomes considerable and the losses increase again. The optimum angle lies between $2\alpha = 4°$ and $8°$, and decreases as Re increases.

The efficiency η_D depends strongly on the boundary layer thickness at entry. For $2\alpha = 8°$, $\eta_D \sim 0.9$ for boundary layer thicknesses of the order of 0.5 percent of $D_1/2$, and decreases to $\eta_D \sim 0.7$ for thicknesses of the order of 5 percent of $D_1/2$. For draft tubes specifically, it has been found that for angles of divergence, 2α, somewhat larger than the optimum, an improvement in efficiency for Kaplan and propeller turbines is obtained by designing for a small swirl component at the entrance to the diffuser in the same direction as the runner rotation. This permits somewhat larger divergence angles and reduces the excavation depth.

For *curved diffusers* the efficiency decreases markedly with deflection angle. Generally, it is best to turn the flow without an area increase, and obtain the pressure recovery in straight-diffuser sections.

Lift

Unless a body is symmetric relative to the direction of a stream, a circulation around it is generated, giving rise to a lift force component normal to the undisturbed stream velocity. Such lift is the basis of performance for airplane wings or for the blades of turbines or propellers, and efforts have long been made to develop efficient profiles exhibiting minimum drag and maximum lift. Abundant information exists for a wide variety of profile shapes, which is useful in the design of axial-flow turbines.

To understand the generation of the circulation, one can consider an aerofoil with a sharp trailing edge, and start the motion suddenly from rest. Immediately after the aerofoil begins to move, the fluid motion is irrotational everywhere because the rate of transport of vorticity away from the aerofoil surface by diffusion, and subsequently by convection as well, is finite. Circulation around the aerofoil is zero, and there is a rear stagnation point some distance away from the trailing edge. The very high velocities at the trailing edge associated with this initial rotational flow immediately cause the formation of a starting vortex originating at the trailing edge, which develops in such a way that the rear stagnation point moves toward, and finally vanishes at, the trailing edge. The shed vorticity is carried away by the flow, which now leaves the trailing edge tangentially, and a circulation Γ exactly equal and opposite that of the shed vortex is established around the aerofoil. This produces a high-pressure (low-velocity) region on the underside of the aerofoil, and a low-pressure (high-velocity) region on the upper side of it, yielding a net force in a direction perpendicular to that of the motion. A theorem of Kutta and Joukowski shows that

$$F_L = \rho V_o \Gamma \tag{6.19}$$

For *Joukowski foils* of a very small thickness, the circulation is approximately related to angle of attack and camber by

$$\Gamma = \pi c V_o \sin(\alpha + \beta) \tag{6.20}$$

where F_L is the lift force per unit length of span, c the chord length, α the angle of attack relative to the chord, and $\beta \approx 2h/C$, where h is the maximum camber. The lift coefficient $C_l = F_L/(\frac{1}{2}\rho V_o^2\, c)$ then becomes theoretically

$$C_L = 2\pi \sin(\alpha + \beta) \tag{6.21}$$

In turbomachinery applications, corrections to tabulated aerofoil data are needed to take into account that we are dealing with a series of blades (cascade) and not with a single profile in infinite fluid.

Similarity Considerations

Similitude Theory

The grouping of parameters brought about by dimensional or inspectional analysis permits the writing of any physical relationship in terms of fewer dimensionless quantities (π numbers) representing ratios of significant forces for the problem. This provides a method to extrapolate model test data to prototype situations by equating corresponding dimensionless numbers.

As applied to hydraulic machinery, similarity considerations provide, furthermore, an answer to the following important question: "Given test data on the performance characteristics of a certain type of machine under certain operating conditions, what can be said about the performance characteristics of the same machine, or of a geometrically similar machine, under different operating conditions?" Similarity considerations provide in addition a means of cataloguing machine types and thus aid in the selection of the type suitable for a particular set of conditions.

The problem of similarity of flow conditions can be summarized as follows: "Under what conditions will geometrically similar flow patterns with proportional velocities and accelerations occur around or within geometrically similar bodies?" Obviously the forces acting on corresponding fluid masses must be proportionally related, as are the kinematic quantities, so as to ensure that the fluid will follow geometrically similar paths. An answer to this question can be obtained by examining the fundamental laws of motion and identifying the relevant forces. While these laws cannot yet be used to predict theoretically the flow conditions in a machine with unknown performance characteristics, the information they provide on forces and boundary conditions enables the determination of an answer to the similitude problem.

Similarity of the velocity diagrams at the entrance to the runner is a necessary requirement. Referring back to Fig. 6.1, assuming equal angles α_1 in model and prototype, the ratio V_1/u_1 must be held constant. If V_n denotes the radial component of the velocity (normal to the flow passages), we have

$$Q_e = f_b \pi \frac{B}{D_1} V_n D_1^2 \tag{6.22}$$

where f_b represents the fraction of free space in the inlet passages of the runner ($f_b \sim 0.95$), D_1 is the inlet diameter, and B is the width of the passages. With

$$V_n = V_1 \sin \alpha_1 = V_{R_1} \sin \beta_1 \tag{6.23}$$

Eq. (6.22) becomes

$$Q_e = \left(f_b \pi \frac{B}{D_1} \sin \alpha_1 \right) V_1 D_1^2$$

Since

$$u = \Omega r = \frac{\pi n}{60} D \tag{6.24}$$

we have

$$\begin{aligned}\frac{V_1}{u_1} &= \frac{1}{\left(f_b\pi \dfrac{B}{D_1} \sin \alpha_1\right)} \frac{Q_e}{u_1 D_1^2} \\ &= \frac{1}{\left(f_b\pi \dfrac{B}{D_1} \sin \alpha_1\right)} \frac{60}{\pi} \frac{Q_e}{n D_1^3}\end{aligned} \tag{6.25}$$

Constancy of the ratio Q_e/nD^3 or $Q_e/\Omega D^3$ (D = reference diameter) is then a necessary condition for similarity. In the rotating coordinate system this ratio represents also the ratio between inertia forces (proportional to $\rho V_R^2/r$) and centrifugal forces (proportional to $\rho\Omega^2 r = u^2/r$), or equivalently the ratio between kinetic energy and the centrifugal-force potential energy in Eq. (6.5).

If we now make the assumption that viscous forces are small relative to inertia forces and thus can be neglected in first approximation, and that furthermore the fluid does not change its physical properties as it passes through the machine (which excludes compressibility effects and cavitation, to be dealt with later), the only other forces that appear in the fundamental equations of motion are the pressure forces. Their ratio to inertia forces is proportional to $\Delta p/\rho V^2$ or, if the head H_u utilized by the runner is introduced, to gH_u/V^2. Under the assumptions made, the condition

$$\frac{Q_e}{\Omega D^3} = \text{constant} \tag{6.26}$$

is sufficient for similarity. The condition

$$\frac{gH_u}{V^2} = \text{constant} \qquad \text{or} \qquad \frac{gH_u}{\Omega^2 D^2} = \text{constant} \tag{6.27}$$

follows from the basic laws and permits calculation of the head H_u for similar operating conditions. The equality of the ratio gH_u/V^2 for model and prototype also follows from inspection of Euler's equation (6.10) under the assumption of negligible viscous effects.

Velocity coefficients ϕ, C_1, and C_2 are customarily introduced as

$$u_1 = \phi\sqrt{2gH} \qquad V_1 = C_1\sqrt{2gH} \qquad V_2 = C_2\sqrt{2gH} \tag{6.28}$$

In terms of these coefficients, the hydraulic efficiency η_h defined by Eq. (6.12) can be written with the use of Eq. (6.10) as

$$\begin{aligned}\eta_h &= \frac{u_1 V_1 \cos \alpha_1 - u_2 V_2 \cos \alpha_2}{gH} \\ &= 2\phi\left(C_1 \cos \alpha_1 - \frac{D_2}{D} C_2 \cos \alpha_2\right)\end{aligned} \tag{6.29}$$

If the viscous losses embodied in η_h can be assumed to occur under hydrodynamically rough conditions, in the sense that the loss coefficients [like f in Eq. (6.16), $1 - \eta_d$ in Eq. (6.18), or the various local energy-loss coefficients] are independent of Re and depend only on geometric ratios and relative roughness (Re must be high enough for the losses to be purely turbulence-controlled), then η_h must be the same in model and prototype, provided that the relative roughnesses are the same and the geometry is faithfully reproduced. Under these conditions Eq. (6.27) becomes

$$\frac{gH}{\Omega^2 D^2} = \text{constant} \tag{6.30}$$

Analogous consideration can be made about the volumetric efficiency η_v. Here Reynolds number effects may be more significant due to the smallness of the leakage-flow passages. But if one can assume Re independence (and strict geometric similarity, including surface roughnesses and running clearances), η_v must be the same in model and prototype and Eq. (6.26) becomes

$$\frac{Q}{\Omega D^3} = \text{constant} \tag{6.31}$$

Equations (6.30) and (6.31) permit calculation of the net head H and total flow rate Q for similar operating conditions.

Scale Effects

The foregoing assumptions are reasonably accurate for turbines of fairly large dimensions operating under noncavitating conditions. In particular, relative roughnesses and running clearance ratios are the same if similarity considerations are applied to the same machine. For large differences in the size of two geometrically similar machines, such as between model and prototype, roughnesses and clearances cannot be geometrically scaled because of fabrication limitations. Certain formulas have been developed to correlate model and prototype data, all of them containing a strong dose of empiricism.

The Moody step-up equation[3]

$$\frac{1 - \eta_1}{1 - \eta_2} = \left(\frac{D_2}{D_1}\right)^n \tag{6.32}$$

has been found to give satisfactory results for turbine flows. The basic assumption in Moody's derivation is Reynolds number independence (hydrodynamically rough flow) and the same degree of surface finish in model and prototype. The derivation is based on assuming losses of the form of Eq. (6.16) for pipe flow, with $f = A(k/D)^n$, where A is constant. Since k is assumed to be the same for model and prototype, it disappears from the final result. The empirical exponent n is based on turbine test results: $n \sim \frac{1}{5}$ according to Moody, but it may become appreciably smaller if the formula is used with models with very smooth walls and close running clearances.

It should be emphasized that Moody's formula has been developed solely with regard to the effect of relative roughness. No consideration is given to changes in the relative size of the running clearances, which obviously affect η_v, nor are mechanical losses explicitly taken into account (although these factors certainly affect the empirical n values). The formula is nevertheless used to correlate overall efficiencies.

The power P developed by the turbine is given by

$$P = \eta\gamma QH = \eta_m\eta_v\eta_h\gamma QH \tag{6.33}$$

Thus, even if η_h and η_v are the same in model and prototype, η_m would be different generally because of differences in disk friction losses and in the losses in bearings and stuffing boxes. Obviously, if the mechanical losses are small, changes in η_m would also be small, and could be subsumed in the empirically determined values of the coefficients and/or exponents in the available step-up formulas.

Camerer's formula,[4,5] as does Moody's Eq. (6.32), takes into account solely surface roughness effects. A formula of Ackeret's,[6] a second formula of Moody,[5,6] and a formula of Hutton[7] all incorporate Reynolds number effects, and the net head ratio thus appears explicitly in them, in addition to the size ratio of model and prototype. All are based on derivations that ignore volumetric losses and mechanical losses, although the selection of the empirical coefficients is based on actual turbine test data.

It may be noted here that Euler's Eqs. (6.8) or (6.10), or the basic relationships of Eqs. (6.4) and (6.5), as applied to reaction turbines, imply simplifications which require in actual calculations the use of experimentally determined coefficients. Thus, for example, fluid particles in different streamlines generally have different velocities and, furthermore, their radial distances to the axis of rotation at entry to or exit from the runner are also different because the entrance or exit edges of the vanes are not always parallel to the turbine axis. Despite these problems, the theory is useful in many ways: It shows the nature of the performance curves of a given machine, it permits identification of each separate factor affecting the performance, and it shows how changes in design should be made to alter the characteristics of a machine as obtained from experimental testing. It also sheds light on the nature of the similarity laws and possible scale effects.

Similar considerations can be made regarding the theory of impulse (Pelton) wheels (to be dealt with later) which is based on a simplified version of Eq. (6.9). Essentially, the same results are obtained regarding similarity relationships. For Pelton wheels, however, the efficiency is nearly independent of size and Eq. (6.32) or similar equations do not apply. As the size of a Pelton wheel increases, there is a deterioration in the smoothness of the jet before it strikes the buckets, which nullifies any benefits from reduced friction losses. Furthermore, there are no leakage losses to make a difference.

The similarity results in these sections can be used, with due regard to the approximations involved, to predict the performance of a machine under operating conditions different from those of available experimental data and the performance of geometrically similar machines if performance characteristics are available for one of them. Some examples of application are given later.

Specific Speed

Similar flow conditions are ensured by the constancy of the ratio $Q/\Omega D^3$, which implies constancy of the ratio $gH/\Omega^2 D^2$. In other words,

$$\frac{gH}{\Omega^2 D^2} = f\left(\frac{Q}{\Omega D^3}\right) \tag{6.34}$$

This relationship can also be written in terms of a third dimensionless number which does not involve the representative dimension D of the machine and which can replace either of the two arguments in Eq. (6.34). Such a number can be obtained by appropriate multiplication of powers of the dimensionless numbers of Eq. (6.34):

$$N_{s_Q} = \left(\frac{Q}{\Omega D^3}\right)^{1/2} \left(\frac{\Omega^2 D^2}{gH}\right)^{3/4} = \frac{\Omega Q^{1/2}}{(gH)^{3/4}} \tag{6.35}$$

This dimensionless number is called the specific speed. For hydraulic turbines, however, the definition of the specific speed is based on the power P delivered by the turbine as a variable, instead of the flow rate Q. The corresponding dimensionless number P is $P/\rho\Omega^3 D^5$, which is a function of $gH/\Omega^2 D^2$. Eliminating D between these two numbers one gets

$$N_s = \frac{\Omega (P/\rho)^{1/2}}{(gH)^{5/4}} \tag{6.36}$$

The two specific speeds are related by

$$N_s = \sqrt{\eta} N_{s_Q} \tag{6.37}$$

which is obtained making use of Eq. (6.33). If we choose N_s as the independent variable in these relationships, then all other dimensionless combinations can be expressed as functions of N_s. These include also the dimensionless torque $T/\rho\Omega^2 D^5$ and the efficiencies, under the assumption of negligible scale effects.

The specific speed describes a specific combination of operating conditions that ensures similar flows in geometrically similar machines. It has thus attached to it a specific value of the efficiency η (assumed approximately constant for similar flow conditions regardless of size). It is then customary to label each series of geometrically similar turbines by the value of N_s which gives maximum η for the series. Unless otherwise stated, this is the N_s value referred to when the terminology specific speed is used. The value of N_s thus defined permits the classification of turbines according to efficiency. Each geometric design has a range of N_s values where it can be used with only one value corresponding to peak efficiency. In subsequent sections this idea will be used to classify turbine designs.

The N_s as defined is dimensionless. It is common in practice to drop g and ρ from the definition and define n_s as

$$n_s = \frac{n\sqrt{P}}{H^{5/4}} \tag{6.38}$$

with n in r/min. In USCS units, the units of P are horsepower, and the units of H are feet. In SI units, the unit of P is either the metric horsepower or the kilowatt, and the unit of H is the meter. The relationship of these three definitions of n_s to the dimensionless N_s are:

$$\begin{aligned} n_s &= 43.5\ N_s \text{ (USCS units)} \\ n_s &= 193.1\ N_s \text{ (SI using metric horsepower)} \\ n_s &= 166\ N_s \text{ (SI using kW for power)} \end{aligned} \tag{6.39}$$

Cavitation

Cavitation can be defined as the formation of the vapor phase in a liquid flow when the hydrodynamic pressure falls below the vapor pressure of the liquid. It is distinguished from *boiling,* which is due to the vapor pressure being raised above the hydrodynamic pressure by heating.

In its initial stages, cavitation is in the form of individual bubbles which are carried out of the minimum pressure region by the flow and collapse in regions of higher pressure. Calculations, as well as sophisticated laboratory experimentation, indicate that collapsing bubbles create very high impulsive pressures. This results in substantial noise (a cavitating turbine sounds like gravel is passing through it). More important, the repetitive application of the shock loading due to bubble collapse at liquid-solid boundaries results in pitting of the material.

As the process continues, cracks form between the pits, and solid material is spalled out from the surface. The mechanical effects of cavitation are enhanced by the high temperature created by collapsing bubbles and the presence of oxygen-rich gases which come out of the solution. The details of the erosion process are complex, but the results are of practical significance. Many components of a turbine are susceptible to extensive damage as illustrated in Fig. 6.4.

In more developed forms of cavitation, large vapor-filled cavities remain attached to the boundary. Each cavity or pocket is formed by the liquid flow

FIG. 6.4 Cavitation erosion on a turbine runner.

detaching from the rigid boundary of an immersed body or flow passage. The maximum length of a fixed cavity depends on the pressure field. Termination may occur by reattachment of the liquid stream at a downstream position on the solid surface, or the cavity may extend well beyond the body. The latter case is known as *supercavitation*. Under these circumstances, the pressure distribution on the boundary can be substantially altered. If developed cavitation occurs on the runner or wicket gates of the turbine, the performance is changed. Cyclical growth and collapse of the cavities can also occur, producing vibration. Thus cavitation can degrade performance and produce vibration, as well as reduce the operational lifetime of the machine through erosion.

The Cavitation Index

The fundamental parameter in the description of cavitation index

$$\sigma = \frac{P_0 - P_v}{\frac{1}{2}\rho V_0^2} \tag{6.40}$$

The state of cavitation is assumed to be a unique function of σ for geometrically similar bodies. If σ is greater than a critical value, say σ_c, there is no cavitation and the various hydrodynamic parameters are independent of σ. When σ is less than σ_c, various hydrodynamic parameters such as the lift and drag of various components and the power and efficiency of a turbine are functions of σ. Noise, vibration, and erosion also scale with σ. It should be emphasized that the value of σ where there is a measurable change in performance is not the value of σ where cavitation can first be determined visually or acoustically. The critical σ_c can be thought of as a performance boundary such that

$\sigma > \sigma_c$ no cavitation effects

$\sigma < \sigma_c$ cavitation effects: performance degradation, noise, and vibration

There are two ways of defining the critical value of sigma. The *incipient cavitation number* σ_i is normally determined in a test facility by lowering the static pressure at constant velocity until cavitation first occurs. A more repeatable parameter is the *desinent cavitation index* σ_d which is based on the static pressure necessary to extinguish cavitation after the pressure has been lowered sufficiently for cavitation to occur. The precise value of σ_c defined by *inception* or *desinence* is normally only determined in the laboratory. It should not be confused with more pragmatic definitions such as the value of the cavitation index at a measurable change in hydraulic performance expressed by power, capacity, or efficiency, or at a measurable change in vibration level.

Cavitation Inception

Cavitation inception is assumed to occur when the local pressure is equal to the vapor pressure. For steady flow over a streamlined body, this implies

$$\sigma_c = -C_{pm} \tag{6.41}$$

where C_{pm} is the minimum value of the pressure coefficient defined by $C_p \equiv (p - p_0)/(\frac{1}{2}\rho U_0^2)$, where p is the pressure at a given position on the surface of the body. In the case of a blade section or a strut, the value of C_{pm} depends on both the shape and angle of attack α.

Inception on hydrofoils has been studied by numerous investigators.[8,9,10] Typical data are shown in Fig. 6.5. Similar trends in cavitation characteristics are evident in the cavitation data for a turbine runner when properly interpreted. For a fixed wicket gate setting, the relative angle of attack of the runner blades of a turbine is a function of the flow coefficient $Q/\Omega D^3$. Cavitation inception for a propeller turbine at a fixed wicket gate setting is shown schematically in Fig. 6.6. It is no accident that Figs. 6.5 and 6.6 are qualitatively similar.

Cavitation inception in separated flows is a more complex process. The minimum pressure is not at a flow boundary as is the case for a streamlined body. Thus σ_c is greater than $-C_{pm}$ measured at the surface. This is an important consideration for cavitation inception on bluff bodies. An example for a sharp-edged disk is shown in Fig. 6.7. The process is highly complex. Cavitation occurs in the turbulent eddies formed in the separated flow. The turbulence is

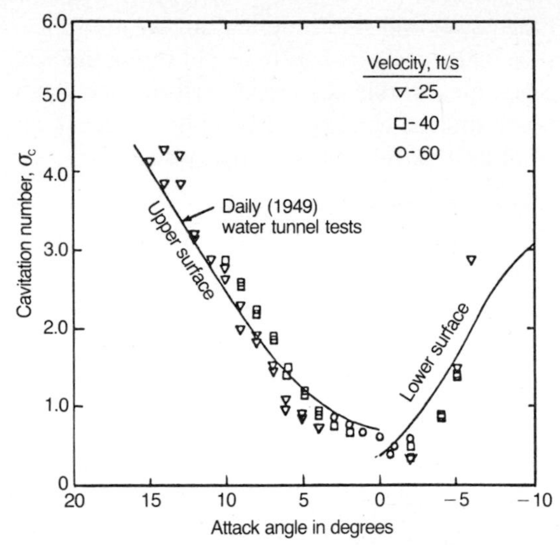

FIG. 6.5 Cavitation inception characteristics of an NACA 4412 hydrofoil. *(Adapted from Ref. 8.)*

found to be most intense along the separation streamline which is the dividing line between the essentially quiescent outer flow and the wake, as illustrated in the sketch accompanying Fig. 6.7. Further details can be found in Arndt.[11]

Vortex cavitation is also an important phenomenon. Cavitation can occur in the vortices formed in the clearance passages of propeller turbines, or at the hub or in the draft tube of propeller and Francis turbines. The minimum pressure is governed by the circulation and core radius:

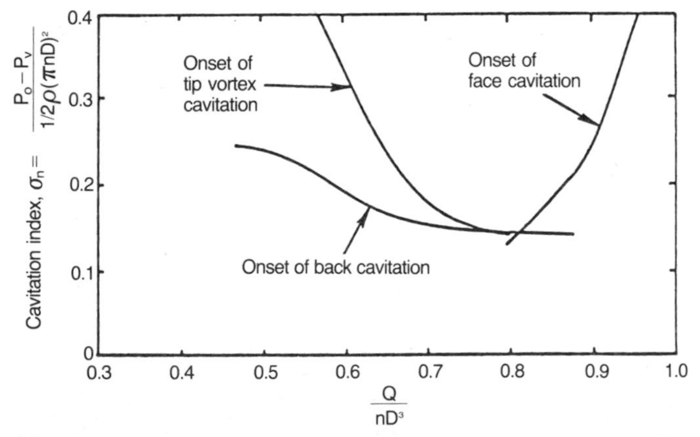

FIG. 6.6 Cavitation inception characteristics of a typical propeller turbine.

$$C_{pm} = -2\left(\frac{\Gamma}{2\pi r_c V_0}\right)^2 \tag{6.42}$$

Unfortunately, it is difficult to quantify the circulation Γ and the core radius r_c. In the case of a hub vortex, Γ is controlled by the operation of the turbine and is proportional to the amount of swirl in the discharge. In the case of a tip vortex, Γ is proportional to blade loading. The core radius is related to the boundary layer thickness at the point of separation. Because of this factor, the pressure field in a vortex is very sensitive to variations in Reynolds number. This is illustrated in Fig. 6.8.

The experimental situation under which these data were collected is an idealized hub vortex, the amount of swirl being controlled by the angle of attack of a series of fixed blades as shown in Fig. 6.8. This clearly shows that

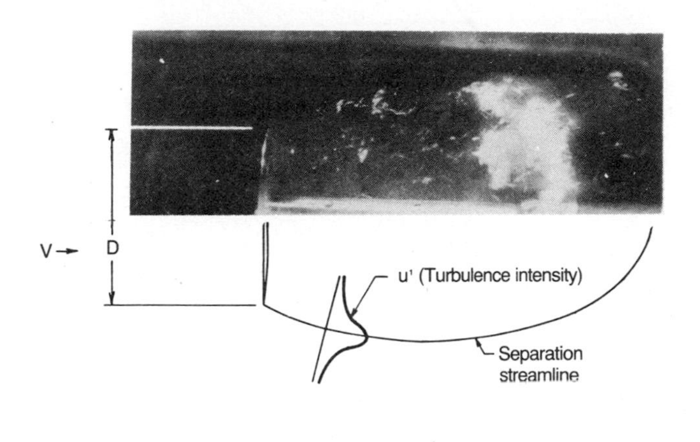

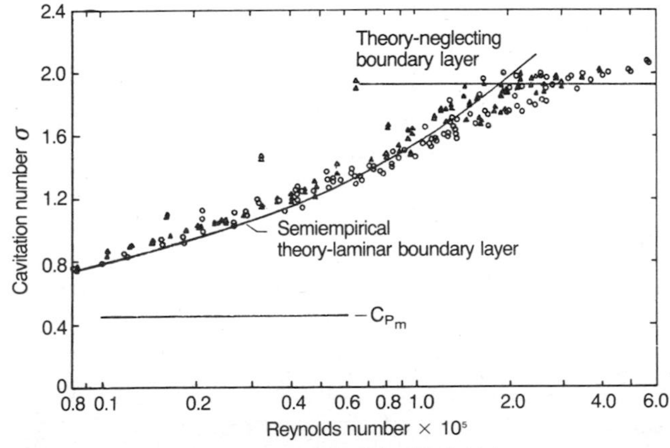

FIG. 6.7 Incipient cavitation on a disk. *(Ref. 11.)*

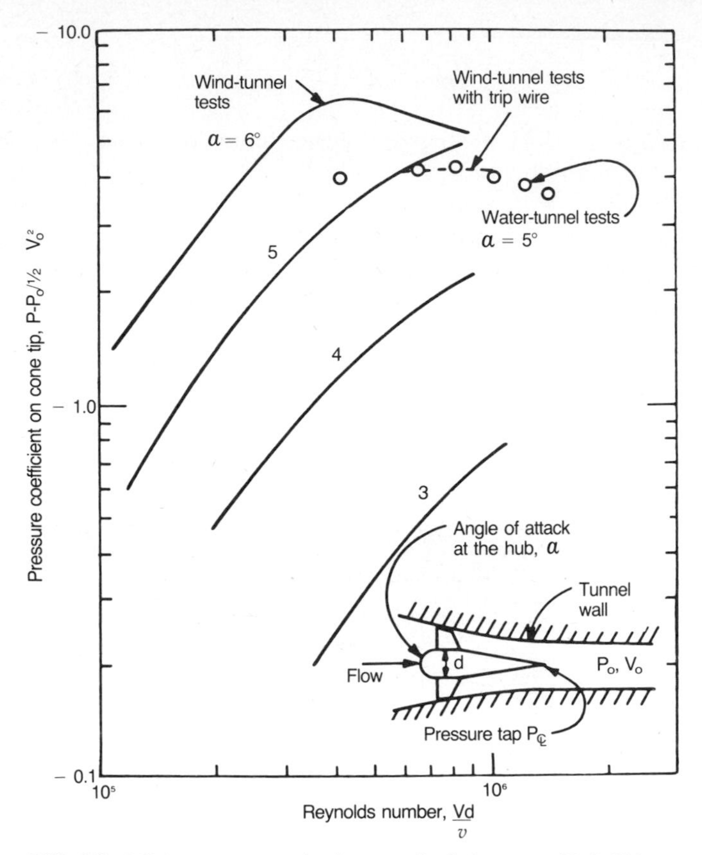

FIG. 6.8 Minimum pressure in the core of a hub vortex. *(Ref. 12.)*

observations of vortex cavitation in a model will not extrapolate to the prototype because of the differences in Reynolds number.

The examples of cavitation in separated flows and vortex flows already illustrate that σ is not a unique parameter for the correlation of cavitation phenomena. In fact, there are a myriad of "scale effects" which preclude using cavitation index alone as the scaling parameter for cavitation phenomena, just as the affinity laws for turbine performance are modified by viscous effects. The problem of scaling cavitation is more complex since the basis of the cavitation index is not only that pressure scales exactly with velocity squared but also that cavitation occurs when the hydrodynamic pressure at a given point is equal to the vapor pressure. The second assumption is also not completely valid; in fact, cavitation inception is also influenced by the level of dissolved gas in the flow as well as by the number, density, and size distribution of microbubbles or nuclei in the flow.[13,14] In addition, surface roughness will increase the susceptibility to cavitate.[15] All these details are of practical importance but are beyond the scope of this book.

Developed Cavitation

When the value of σ is less than σ_c, fixed or attached cavities can form on the suction side of a lifting surface. The minimum pressure is the vapor pressure, independent of upstream velocity and pressure; hence

$$C_{pm} = -\sigma \qquad \sigma < \sigma_c \tag{6.43}$$

Assuming the pressure distribution on the pressure side to be uninfluenced by cavitation on the suction side, it is easily seen that the lift coefficient C_L should be roughly proportional to C_{pm}. This is shown in Fig. 6.9. At each value of the angle of attack α, there is a value of σ above which C_L is independent of this parameter. At lower values of σ, C_L decreases with decreasing σ. Note that as angle of attack increases, C_L increases and "cavitation stall" occurs at increasingly higher values of σ. In a turbomachine the picture is qualitatively the same. As previously mentioned, the angle of attack is proportional to flow coefficient at a fixed wicket gate setting. Obviously the flow is more complicated, but the analogy between a hydrofoil and the blade section of a propeller turbine can be seen.

Cavitation Erosion

A detailed explanation of cavitation erosion is beyond the scope of this book. Most of the research in this area has been directed toward the mechanics of bubble collapse and the associated impulsive pressures as well as toward a quantification of those material properties that are of importance in resistance to cavitation. Little has been done to correlate cavitation erosion with the properties of a given flow field. However, it is important to have in mind that cav-

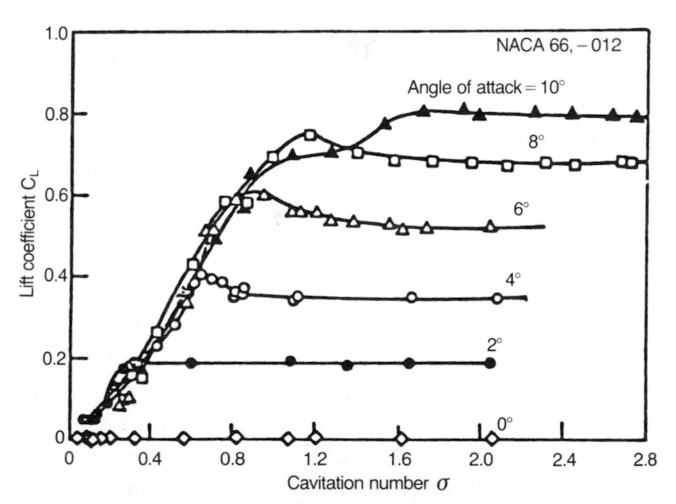

FIG. 6.9 Variation in lift coefficient with cavitation number. *(Ref. 9.)*

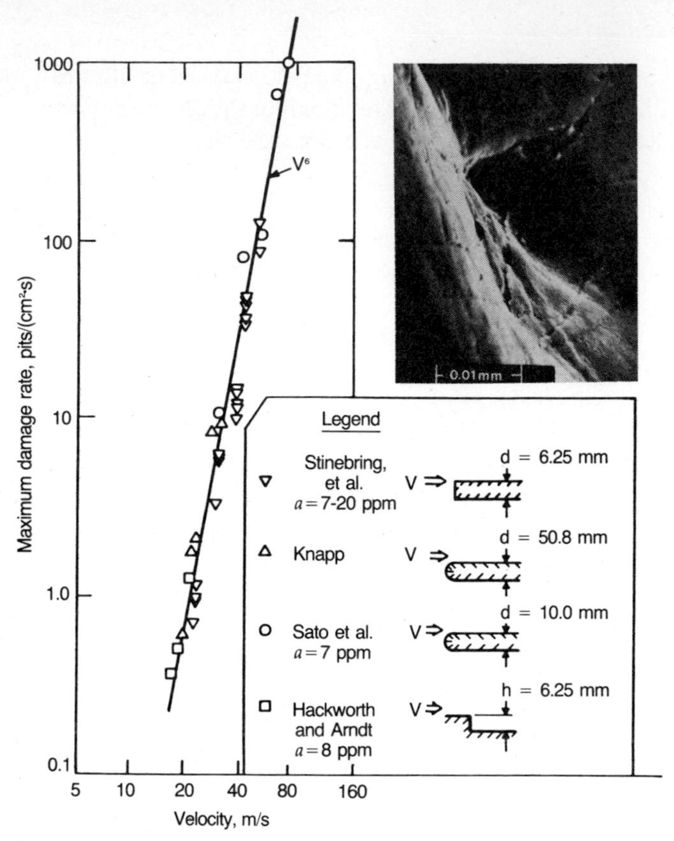

FIG. 6.10 Cavitation erosion data. Constant cavitation number. Microphotograph of a typical pit is shown in the insert. *(Adapted from Ref. 14.)*

itation erosion will scale with a high power of velocity at a given cavitation number and that cavitation erosion does not necessarily increase with a decrease in the cavitation index. Figure 6.10 is a compilation of data obtained in various laboratories for cavitation erosion on various axisymmetric bodies as well as two-dimensional flow over a backward facing step. For the purposes of this illustration, erosion rate is defined as the pitting rate of soft aluminum in number of pits per unit area per unit time, and corrections have been made for variations in dissolved gas. Very little quantitative information is available, but it has been observed that the pitting rate is measurably reduced with an increased concentration of gas.[16] A typical microphotograph of a pit is also shown in the figure. The important feature is that at constant σ the pitting rate scales with the sixth power of velocity. The pitting rate is *not* quantitatively equivalent to the measured weight loss observed in specially designed erosion test apparatus. However, Stinebring et al.[16] have been able to measure the energy absorbed per pit. This is directly proportional to pit volume which is found to scale with velocity to the fifth power. Thus, energy absorbed per unit area is

$$\frac{\text{Energy}}{\text{Unit area time}} = \frac{\text{pits}}{\text{area time}} \times \frac{\text{energy}}{\text{pit}} \tag{6.44}$$
$$= kV^6 \times V^5 = kV^{11}$$

Obviously, the erosion rate is very sensitive to velocity in the initial stages of cavitation. This implies that the magnitude of the erosion problem increases very rapidly with the head.

In many cases cavitation erosion occurs at the trailing edge of an attached cavity. The number of bubbles that collapse in this region in a given period of time will be a function of the geometry of the cavity which in turn is a function of the cavitation index. This is illustrated in Fig. 6.11. This is admittedly for an idealized situation, namely cavitation on an axisymmetric body under carefully controlled laboratory conditions. However, it is important for every hydraulic engineer to recognize that maximum erosion intensity does not occur at the lowest possible cavitation number.

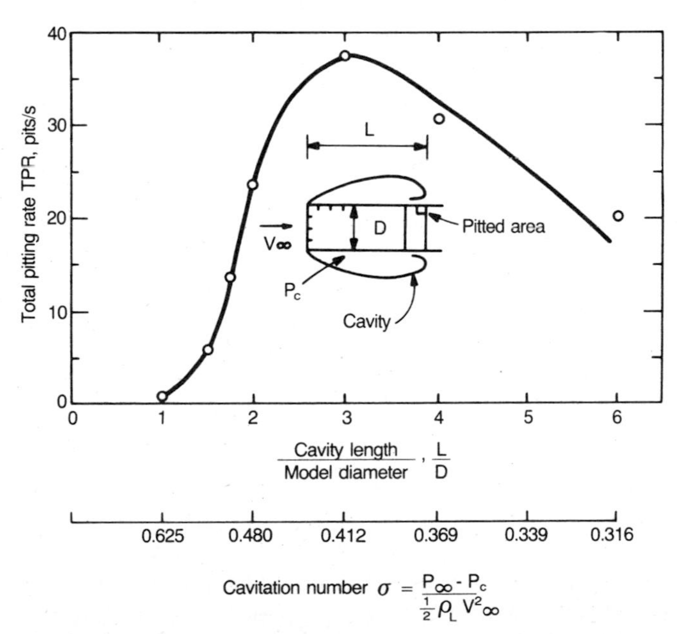

FIG. 6.11 Variation of cavitation erosion with cavity length and cavitation number. *(Adapted from Ref. 16.)*

Thoma's Sigma

The introductory material on cavitation was based on simple geometric shapes and an easily defined cavitation parameter. The flow in a turbine is obviously more complex and less easily quantified. There still is, however, a definite need to define operating conditions with respect to cavitation. For example, it is sometimes necessary to specify under what conditions the degree of cavitation will be the same for the same machine operating under different heads and speeds, or for two machines of similar design but different heads and speeds, or for two machines of similar design but different size, e.g., a model and a prototype. The accepted parameter for this purpose is the Thoma sigma σ_T. This is defined as

$$\sigma_T = \frac{H_{sv}}{H} \tag{6.45}$$

where H_{sv} is the net positive suction head. Referring back to Fig. 6.3, this is defined as

$$H_{sv} = H_a - z - H_v + \frac{V_e^2}{2g} + H_l \tag{6.46}$$

where H_a is the atmospheric pressure head, z is the elevation of the critical location for cavitation above the tailwater elevation, V_e the average velocity in the tailrace, and H_l the head head loss in the draft tube.

If we neglect the draft-tube losses and the exit-velocity head, Thoma's sigma is

$$\sigma_T = \frac{H_a - H_v - z}{H} \tag{6.47}$$

Each type of turbine will cavitate at a given value of σ_T. Clearly cavitation can only be avoided if the installation is such that σ_T is greater than this critical value. The value of σ_T for a given installation is known as the plant sigma. For a given turbine operating under a given head, the only variable is the turbine setting z. The critical value of Thoma's sigma, σ_{TC}, controls the allowable setting above tailwater:

$$z_{\text{allow}} = H_a - H_v - \sigma_{TC} H \tag{6.48}$$

It must be borne in mind that H_a varies with elevation. As a rule of thumb, H_a decreases from the sea level value of 10.3 m (33.8 ft) by 1.1 m (3.6 ft) for every 1000 m (3280 ft) above sea level. Thus a turbine sited at Leadville, Colorado, or Quito, Ecuador, for example, would have an allowable turbine setting that is 3 m (9.84 ft) less than that at sea level. In fact, z_{allow} could easily be negative, implying a required turbine setting below the tailwater elevation.

The determination of σ_{TC} is usually done by a model test. A schematic of the correlation between performance breakdown and σ_T is shown in Fig. 6.12.

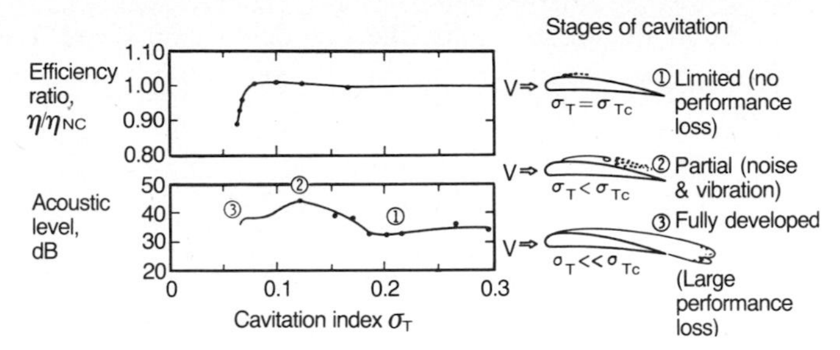

FIG. 6.12 Schematic of the correlation between performance breakdown, noise, and erosion with cavitation index.

This figure is based on information presented by Deeprose et al.[17] Note the similarity in the trend of performance with σ_T and the correlation of lift coefficient with σ as shown in Fig. 6.9. As has already been emphasized, a measurable drop in efficiency occurs at a value of σ_T that is well below the value corresponding to the detection of cavitation inception acoustically. Note also that maximum noise and presumably maximum erosion rate occur at a value of σ_T intermediate between the value at inception and that at performance breakdown. A slight rise in efficiency is often also noted at intermediate values of σ_T.

Suction Specific Speed

The critical value of σ_T is a function of the type of turbine involved, i.e., the specific speed of thc machine. A cavitation scaling parameter often used in pump application is the suction specific speed

$$S = \frac{\Omega\sqrt{Q}}{(gH_{sv})^{3/4}} \tag{6.49}$$

The suction specific speed is a natural consequence of considering dynamic similarity in the low perssure region of a turbomachine. The dynamic relations are

$$\frac{gH_{sv}D_e^4}{Q^2} = \text{constant} \qquad \frac{gH_{sv}}{\Omega^2 D_e^2} = \text{constant} \tag{6.50}$$

where D_e is the eye or throat diameter.

These relations hold when the kinematic condition for similarity of flow in the low-pressure region of the machine is satisfied:

$$\frac{Q}{\Omega D_e^3} = \text{constant} \tag{6.51}$$

Elimination of D_e in Eq. (6.50) yields the suction specific speed. Using Eq. (6.33) for the power developed by a turbine, the relationship between σ_T, N_s, and S is given by

$$\sigma_T = \frac{1}{\eta^{2/3}}\left(\frac{N_s}{S}\right)^{4/3} \tag{6.52}$$

If S can be assumed to be constant, then Eq. (6.52) produces a relationship between σ_T and N_s. Allowable values of S do vary, but an acceptable conservative value in nondimensional units is 3.

A comparison between Eq. (6.52), assuming $2 < S < 4$, and actual turbine experience as quoted by Moody and Zowski,[3] is shown in Fig. 6.13. An efficiency of 0.9 is assumed. The allowable S for turbines appears to be higher than for equivalent pumps. Note also that the trend of limiting σ_T for turbines has a steeper slope than the constant S lines. This could imply that different specific speed designs are not equally close to the optimum with regard to cav-

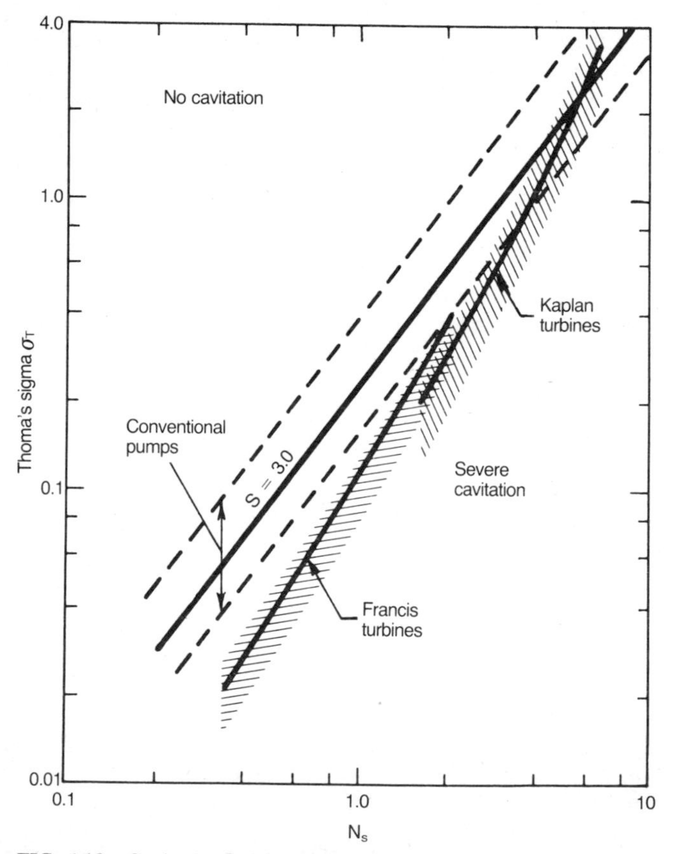

FIG. 6.13 Cavitation index as a function of specific speed.

itation or that the factor S cannot be considered a constant. It should be kept in mind that since the flow direction for a pump and a turbine are in opposite directions, only the inception point should be similar. Under developed cavitating conditions, the flow situation could be quite different with cavity closure occurring on the runner in the case of a pump, whereas it would occur downstream of the runner in the case of the equivalent turbine.

Figure 6.13 is a useful chart for estimating the turbine setting for various types of turbines in conjunction with Eq. (6.48). This is a useful procedure for preliminary design and comparison between different types of turbines for the same installation. However, the manufacturer's recommendation should be followed in the final design.

6.4 TURBINE TECHNOLOGY

Overall Description of a Hydropower Installation

The hydraulic components of a hydropower installation consist of an intake, penstock, guide vanes or distributor, turbine, and draft tube. The intake is designed to withdraw flow from the forebay as efficiently as possible, with no, or minimal, vorticity. Trashracks are commonly provided to prevent ingestion of debris into the turbine. Intakes usually require some type of shape transition to match the passageway to the turbine and also incorporate a gate or some other means of stopping the flow in case of an emergency or turbine maintenance.

Some types of turbines are set in an open flume; others are attached to a closed-conduit penstock. In all cases, efforts should be made to provide uniformity of the flow, as this uniformity has an effect on the efficiency of the turbine. For low-head installations, the diameter of a closed penstock must be quite large to accommodate the large discharges necessary for a given power output. Its size is a compromise between head loss and cost. The selection of the actual penstock configuration is dependent on the location of the powerhouse with respect to the dam.

For some types of reaction turbines, the water is introduced to the turbine through the casings or flumes which vary widely in design. The particular type of casing is dependent on the turbine size and head. For small heads and power output, open flumes are commonly employed. Steel spiral casings are used for higher heads, and the casing is designed so that the tangential velocity is essentially constant at consecutive sections around the circumference. This requirement necessitates a changing cross-sectional area of the casing.

Some examples of intakes and casings are shown in Fig. 6.14, where dimensions are given in terms of the runner diameter. As the inflow has an effect on the turbine efficiency, the design of the special casing is carried out by the turbine manufacturer.

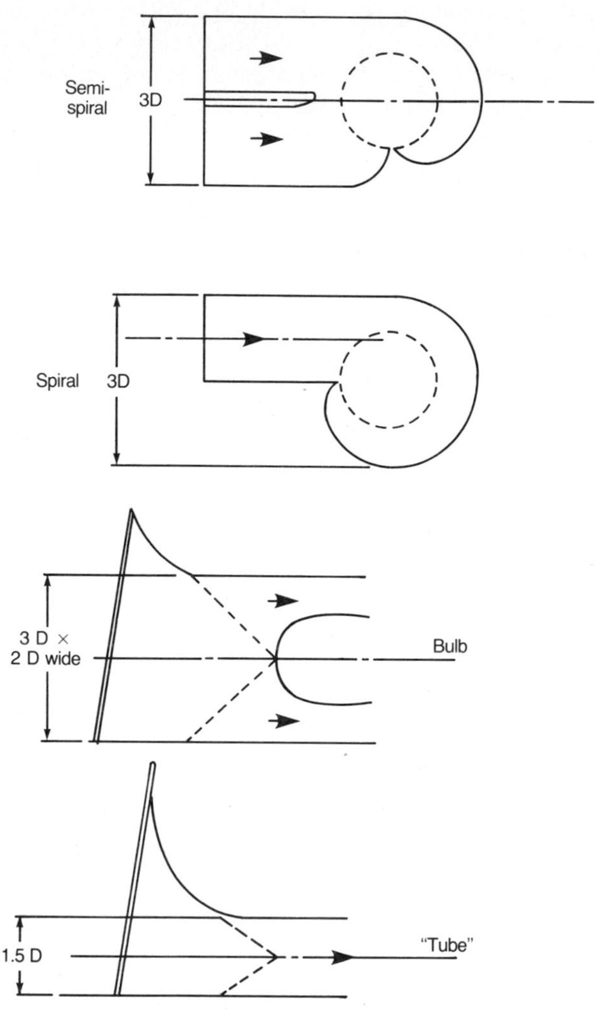

FIG. 6.14 Typical intake and case dimensions. *(Reprinted with permission from Ref. 18.)*

Discharge control for some types of reaction turbines is provided by means of adjustable guide vanes or wicket gates around the outer edge of the turbine runner. The vanes are tied together with linkages and their positioning is regulated by a governor. The adjustable vanes are shown schematically in Fig. 6.1. The flow areas can be readily varied from zero to a maximum by rotation of the vanes. In addition, the velocity diagrams at the entrance and exit are a function of the guide-vane position, and therefore the efficiency of the turbine also changes. Wicket gates can also be used to shut off the flow to the turbine in emergency situations. Various types of valves are installed upstream of the turbine for this purpose for turbines without wicket gates.

One purpose of the draft tube is to reduce the kinetic energy of the water exiting the turbine runner. Within limits a well-designed draft tube will permit installation of the turbine above the tailwater elevation without losing any head. Different designs of a draft tube are common, ranging from a straight conical diffuser to configurations with bends and bifurcations. Some typical shapes and relative dimensions are shown in Fig. 6.15, where the dimensions again are given in terms of the runner diameter.

The simplest form of draft tube is the *straight conical diffuser*. Efficiency of energy conversion is dependent on the angle of the diverging walls, as discussed in Sec. 6.3 under Turbine Efficiency and Losses. Small divergence angles require long diffusers to achieve the area necessary to reduce the exit velocity. Long diffusers increase construction costs; therefore the angle in some cases may be increased up to about 15° from the typical optimum value of about 7°. In addition to the increased loss through a large-angle diffuser, flow separation can lead to unstable flow. Flow instability is to be avoided, as it has an adverse effect on turbine performance.

For some types of turbine installations, such as a vertical-axis turbine, the flow must be turned through a 90° angle after leaving the turbine. This is accomplished by adding an elbow between the turbine and draft tube, which has an influence on the draft tube performance (as discussed briefly in Sec. 6.3 under Turbine Efficiency and Losses) and requires careful design.

Experimental data on diffusers are available in the literature. However, the flow leaving the turbine runner can have a swirl component of velocity which has an effect on the draft-tube efficiency. The magnitude of the swirl is dependent on the type of turbine and operating conditions. Excessive swirl can result in surging in the draft tube, as well as load fluctuations and pressure fluctuations that can cause mechanical vibrations of severe magnitude. However, a small swirl component has been found to be beneficial, as pointed out in Sec. 6.3 under Turbine Efficiency and Losses.

A draft-tube design adequate for one type of runner may not be satisfactory for another. Therefore, the

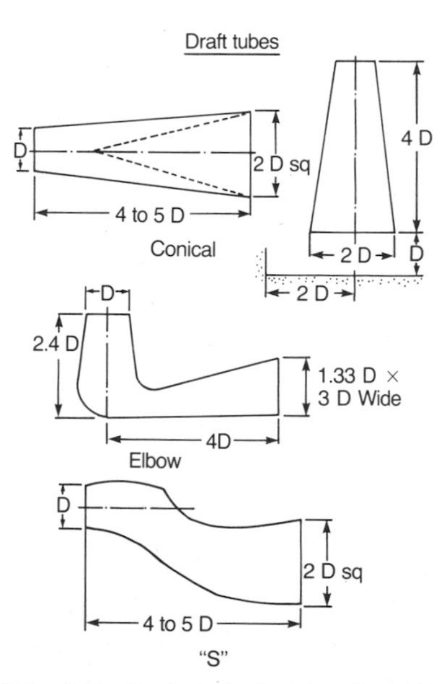

FIG. 6.15 Typical draft tube dimensions. *(Reprinted with permission from Ref. 18.)*

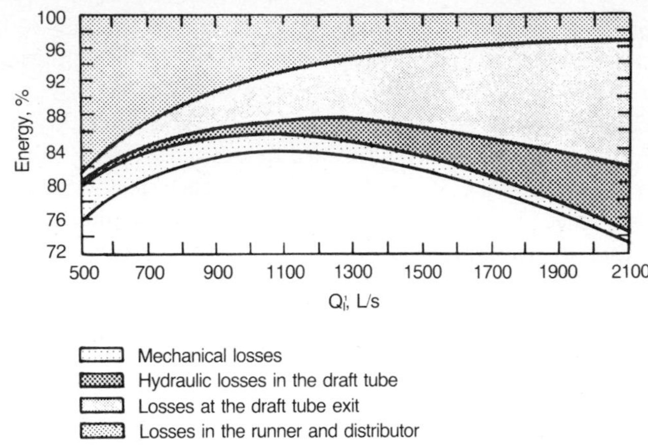

FIG. 6.16 Energy balance for a medium-speed Kaplan turbine model as a function of unit discharge $Q_i = Q/D_o\sqrt{H}$ for a 1-m (3.28-ft) diameter and 1-m (3.28-ft) head. *(Reprinted with permission from Ref. 19.)*

draft tube is considered an essential part of the turbine, and its design is carried out by the turbine manufacturer.

An example of some typical losses and their sources are shown in Fig. 6.16 for a Kaplan turbine. For small discharges, major losses occur in the runner and distributor. This is typical for a turbine operating at off-design conditions and is associated with the shock losses in the runner. As the discharge increases, the runner and distributor losses decrease to relatively small values. The draft-tube losses increase, but the largest increase is the losses at the draft-tube exit. It becomes obvious that efforts should be made to reduce this loss, which can be accomplished by enlarging the draft-tube exit area. However, as such enlargement increases the construction cost, compromises must again be made.

Turbine Classification and Description

There are two basic types of turbines, denoted as *impulse* and *reaction.* In an impulse turbine the available head is converted to kinetic energy before entering the runner, the power available being extracted from the flow at atmospheric pressure. In a reaction turbine the runner is completely submerged and both the pressure and the velocity decrease from inlet to outlet. The velocity head at the inlet to the turbine runner is typically less than 50 percent of the total head available. In either machine the torque is equal to the rate of change of angular momentum through the machine as expressed by the Euler equation.

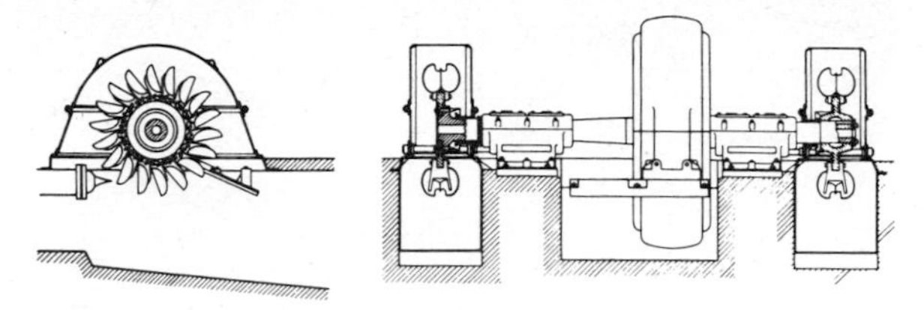

FIG. 6.17 Double-overhung impulse wheel. *(Reprinted with permission from Ref. 20.)*

Impulse Turbines

Modern impulse units are generally of the Pelton type and are restricted to relatively high head application (Fig. 6.17). One or more jets of water impinge on a wheel containing many curved buckets. The jet stream is directed inwardly, sideways, and outwardly thereby producing a force on the bucket which in turn results in a torque on the shaft. All the available head is converted to kinetic energy at the nozzle. Any kinetic energy leaving the runner is "lost." It is essential that the buckets are designed in such a manner that exit velocities are a minimum. No draft tube is used since the runner operates under essentially atmospheric pressure and the head represented by the elevation of the unit above tailwater cannot be utilized.* Since this is a high-head device, this loss in available head is relatively unimportant. As will be shown later, the Pelton wheel is a low-specific-speed device. Specific speed can be increased by the addition of extra nozzles, the specific speed increasing by the square root of the number of nozzles.

Specific speed can also be increased by a change in the manner of inflow and outflow. As shown in Fig. 6.18, a Turgo turbine can handle relatively larger quantities of flow at a given speed and runner diameter by passing the jet obliquely through the runner in a manner similar to a steam turbine. The jet impinges on several buckets continuously, whereas only a single bucket per jet is effective at any instant in a Pelton wheel. The Banki-Mitchell turbine illustrated in Fig. 6.19 is a variation on this theme. The flow passes through the blade row twice, first at the upper portion of the wheel and again at the lower portion. The flow exits the blade in the opposite direction from the first pass and hence this configuration tends to be self-cleaning since debris impinging on the periphery of the runner at the top dead center is removed by the flow on the second pass at essentially bottom dead center.

Most Pelton wheels are mounted on a horizontal axis, although newer vertical-axis units have been developed. Because of physical constraints on orderly outflow from the unit, the maximum number of nozzles is generally limited to

*In principle, a draft tube could be used, which requires the runner to operate in air under reduced pressure. Attempts at operating an impulse turbine with a draft tube have not met with much success.

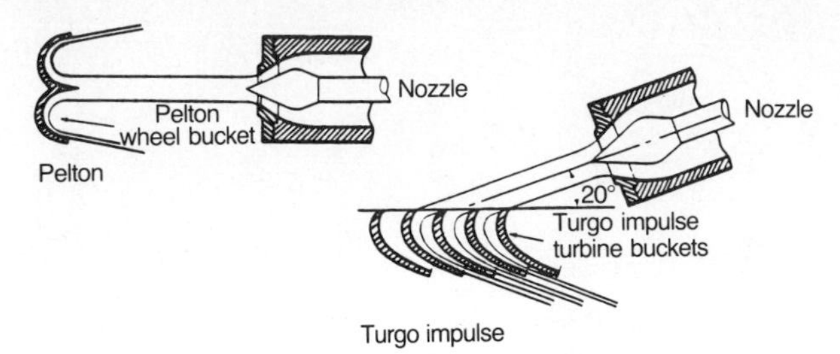

FIG. 6.18 Turgo and Pelton wheels contrasted. The jet on the Turgo strikes three buckets continuously, whereas on the Pelton it strikes only one. A similar speed-increasing effect can be had on the Pelton by adding another jet or two.

six or less. While the power of a reaction turbine is controlled by the wicket gates, the power of the Pelton wheel is controlled by varying the nozzle discharge by means of an automatically adjusted needle, as illustrated in Fig. 6.20. Jet deflectors, Fig. 20*a*, or auxilliary nozzles arranged as in Fig. 20*b* are provided for emergency unloading of the wheel.

Additional power can be obtained by connecting two wheels to a single generator or by using multiple nozzles. Since the needle valve can throttle the flow while maintaining essentially constant jet velocity, the relative velocities at entrance and exit remain unchanged, producing nearly constant efficiency over

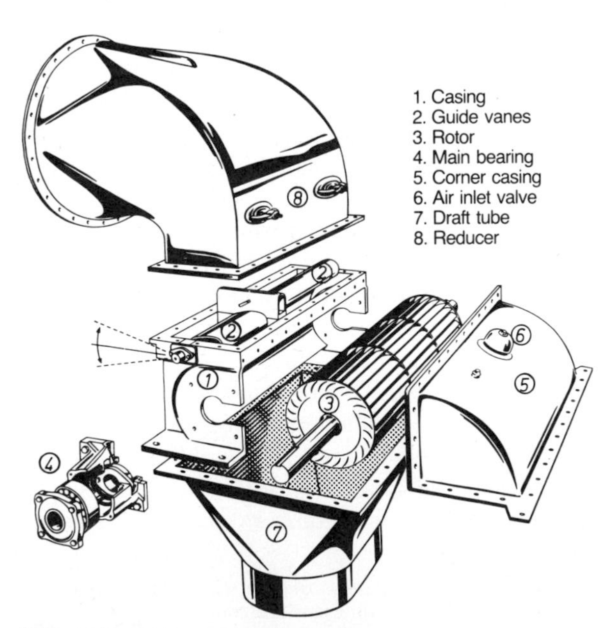

FIG. 6.19 Ossberger crossflow turbine.

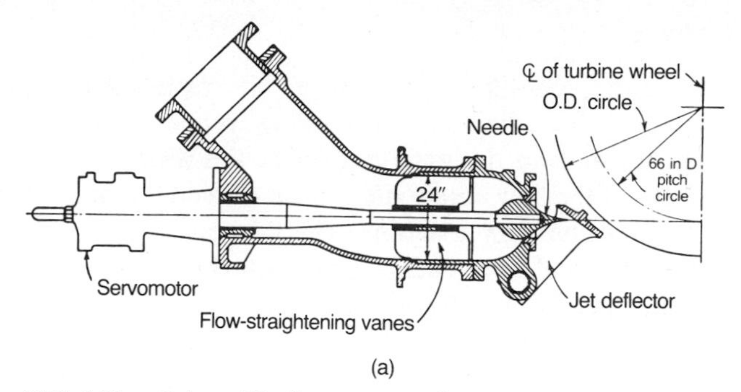

FIG. 6.20*a* Pelton 45° elbow-type needle nozzle with jet deflector. *(Reprinted with permission from Ref. 20.)*

a wide range of power output. This is a desirable feature of Pelton and Turgo wheels.

Throttling of the Banki-Mitchell turbine is accomplished differently, as illustrated in Fig. 6.19. An adjustable guide vane is used which functions in a manner similar to the wicket gates in a reaction turbine. If operating conditions require, the guide vanes can be divided into two separately controlled sections. For most installations, the lengths of the two guide-vane sections are in the ratio 1:2, allowing for utilization of ⅓, ⅔, or the entire runner, depending on the flow conditions. This combination provides a relatively flat efficiency curve over the power range 15 to 100 percent.

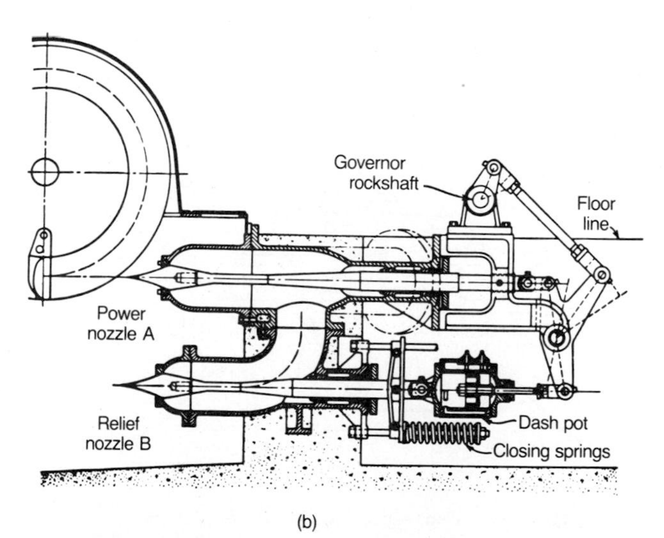

FIG. 6.20*b* Pelton nozzle with auxiliary relief nozzle. *(Reprinted with permission from Ref. 20.)*

Reaction Turbines

Reaction turbines are classified according to the variation in flow direction through the runner. In radial- and mixed-flow runners, the flow exits at a radius different (in modern designs the inlet flow is always inward) than the radius at the inlet. If the flow enters the runner with only radial and tangential components, it is a radial-flow machine. The flow enters a mixed-flow runner with both radial and axial components. Francis turbines are of the radial- and mixed-flow type, depending on the design specific speed. Two Francis turbines are illustrated in Fig. 6.21. The radial-flow runner (Fig. 21*a*) is a low-specific-speed design, whereas the mixed-flow runner (Fig. 21*b*) achieves peak efficiency at considerably higher specific speed.

Axial-flow propeller turbines are generally either of the fixed-blade or

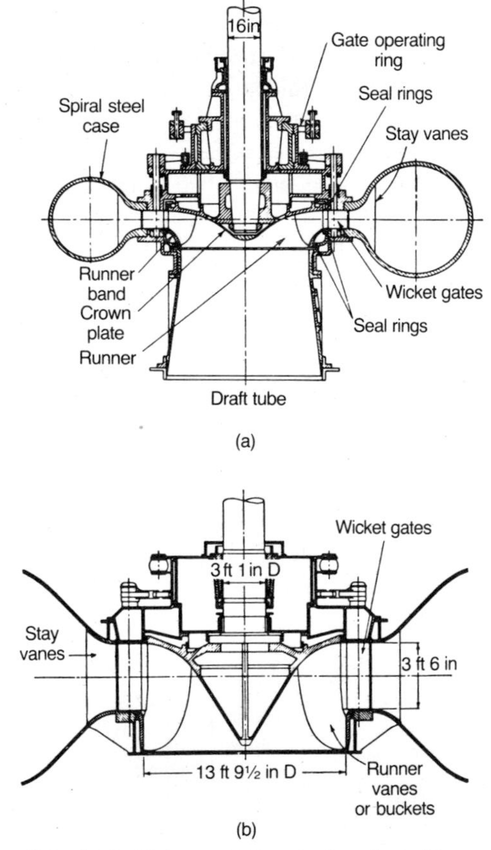

FIG. 6.21 Two examples of Francis turbine. *(Reprinted with permission from Ref. 20.)*

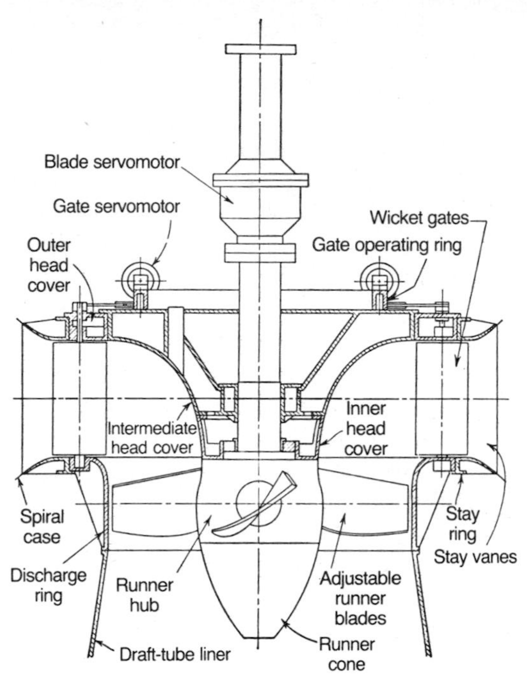

FIG. 6.22 Smith-Kaplan axial-flow turbine with adjustable-pitch runner blades, $N_s \sim 3.4$. *(Reprinted with permission from Ref. 20.)*

Kaplan (adjustable-blade) variety. The "classical" propeller turbine, illustrated in Fig. 6.22, is a vertical-axis machine with a scroll case and a radial-wicket-gate configuration that is very similar to the flow inlet for a Francis turbine. The flow enters radially inward and makes a right-angle turn before entering the runner in an axial direction. The Kaplan turbine has both adjustable runner blades and adjustable wicket gates. The control system is designed so that the variation in blade angle is coupled with the wicket-gate setting in a manner which achieves best overall efficiency over a wide range of flow rate.

The classical design does not take full advantage of the geometric properties of an axial-flow runner. The flow enters the scroll case in a horizontal direction, issues radially inward from the guide case, where it forms a vortex and discharges into the draft tube in a vertical direction with very little whirl component remaining. The flow must then again be turned through 90° to discharge into the tailwater in a horizontal direction. From a design point of view, this is less than desirable for many reasons. The flow field entering the runner is highly complex, and it is difficult to design the proper pitch distribution from hub to tip for minimal shock losses. There are additional losses in the elbow, and the tortuous flow path required from inlet to outlet requires additional civil works.

More modern designs take full advantage of the axial-flow runner; these include the tube, bulb, and Straflo types illustrated in Fig. 6.23. The flow enters and exits the turbine with minor changes in direction. A wide variation in civil works design is also permissible. The tube-type can be fixed-propeller, semi-Kaplan, or fully adjustable. An externally mounted generator is driven by a shaft which passes through the flow passage either upstream or downstream of the runner.

The bulb turbine was originally designed as a high-output low-head unit. In large units, the generator is housed within the bulb and is driven by a variable-pitch propeller at the trailing end of the bulb. Smaller units are available in which an externally mounted generator is driven by a right-angle drive which is housed within the bulb (Fig. 6.24). Because of the simplicity of installation, the various modern axial-flow machines are of considerable interest for low-head applications.

In addition to the radial-flow and mixed-flow-Francis and axial-flow propeller units, there is the Deriaz turbine which is a mixed-flow propeller unit of the Kaplan type. This turbine was originally developed for pumped storage applications, but shows great promise for applications in the medium-head range. The turbine consists of a series of controllable-pitch blades mounted on a conical hub. The turbine can have either a conventional scroll case and gate apparatus or a more specialized flap system for controlling the inlet flow. This unit provides the same flat efficiency curve over a wide range of power as the standard Kaplan propeller units, but because of the mixed-flow design, it is applicable to higher head applications. However, in 1981, there did not appear to be any units available of a small size (less than 1-MW capacity).

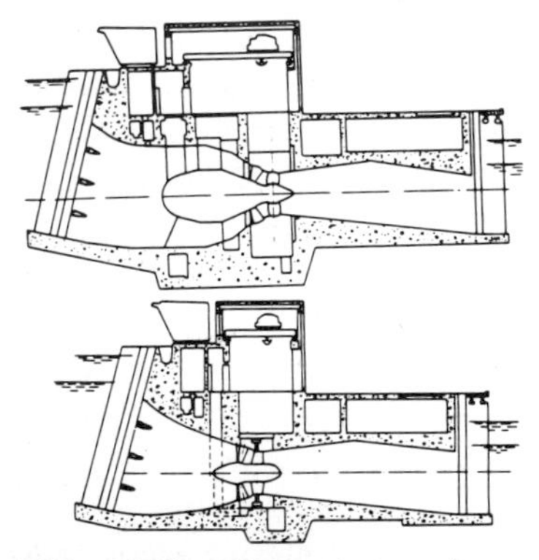

FIG. 6.23*a* Comparison of structures required for Straflo vs. bulb turbine with same output and head. *(U.S. Department of Energy Report IDO-1962-1, April 1978.)*

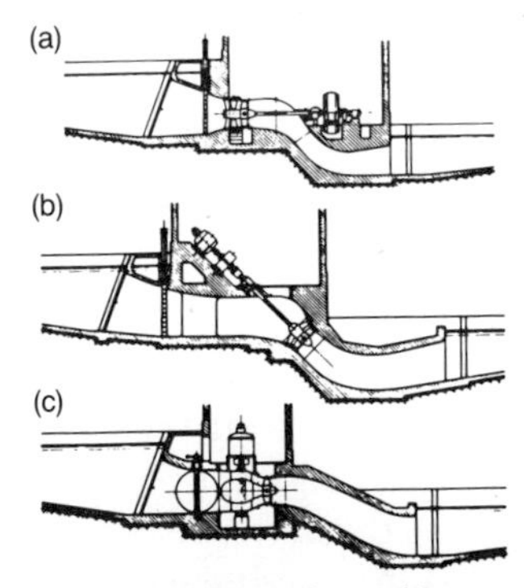

FIG. 6.23*b* Various tube turbine arrangements. *(U.S. Department of Energy Report IDO-1962-1, April 1978.)*

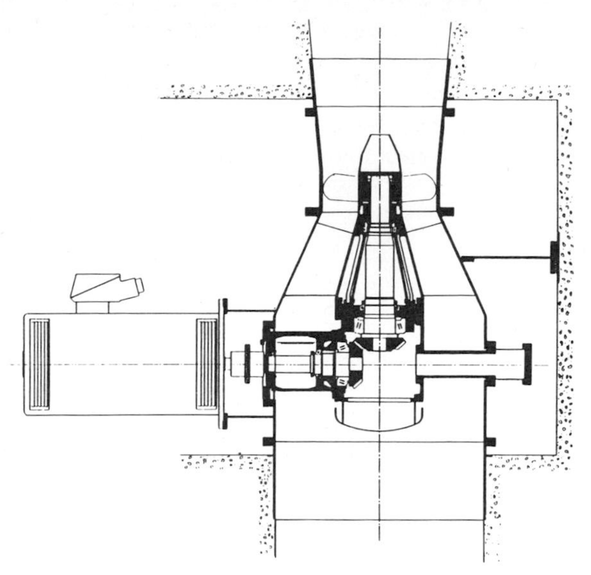

FIG. 6.24 Right-angle drive bulb turbine *(Neyrpic.)*

Performance Characteristics

Comparative Performance of Impulse and Reaction Turbines

The two basic types of turbines tend to operate at peak efficiency over different ranges of specific speed. This is due to geometric and operational differences. In order to give the reader a perspective of the operational characteristics of each type, a brief discussion of operational principles is presented, followed by a summary of the performance of commercially available equipment in subsequent sections.

Impulse Wheels: Typical types of impulse wheels are illustrated in Figs. 6.17 to 6.19. For a given pipeline there is a unique jet diameter that will deliver maximum power to a jet. Denoting the jet diameter by d_j the power is given by

$$P_j = \gamma Q \frac{V_j^2}{2g} = \gamma \frac{\pi}{8g} d_j^2 V_j^3 \tag{6.53}$$

Let Δh denote the difference between the reservoir surface and the nozzle elevation. Neglecting losses at the entrance to the pipe and in the nozzle, Eq. (6.53) yields

$$\frac{V_j^2}{2g} + f \frac{L}{d_p} \frac{V_p^2}{2g} = \Delta h \tag{6.54}$$

where V_j is the jet velocity, V_p is the velocity in the pipe, and d_p and L denote, respectively, the pipe diameter and length. As the size of the nozzle opening is increased, the flow rate Q gets larger while the jet velocity V_j gets smaller, since the losses in the pipeline increase with Q. Using $V_p = V_j(d_j/d_p)^2$ and Eq. (6.54), it can be shown that maximum power is obtained when

$$\Delta h = 3f\frac{L}{d_p}\frac{V_p^2}{2g} \tag{6.55}$$

$$H = \frac{V_j^2}{2g} = \frac{2}{3}\Delta h \tag{6.56a}$$

and

$$dj = \left(\frac{d_p^5}{2fL}\right)^{1/4} \tag{6.56b}$$

Thus, for a given penstock geometry and Δh, the maximum power available to the turbine can be calculated. From Eqs. (6.33) and 6.56a), it should be noted that the maximum possible plant efficiency is ⅔ for this case.

For the usual design problem in which Q and Δh are given, the maximum possible plant efficiency could theoretically be unity if d_p is so large that the loss term in Eq. (6.54) is negligible. The jet velocity V_j would then be given by $V_j = \sqrt{2g\,\Delta h}$ and the jet diameter by $\pi d_j^2/4 = Q/\sqrt{2g\,\Delta h}$. This is a minimum jet diameter. Note that selection of d_p and d_j in specific cases requires an economic analysis and depends strongly on the characteristics of the site and also on the turbines that are available.

Of the head available at the nozzle inlet, a small portion is lost to friction in the nozzle and to friction on the buckets. The rest is available to drive the wheel. The actual utilization of this head depends on the velocity head of the flow leaving the turbine and the setting above tailwater. Optimum conditions corresponding to maximum utilization of the head available dictate that the flow leaves at essentially zero velocity. Under ideal conditions, this occurs when the peripheral speed of the wheel is one-half the jet velocity. In practice, optimum power occurs at a speed coefficient, $\phi = u_1/\sqrt{2gH}$, somewhat less than 0.5. In fact, it can be shown that best efficiency will occur when $\phi = \frac{1}{2}C_v \cos \alpha_1$, where C_v is the velocity coefficient for the nozzle,

$$C_v = \frac{V_j}{\sqrt{2gH}} \tag{6.57}$$

and α_1 represents the effective angle between the jet velocity and the peripheral velocity of the runner at entrance to the bucket. This is illustrated in Fig. 6.25.

Since maximum efficiency occurs at fixed speed for fixed H, V_j must remain constant under varying flow conditions. Thus the flow rate Q is regulated with an adjustable nozzle. There is some variation in C_v and α_1 with regulation and maximum efficiency occurs at slightly lower values of ϕ under partial power settings. Present nozzle technology is such that the discharge can be regulated over a wide range at high efficiency.

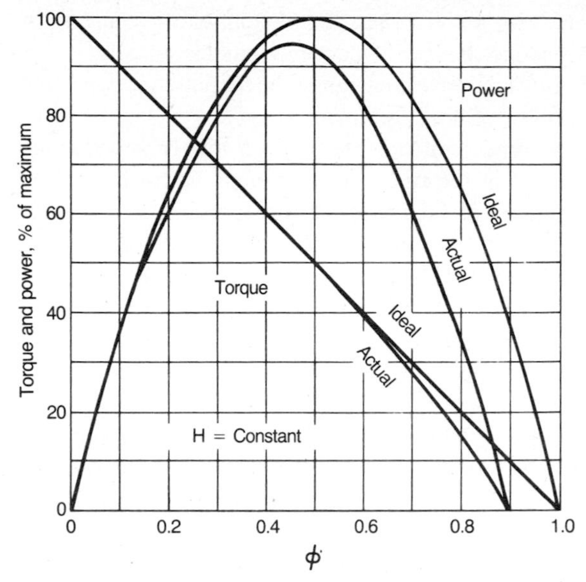

FIG. 6.25 Ideal and actual variable-speed performance for an impulse turbine. *(Adapted with permission from Ref. 20.)*

A given head and penstock configuration establishes the optimum jet velocity and diameter. The size of the wheel determines the speed of the machine. For a wheel of diameter D, the speed in radians per second is

$$\Omega = 2 \frac{u_1}{D} = \frac{2\phi}{D} \sqrt{2gH} \tag{6.58}$$

Using Eq. (6.33) for the power P and the equation

$$Q = V_j \frac{\pi d_j^2}{4} = C_v \sqrt{2gH} \frac{\pi d_j^2}{4} \tag{6.59}$$

for the flow rate Q, one obtains for the specific speed of the machine the relationship

$$N_s = 2^{1/4} \sqrt{2\pi\eta C_v \phi} \frac{d_j}{D} \tag{6.60}$$

or approximately

$$N_s = 1.3 \frac{d_j}{D} \tag{6.61}$$

Practical values of d_j/D for Pelton wheels to ensure good efficiency are in the range 0.04 to 0.1, corresponding to N_s values in the range 0.05 to 0.13 (10 to 25 in metric units using the metric horsepower). In Turgo turbines the relative wheel diameter can be half that of a Pelton wheel resulting in specific speeds approximately twice that of the conventional design. Higher specific speeds are possible with multiple-nozzle designs. The increase is proportional to the square root of the number of nozzles.

Crossflow turbines can operate at even higher specific speed ($N_s = 0.6$) because the length of the runner can be much larger than the diameter, which permits large values of flow through a relatively small diameter runner. This is possibly one of the reasons why the crossflow turbine has seen application over such a wide range of head and power (Fig. 6.34). However, in considering an impulse unit, one must remember that efficiency is based on net head, and the net head for an impulse unit is generally less than the net head for a reaction turbine at the same gross head because of the lack of a draft tube.

Reaction Turbines: The main difference between impulse wheels and reaction turbines is the fact that a pressure drop takes place in the rotating passages of the reaction turbine. This implies that the entire flow passage from the turbine inlet to the discharge at the tailwater must be completely filled. A major factor in the overall design of modern reaction turbines is the draft tube.

This was not always the case. In earlier days, when low-speed large-diameter Francis turbines were installed under low heads, the lack of a draft tube or the use of a very short conical tube resulted in a nominal velocity head loss from the runner. This was not particularly critical for installations which are underdeveloped based on current standards since water was spilled over the dam much of the year. However, since it is now desirable to reduce the overall equipment and civil construction costs by using high-specific-speed propeller runners, the draft tube is extremely critical from both a flow stability and an efficiency viewpoint. Since the runner diameter is relatively small, a substantial percentage of the total energy is in the form of kinetic energy leaving the runner. To recover this efficiently, considerable emphasis should be placed on the draft-tube design.

The practical specific-speed range for reaction turbines is much broader than for impulse wheels. This is due to the wider range of variables which control the basic operation of the turbine. As an illustration of the design and operation of reaction turbines for constant speed, refer again to Fig. 6.1. The pivoted guide vanes allow for control of the magnitude and direction of $\mathbf{V}_1$, that is, V_1 and α_1. The relationship between blade angle, inlet velocity, and peripheral speed for shock-free entry can be obtained from Eq. (6.7) as

$$\cos \beta_1 = \frac{V_1 \cos \alpha_1 - u_1}{V_1 \sin \alpha_1} \tag{6.62}$$

Without the ability to vary the blade angle, it is obvious that shock-free entry cannot be completely satisfied at partial flow. This is the distinction between the power efficiency of fixed-propeller and Francis types at partial loads and the fully adjustable Kaplan design.

Referring to Eq. (6.29), optimum hydraulic efficiency would occur when α_2 is equal to 90°. However, overall efficiency of the turbine is dependent on the optimum performance of the draft tube which occurs with a little swirl in the flow. Thus, best overall efficiency occurs with $\alpha_2 \cong 75°$ for high specific speed turbines. The hydraulic efficiency (Eq. 6.29) is approximately

$$\eta_h = 2\phi C_1 \cos \alpha_1 \tag{6.63}$$

With α_1 in the range of 10 to 25° and $C_1 \cong 0.6$, the speed coefficient ϕ, is approximately 0.8 compared with a little less than 0.5 for an impulse turbine. Note also that $C_1 \cong 0.6$ implies that only 40 percent of the available head is converted to velocity head at the turbine inlet compared with 100 percent for the impulse wheel.

The determination of optimum specific speed in a reaction turbine is more

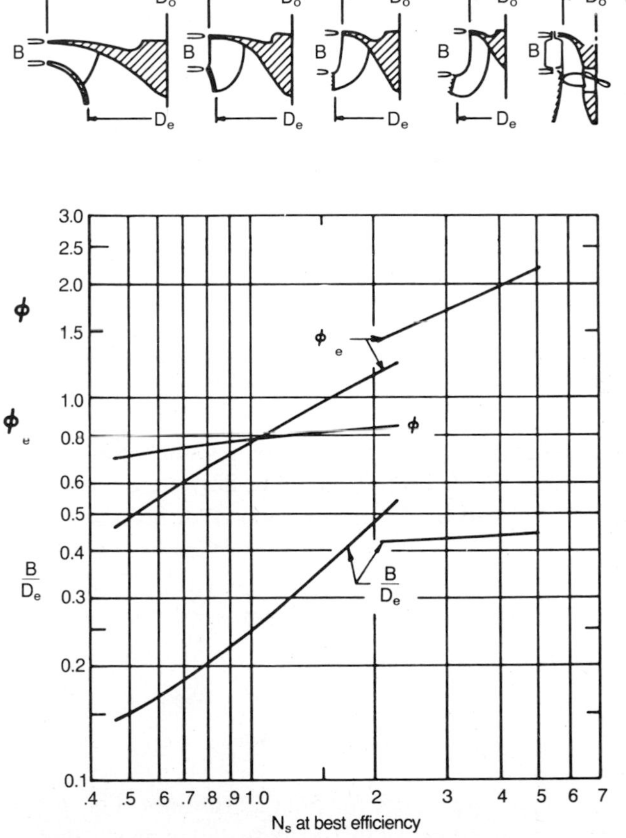

FIG. 6.26 Empirical design constants for reaction turbines. *(Adapted with permission from Ref. 20.)*

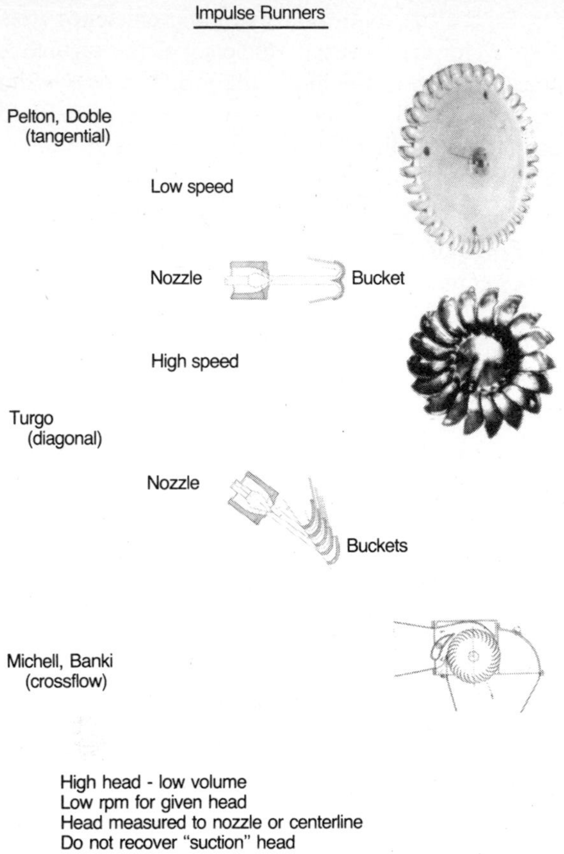

Impulse Runners

Pelton, Doble
(tangential)

Low speed

Nozzle
Bucket

High speed

Turgo
(diagonal)

Nozzle

Buckets

Michell, Banki
(crossflow)

High head - low volume
Low rpm for given head
Head measured to nozzle or centerline
Do not recover "suction" head

FIG. 6.27 Physical characteristics of various turbine runners compared. *(Adapted with permission from Ref. 18.)*

complex since there are more variables. For a radial-flow machine (refer to Fig. 6.1), a relatively simple expression can be derived. Combining Eqs. (6.22), (6.23), (6.28), (6.33), and (6.58), the expression for specific speed, Eq. (6.36), is

$$N_s = 2^{5/4}\left(2\pi\eta f_b C_1 \sin \alpha_1 \frac{B}{D}\right)^{1/2} \phi \tag{6.64}$$

or approximately

$$N_s = 5.5\left(C_1 \sin \alpha_1 \frac{B}{D}\right)^{1/2} \phi \tag{6.65}$$

Using standardized design charts for Francis turbines (Fig. 6.26), N_s is normally found to be in the range 0.3 to 2.5 (58 to 480 in metric units using metric horsepower). Note that the actual values of the reference diameters used in

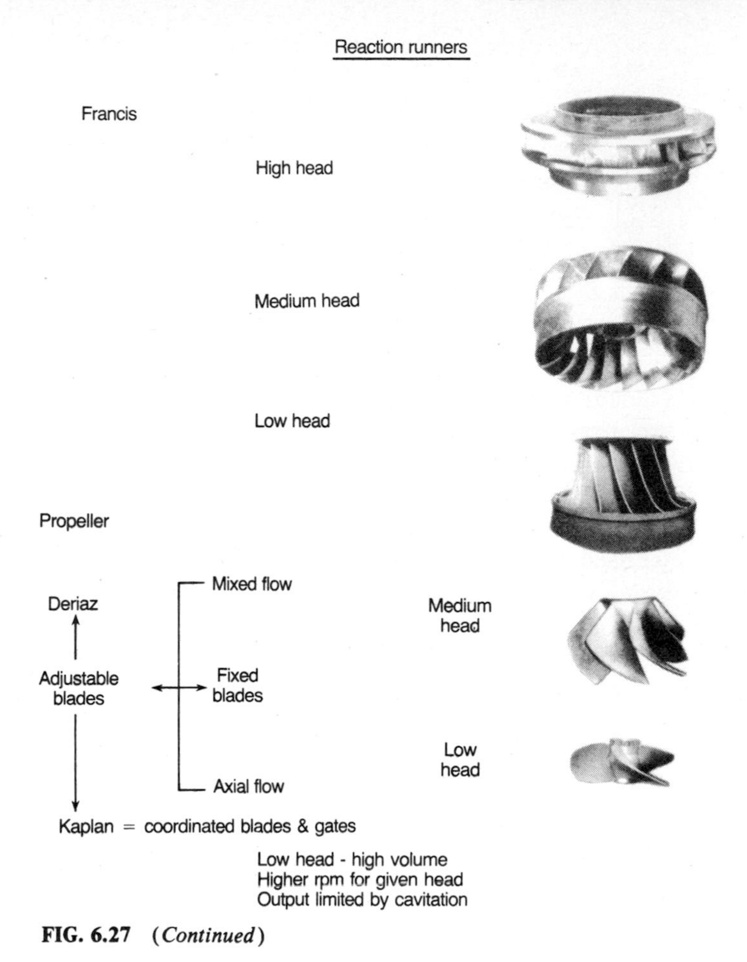

FIG. 6.27 (*Continued*)

Fig. 6.26 are defined in the figure itself and do not necessarily correspond to D_1 and D_2 in Fig. 6.1.

Performance Comparison: The physical characteristics of various runner configurations are summarized in Fig. 6.27. It is obvious that the configuration changes with speed and head. This can be expressed in terms of peak efficiency versus specific speed, as illustrated in Fig. 6.28. As already discussed, impulse turbines are efficient over a relatively narrow range of specific speed, whereas Francis and propeller turbines have a wider useful range.

Variable geometry is an important consideration when a turbine is required to operate over a wide range of load. Pelton wheels and Turgo wheels tend to operate efficiently over a wide range of power loading because the needle valve is capable of metering flow at a constant discharge velocity. Thus, the relative velocities through the runner remain fixed in magnitude and direction which

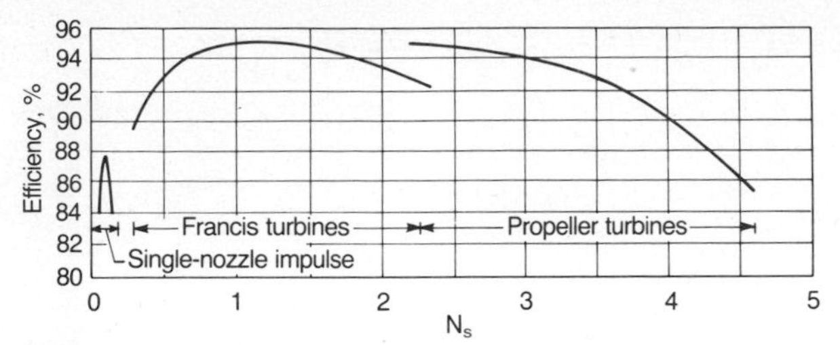

FIG. 6.28 Efficiency of various types of turbines as a function of specific speed.

allows for maximum runner efficiency independent of flow rate. A comparison of efficiency variation with load as a function of the level of sophistication of a tube turbine is illustrated in Fig. 6.29. Fixed gates and blade settings result in peak efficiency at 100 percent load, and the efficiency drops off rapidly with changes in load.

On the other hand, a Kaplan-type tube turbine can maintain efficiency over a relatively broad range of conditions. Also illustrated in Fig. 6.29 is the variation of efficiency for an impulse wheel. Although its peak efficiency is less than the high-speed tube turbine, the impulse unit is able to maintain a relatively high efficiency over a wide range of conditions. Both Francis and Deriaz units are designed to operate at medium specific speed. The efficiency of these two turbines is compared in Fig. 6.30. Again, the advantages of controllable-pitch blades are evident in this comparison. The decision of whether to select a simple configuration with a relatively "peaky" efficiency curve or to go to the addi-

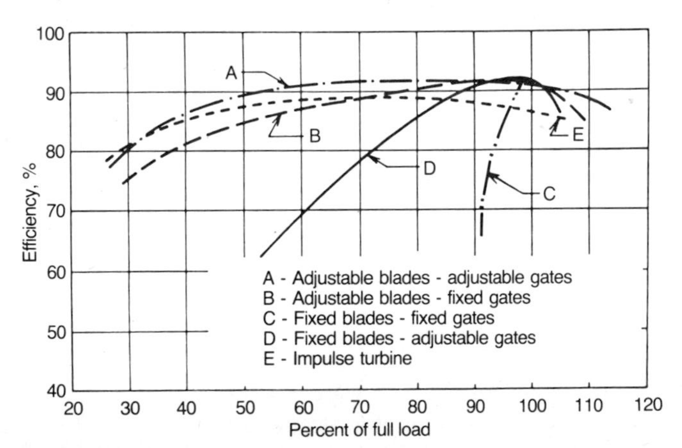

FIG. 6.29 Turbine efficiency as a function of load. Comparison between an impulse turbine and various configurations of a high-speed propeller unit.

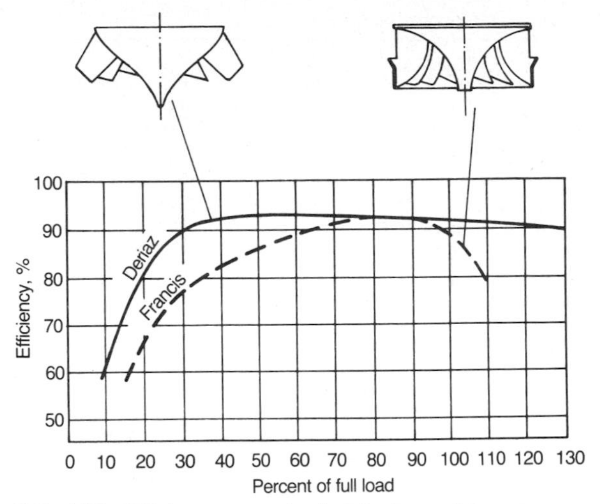

FIG. 6.30 Efficiency versus load of Deriaz and Francis turbines of equivalent specific speed and size.

tional expense of installing a more complex machine with a broad efficiency curve will depend on the expected operation of the plant. If the head and flow are relatively constant, then the less expensive choice is justified. On the other hand, many run-of-river plants may be more economical with the installation of Kaplan or Deriaz units. Table 6.1 summarizes the range of head, flow, and power of various types of turbines.

Detailed performance maps that are typical of Francis- and Kaplan-type turbines are shown in Figs. 6.31 and 6.32. Figure 6.31 clearly shows the narrower operating range of a fixed-geometry Francis turbine when compared to the fully adjustable Kaplan type. Note that maximum efficiency does not occur at maximum gate opening but at an opening of about 75 percent, where 85 percent power is produced under design head. Figure 6.32 makes quite clear the operating advantages which accrue with the increased complexity of the Kaplan design.

Because various types of turbines tend to operate best over different specific-speed ranges, the head and power available at a given site dictate what options are practical. This is illustrated in Fig. 6.33, where the various types of turbines that would be useful at various combinations of head and desired power output are plotted over a range of head and power from 2 to 400 m (about 6 to 1,300 ft) and 10 to 20,000 kW. Figure 6.33 is constructed with the following assumptions: speed in the range 600 to 3,600 r/min,* direct drive, and specific speed

*The speed range is based on the assumption of 60-Hz current and a maximum of 12 poles in the generator. (Number of poles equals 7,200/*n*.)

TABLE 6.1 Turbine Performance Characteristics*

Turbine type	Rated head† H_R, ft (m)	Min/max head as % of H_R, %	Capacity, MW	Min/max capacity as % of rated power, %
1. Vertical fixed-blade propeller	7–65.6 & over (2–20) & over	55–125‡	0.25–15 & above	30–115
2. Vertical Kaplan (adjustable-blade propeller)	7–65.6 & over (2–20) & over	45–150	1–15 & above	10–115
3. Vertical Francis	25–65.6 & over (8–20) & over	50–125‡ & over	0.25–15	35–115
4. Horizontal Francis	25–65.6 & over (8–20) & over	50–125‡	0.25–2	35–115
5. Tubular (with adjustable blades and fixed gates)	7–59 (2–18)	65–140	0.25–15	45–115
6. Tubular (fixed-blade runner with wicket gates)	7–59 (2–18)	55–140	0.25–15	35–115
7. Bulb	7–66 (2–20)	45–140 & over	1–15	10–115
8. Rim	7–30 (2–9)	45–140	1–8	10–115
9. Right-angle-drive propeller	7–59 (2–18)	55–140	0.25–2	45–115
10. Open flume	7–36 (2–11)	90–110	0.25–2	30–115
11. Closed flume	7–54.6 & over (2–20) & over	50–140	0.25–3	35–115
12. Crossflow	20–54.6 & over (6–20) & over	50–125	0.25–2	10–115

*From Ref. 22.
†Rated head is defined as the head at which full gate output equals rated generator output.
‡May be operated to 140 percent if proper turbine setting is used.

in the range of optimum efficiency for a given design. At constant n and n_s, the head is related to the power by

$$H \sim \left(\frac{n}{n_s}\right)^{4/5} (P)^{2/5} \tag{6.66}$$

Thus the upper limits for each turbine type represent maximum r/min (if possible without cavitation) and minimum n_s. The lower boundary is determined from the lowest r/min and maximum n_s without cavitation.

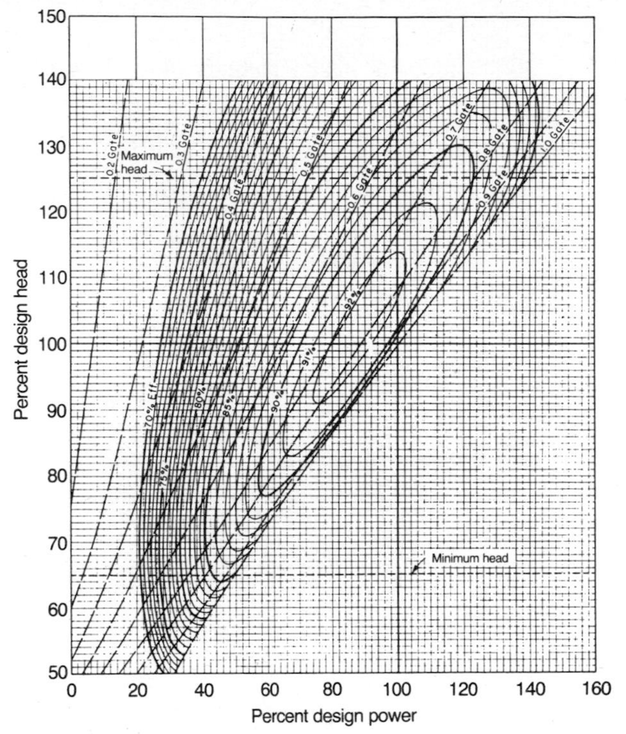

FIG. 6.31 Francis turbine performance—23 to 46 m head range. *(Ref. 21.)*

Cavitation limits are based on a net positive draft head $H_{sv} = H_a - H_v - z$ of 1 atm. For example, the lower curve for the propeller turbine is determined with n of 600 r/min and a maximum n_s of 764 in metric units using kW for power ($N_s = 4.6$; see Fig. 6.28). Since the critical Thoma's sigma is a function of n_s (Fig. 6.13), and the plant sigma is a function of draft head and net head (Eq. 6.47), a point is reached where the two sigmas are equal. This occurs at the break in the lower curve. If it is desired to maintain the same speed at higher heads, n_s must decrease to avoid cavitation. Thus, n_s decreases along the lower curve to the right of the breakpoint.

Survey of Commercially Available Equipment

Recent interest in small-scale hydropower has stimulated the turbine manufacturers to produce turbines suitable for this application. Larger units have been scaled down to match the lower head and power requirements. As cost of equipment has a significant impact on the economic feasibility of a small-scale installation, a major thrust has been made to develop standardized units to reduce cost. Many of these standardized units are supplied complete with the

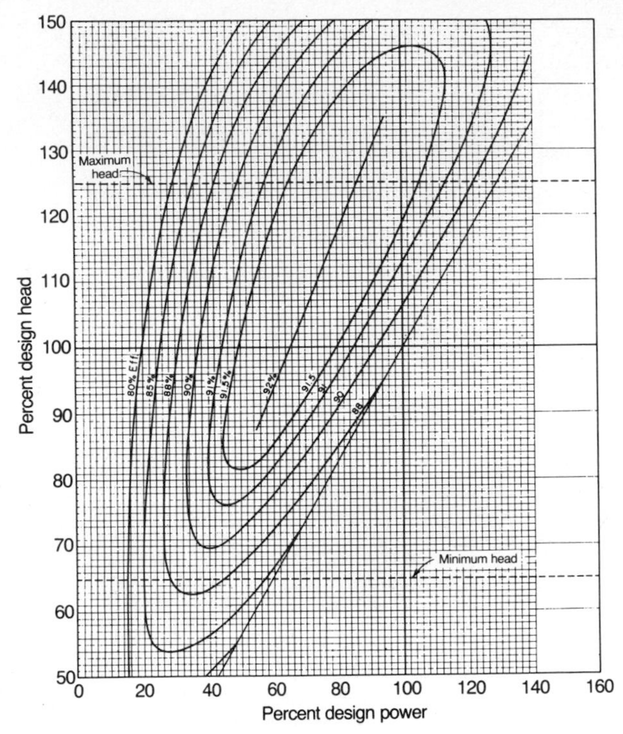

FIG. 6.32 Performance curve for adjustable-blade turbine, N_s = 3.3. *(Ref. 21.)*

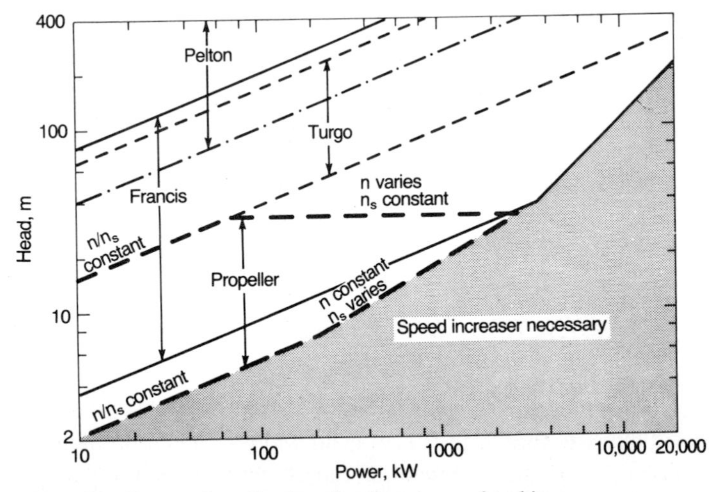

FIG. 6.33 Range of application of various types of turbines.

generator and auxiliary equipment. The larger and well-established equipment manufacturers are adding such equipment as a line item, and the number of smaller manufacturers is rapidly increasing in response to the anticipated demand. The small companies are in general developing equipment for the lower power outputs in the range of less than 20 kW.

A summary graph based on commercially available equipment as a function of head and power output is shown in Fig. 6.34 to indicate the coverage as it currently exists. This graph should be compared with Fig. 6.33. This summary should be used only as a guide to available turbines, as different units are rapidly being offered by the various manufacturers. It can readily be seen that several types of turbines are available for a given head and power output. The crossflow turbine covers a wide range of conditions. The propeller turbine classification includes the vertical Kaplan, bulb, and tube turbine units. The Kaplan turbine is commonly used for the higher heads and power output, and the bulb for essentially the same output at lower heads. The tube-turbine range includes the lower heads and power outputs. Standardized tube turbines are available in this range and may be economically attractive for the mini hydro projects and also in the upgrading or rehabilitation of old hydropower stations.

For the micro hydropower sites, which are low-head and have power outputs of less than 100 kW, some standardized propeller turbines are also available. These small units have been developed specifically for this application, and attempts have been made to simplify the machine and thus lower initial equipment costs. The simplification may result in reduction of efficiency of the unit, and this should be considered in the assessment of economic feasibility.

Accurate performance data are usually not available for smaller turbines.

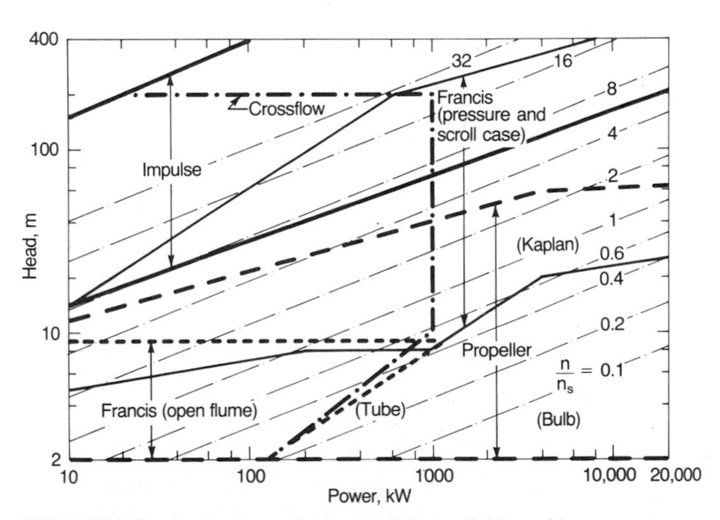

FIG. 6.34 Summary chart of commercially available turbines.

In fact, model tests are often not performed for turbines smaller than about 5 to 10 MW. As an example, consider a 500 r/min 15-MW turbine operating at its design point with an efficiency of 92 percent under a 100-m (328-ft) head. The same design could be scaled down to 500 kW at 1200 r/min and 50-m (164-ft) head. Since the specific speed is constant, the size, speed, and head will vary according to

$$\frac{H_1}{H_2} = \frac{n_1^2 D_1^2}{n_2^2 D_2^2} \tag{6.67}$$

Thus, $D_1/D_2 = 3.39$. The change in efficiency could be estimated by using the Moody formula (Eq. 6.32) in reverse (a Moody step-down equation, if you will). This yields $\eta_2 = 89.9$ percent. Scaling down to even smaller sizes would bring about even more dramatic reduction in efficiency. This is true only for reaction turbines in which leakage and frictional losses are disproportionately higher in the smaller sizes.

Present Trends in Turbine Development

As previously mentioned, attention is being directed toward the development of standardized turbines to cover a wide range of applications. Some turbine manufacturers are exploring the possibility of using pumps operated as turbines. It is expected that continued efforts will be made in this area. For many remote and relatively inaccessible sites, lightweight turbines of small size would be attractive. The use of plastics, etc. for various elements could perhaps reduce cost through mass production techniques as well as reduce the weight of the smaller units. These elements may require more maintenance, but the lower cost of the parts may offset the increased maintenance cost.

6.5 HYDRAULIC STRUCTURES AND OPERATIONAL CONSIDERATIONS

Integration of Turbine with Inlet and Outlet Works

From the discussion in Sec. 6.4, it is apparent that several types of turbines may be appropriate for given flow conditions. It is therefore necessary to make a decision as to which particular unit is most economical. In addition to the cost of the turbine and generator, the cost of the associated civil works must be considered, as this cost can represent a large portion of the total cost of a small-scale hydropower installation.

Some types of turbines require larger civil works than others. Several alternative preliminary layouts should be evaluated, each of which may have different inlet and draft-tube requirements. The necessity of this evaluation has become increasingly evident with the recent interest in retrofitting existing sites for power production. The original turbines either may not exist or may not be capable of repair, or it may be desirable to replace the unit with a modern turbine of different capacity.

The condition and the extent of the existing civil works may influence the type of turbine to be finally selected. For example, if an open-flume Francis turbine was originally installed but is no longer usable, it may prove to be more economical to install a new turbine of the same type or a different design with roughly the same overall dimensions as the original equipment if the rest of the structure is still in good condition. In other cases, it may be most cost effective to abandon the existing structure and select a different type of turbine. It is therefore necessary to analyze each site on an individual basis. In so doing, considerable cost savings may be realized.

Cavitation and Turbine Setting

Another factor that must be considered prior to equipment selection is the evaluation of the turbine with respect to tailwater elevations. As previously discussed, hydraulic turbines are subject to pitting due to cavitation. For a given head, a smaller, lower-cost high-speed runner must be set lower (i.e., closer to tailwater or even below tailwater) than a larger, higher-cost low-speed turbine runner. Also, atmospheric pressure or plant elevation above sea level is a factor along with tailwater-elevation variations and operating requirements. This is a complex subject which can only be accurately resolved by model tests. Every runner design will have different cavitation characteristics; therefore, the anticipated turbine location or setting with respect to tailwater elevations is an important consideration in turbine selection.

Cavitation is not normally a problem with impulse wheels. However, by the very nature of their operation, cavitation is an important factor in reaction-turbine installations. The susceptibility for cavitation to occur is a function of the installation and the turbine design. As already discussed, this can be expressed conveniently in terms of Thoma's sigma. The critical value of σ_T is a function of specific speed, as illustrated in Fig. 6.13. As the specific speed increases, the critical value of σ_T increases dramatically.

For minimization of cavitation problems, the plant σ_T must be in excess of the critical σ_T denoted in Fig. 6.13. This can have important implications for turbine settings and the amount of excavation necessary. The criteria for establishing the turbine setting have already been discussed in Sec. 6.3 under Cavitation. As an example, consider a 500-kW machine operating at 500 r/min under a head of 10 m (32.8 ft). The specific speed is 3.8. This will be an axial-flow turbine having a critical σ_T of about 0.9. At sea level the maximum turbine setting would be

$$z_{B_m} = 10 - 0.09 - (0.9 \times 10) = 1 \text{ m } (3.28 \text{ ft})$$

If the same turbine was installed at Leadville, Colorado, elevation ~ 3000 m (~ 9,840 ft), the maximum turbine setting would be

$$z_B = 6.7 - 0.09 - (0.9 \times 10) = -2.3 \text{ m } (-7.54 \text{ ft})$$

Considerable excavation would be necessary. Thus, cavitation can be an important consideration.

Speed Regulation

The speed regulation of a turbine is an important and complicated problem. The magnitude of the problem varies with size, type of machine and installation, type of electrical load, and whether the plant is tied into an electrical grid. It should also be kept in mind that runaway or no-load speed can be higher than the design speed by factors as high as 2.5. This is an important design consideration for all rotating parts, including the generator.

It is beyond the scope of this section to discuss the question of speed regulation in detail. However, some mention should be made since much of the technology is derived from large units. The cost of standard governors is thus disproportionately high in the smaller sizes. Regulation of speed is normally accomplished through flow control. Adequate control requires sufficient rotational inertia of the rotating parts. When load is rejected power is absorbed, accelerating the flywheel, and when load is applied, some additional power is available from deceleration of the flywheel. Response time of the governor must be carefully selected since rapid closing time can lead to excessive pressures in the penstock.

A Francis turbine is controlled by opening and closing the guide vanes which vary the flow of water according to the load. A powerful governor is required to overcome the hydraulic and frictional forces and to maintain the guide vanes in fixed position under steady load. Impulse turbines are more easily controlled because the jet can be deflected or an auxiliary jet can bypass flow from the power-producing jet without changing the flow rate in the penstock. This permits long delay times for adjusting the flow rate to the new power conditions. The spear or needle valve controlling the flow rate can close quite slowly, say 30 to 60 s, thereby minimizing any pressure rise in the penstock.

Several types of governors are available which vary with the work capacity desired and/or the degree of sophistication of control. These vary from pure mechanical to mechanical-hydraulic and electro-hydraulic. Electro-hydraulic units are sophisticated pieces of equipment and would not be suitable for remote regions. The precision of governing necessary will depend on whether the electrical generator is synchronous or asynchronous (induction type).

There are advantages to the induction-type generator. It is less complex and therefore less expensive, but it has typically slightly lower efficiency. Its frequency is controlled by the frequency of the grid it is feeding into, thereby eliminating the need of an expensive conventional governor. It cannot operate independently but can only feed into a network and does so with lagging power factor which may or may not be a disadvantage, depending on the nature of the load. Long transmission lines, for example, have a high capacitance and in this case the lagging power factor may be an advantage.

Some general features of the overall regulation problem can be demonstrated by examination of the basic equation for a rotating system

$$J\frac{d\Omega}{dt} = T_t - T_L \tag{6.68}$$

where

J = moment of inertia of rotating components
Ω = angular velocity
T_t = torque of turbine
T_L = torque due to load

Three cases may be considered in which T_t is equal to, less than, or greater than T_L.

For the first case, the operation is steady. The other two cases imply unsteady operation, since $d\Omega/dt$ is not constant, and usually a governor is provided so that the turbine output matches the generator load.

Speed regulation is a function of the flywheel effect of the rotating components and the inertia of the water column of the system. The start-up time of the rotating system[21] is given by

$$T_s = \frac{J\Omega^2}{P} = \frac{Jn_0^2}{6818HP_r} \tag{6.69}$$

where

J = flywheel effect of generator and turbine, $kg \cdot m \cdot s^2$
n_0 = normal turbine speed, r/min
HP_r = rated metric horsepower

The starting up time of the water column is given by

$$T_p = \frac{\Sigma LV}{gh_r} \tag{6.70}$$

where

L = length of water column
V = velocity of each component of water column
h_r = rated head

For good speed regulation it is desired to keep $T_s/T_p \leq 8$. Lower values can also be used, although special precautions are necessary in the control equipment. It can readily be seen that higher ratios of T_s/T_p can be obtained by increasing J or decreasing T_p. Increasing J implies a larger generator, which also results in higher costs. The start-up time of the water column can be

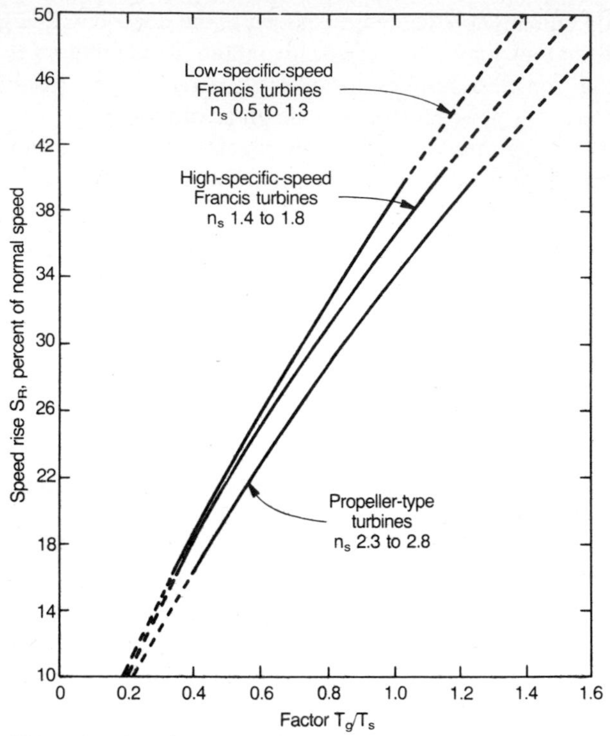

FIG. 6.35 Speed rise for full gate load rejected with no water hammer. *(Ref. 21.)*

reduced by reducing the length of the flow system, by using lower velocities, or by addition of surge tanks, which essentially reduce the effective length of the conduit. A detailed analysis should be made for each installation, as for a given length, head, and discharge the flow area must be increased to reduce T_p, which leads to associated higher construction costs.

A method for determining the speed rise as a result of load rejection is incorporated in Fig. 6.35 for several specific-speed machines. The abscissa is the ratio of T_g/T_s, where T_g is the full closing time of the governor

$$T_g = 0.25 + T_c \tag{6.71}$$

and T_c is the rated governor time in seconds, which generally varies from 3 to 5 s. With the ratio determined, the percent speed rise S_R for no water hammer and a given specific speed can be found in Fig. 6.35. This value should be modified to include the start-up time of the pipeline, or

$$S'_R = S_R\left(1 + \frac{T_p}{T_c}\right) \tag{6.72}$$

It is desired to keep the speed rise for full-load rejection to less than 45 percent, although in some cases higher percentages can be permitted if regulating ability is sacrificed. If the speed rise is excessive, consideration should be given to providing surge tanks. Further discussion is beyond the scope of this section, and the reader is referred to Ref. 21 for more detail.

Emergency and Abnormal Conditions

Emergency conditions can arise if the system experiences a sudden drop in load and the guide vanes remain open as a result of failure in the regulating system. The speed will rise rapidly until a maximum is reached, which is called the *runaway speed.* The runaway speed is dependent on the type of turbine, distributor opening, head, and in the case of a Kaplan turbine, the runner blade angle.

Based on field tests, an equation[21] has been developed for use in predicting the runaway speed, n_r, for various types of turbines. This formulation is

$$\frac{n_r}{n_d} = K_n \left(\frac{H_{\max}}{H_d} \right)^{1/2} \tag{6.73}$$

where

$K_n = 0.28\,N_s + 1.45$
n_d = design speed, r/min
N_s = specific speed
$H_{\max}$ = max. head
H_d = design head

Thus, the ratio of runaway to normal speed is higher for propeller turbines than for Francis turbines and may attain values up to about 2.6. This factor must be considered in the design of the turbine and generating equipment as the increase in centrifugal force can be substantial.

As previously mentioned, for an adjustable-blade runner the runaway speed is a function of the blade angle. If the blade angles are increased from their optimum value, the runaway speed is decreased. At the larger blade angles, the shock losses are higher and equilibrium conditions are reached at lower speeds. However, an increase in blade angle can also result in serious vibration, which can cause damage to the turbine unit.

If the blade angle is decreased, the velocity vectors are changed so that the losses are reduced, and the runaway speed is increased. In fact, the theoretical runaway speed with a closed blade runner approaches infinity. This is not actually realized, however, due to frictional windage losses in the generator.

6.6 EXAMPLES

Two simplified examples of application of the concepts developed in this chapter are given below. The purpose of the examples is to illustrate the use of the charts and similarity laws. Use of multiple turbines has not been considered, and possible load and/or discharge variations have not been included in these simplified cases.

Example 1

Select a turbine for $H = 100$ m (328 ft) and $Q = 2\ m^3/s$ ($\sim$ 70 ft^3/s).

Assuming $\eta = 0.85$, the power is

$$P = \gamma Q H \eta = \frac{1000 \times 2 \times 100 \times 0.85 \times 0.735}{75} = 1667 \text{ kW}$$

From Fig. 6.34, it is seen that either an impulse or Francis turbine may be suitable. The specific speed is related to the rotational speed by

$$n_s = \frac{n\sqrt{P}}{H^{5/4}} = \frac{n\sqrt{1667}}{(100)^{5/4}}$$

or

$$n = 7.75 n_s$$

Several values of n_s will be used to determine n. Velocity coefficients ϕ are taken from Figs. 6.25 and 6.26 for each turbine, and diameters of the runners are calculated from

$$\phi\sqrt{2gH} = \frac{\pi D_0 n}{60}$$

and

$$\phi_e\sqrt{2gH} = \frac{\pi D_e n}{60} \qquad \text{(only for Francis turbine)}$$

The width of the Francis runner B can also be obtained from the ratio B/D_e in Fig. 6.26, and σ_{Tc} taken from Fig. 6.13 for the Francis turbine. The computations are summarized in Table 6.2.

TABLE 6.2 Computations for Example 1.

	n_s	N_s	n, r/min	ϕ	ϕ_e	D_0, m	D_e, m	B/D_e	B, mm	σ_{Tc}
Single	5	0.03	38.8	0.45		9.82				
Jet	10	0.06	77.5	0.45		4.92				
Pelton	20	0.12	155	0.45		2.46				
	30	0.18	232.5	0.45		1.65				
	75	0.45	581	0.7	0.46	1.02	0.67	0.145	97	0.03
Francis	100	0.60	775	0.72	0.55	0.786	0.60	0.165	99	0.05
	150	0.90	1,163	0.78	0.74	0.57	0.54	0.22	119	0.10
	200	1.20	1,550	0.8	0.82	0.437	0.44	0.28	123	0.18

If the practical upper limit for d_j/D_0 is about 0.1, then $D_0 \sim 2.4$ m (~ 7.9 ft) which corresponds to the unit with a specific speed of about 20. For generation of 60 Hz power, the rotational speed n is $7{,}200/p$, where p is the number of poles, preferably an even number divisible by four. With $p = 44$, $n = 164$ r/min, $D_0 = 2.32$ m, and $n_s = 21$ or $N_s = 0.13$; these conditions may be reasonable. Use of a speed increaser would probably be desirable to reduce the size of the generator.

From the tabulated values of the Francis turbine, the physical size of the turbine is quite small, and the rpm high. If a 4-pole generator is considered, $n = 1{,}800$ r/min, corresponding to $n_s = 232$ or $N_s = 1.4$. A preliminary check can be made of the cavitation limits, and the final turbine setting is based on recommendations of the manufacturer. The allowable turbine setting is given by Eq. (6.48)

$$Z = H_a - H_v - \sigma_{Tc} H$$

Assuming $H_a = 8.6$ m (28.2 ft) at an elevation 1500 m (4,920 ft) above sea level and neglecting H_v, the setting for a turbine with $n_s = 150$ and $\sigma_{Tc} = 0.1$ is -1.4 m (-4.6 ft) below tailwater. For $n_s = 100$ and $\sigma_{Tc} = 0.05$, the setting Z is 3.6 m (11.8 ft) above tailwater, a more reasonable value. An 8-pole generator requires $n = 800$ r/min, and $n_s = 103$ or $N_s = 0.62$. This turbine probably is the best selection for this example, as the generator costs would be lower. In any case, the total costs of the turbine, generator, and other civil works must be evaluated in the final selection.

Example 2

A retired dam site with $H = 10$ m (32.8 ft) is being rehabilitated. A machinery broker has a used Francis turbine to be sold at low cost which is rated at $H = 15$ m (49.2 ft), $P = 300$ kW, $n = 450$ r/min, and $\eta = 0.9$. Calculate the discharge Q_1, power P_1, and speed n_1 at which the turbine must operate at the proposed site.

If the specific speed is kept constant, the efficiency should be the same for both cases. The specific speed is

$$n_s = \frac{n\sqrt{P}}{H^{5/4}} = \frac{450\sqrt{300}}{15^{5/4}} = 264$$

and

$$Q = \frac{P}{\gamma H \eta} = \frac{300 \times 75}{0.735 \times 1000 \times 15 \times .9} = 2.27\text{m}^3/\text{s}\ (80.15\ \text{ft}^3/\text{s})$$

From the similarity relationships for constant diameter

$$\frac{H}{n^2} = \frac{H_1}{n_1^2} \qquad \frac{Q}{n} = \frac{Q_1}{n_1} \qquad \frac{P}{n_1^3} = \frac{P_1}{n^3}$$

Therefore

$$n_1 = n\sqrt{\frac{H_1}{H}} = 450\sqrt{\frac{10}{15}} = 367 \text{ r/min}$$

$$Q_1 = Q\frac{n_1}{n} = 2.27 \times \frac{367}{450} = 1.85 \text{ m}^3\text{/s } (65.32 \text{ ft}^3\text{/s})$$

$$P_1 = P\left(\frac{n_1}{n}\right)^3 = 300\left(\frac{367}{450}\right)^3 = 163 \text{ kW}$$

With a 20-pole generator and 60-Hz current, $n = 7200/20 = 360$ r/min. As this n is slightly less than that required for constant specific speed, the efficiency would be slightly lower for the same head of 10 m (32.8 ft). In a situation of this type, one must also be certain that the machine used in the new application is not overstressed, which is not the case here.

6.7 SUMMARY

It has been shown that the head utilized by a turbine runner to produce power can be derived from a suitable form of Euler's equation of motion. The head utilized, and consequently the power developed, is dependent on the velocity vectors of the inlet and exit flow of the runner. The velocity vectors are determined by the operational conditions and the turbine design. Overall efficiency of the turbine is the product of the hydraulic, volumetric, and mechanical efficiencies. Each of these are dependent on various energy losses in the turbine unit, and the origin of these losses has been briefly discussed.

Similarity considerations permit the formulation of dimensionless numbers. These numbers are useful in the extrapolation of test data taken with a model turbine to full-scale conditions and therefore predict performance. One of the most significant dimensionless numbers is the specific speed, which consists of a combination of operating conditions that ensures similar flows in geometrically similar machines. Each type of machine has a value of specific speed that gives maximum efficiency, and it is therefore convenient to classify the various turbine designs by the specific speed at best efficiency.

Cavitation must be avoided in turbines, as it results in loss of performance and can cause erosion damage to the runner and possibly other parts of the structure. Each particular turbine type has its own cavitation limits which are determined from model tests. High-specific-speed turbines are more susceptible to cavitation than low-specific-speed units. The setting of the turbine with respect to the tailwater elevation must be carefully considered to ensure cavitation-free operation and is based on the manufacturer's recommendations.

Turbines can be classified in two broad groups, impulse and reaction turbines. An impulse turbine is driven by a high-velocity jet impinging on buckets around the periphery of the wheel, whereas the reaction turbine requires that the flow passages be completely filled. Reaction turbines can be subdivided further into radial-, mixed-, or axial-flow types. The radial- and mixed-flow types have fixed runner blades, except for the Deriaz turbine, and the axial-flow machine may have either fixed or adjustable blades. In addition to differences in blade geometry, each type of reaction turbine has different requirements for a draft tube. The draft tube is considered part of the turbine, and its energy losses are charged to the turbine performance. Some draft-tube configurations may require a large amount of excavation to achieve the desired turbine setting.

The overall efficiency of an impulse turbine is quite constant over a broad range of operating conditions, which is achieved by throttling of the flow at the nozzle. Fixed-blade reaction turbines have a more peaky efficiency curve whereas the efficiency curve for adjustable-blade units is relatively flat. The latter unit is particularly suited for installations subject to a wide variation in flow conditions. However, an economic analysis must be made to justify the higher cost of the fully adjustable turbine.

A wide variety of turbines is becoming commercially available for application to small-scale hydropower sites. Standardized units are offered by manufacturers in a range of sizes to reduce equipment costs. With increased demand for lower cost units to make marginal sites feasible, it is expected that further developments in standardization will be made. The manufacturers should be asked for their recommendations.

Several operational conditions are also significant. Most turbines are designed to operate at a constant rotational speed which is controlled by a governor. Some general guidelines are given concerning speed regulation. With a sudden loss in electrical load, the turbine will reach a runaway speed that is considerably greater than the normal speed. The turbine and generator must be designed to tolerate the additional centrifugal forces. Provisions should also be made for emergency shutdown of the flow to the turbine, either by wicket gates or appropriate valves.

6.8 REFERENCES

1. H. Rouse and S. Ince, *History of Hydraulics,* Dover, New York, 1963.
2. N. Smith, "The Origins of the Water Turbine," *Scientific American,* vol. 242, no. 1, January 1980.
3. L. F. Moody and T. Zowski, "Hydraulic Machinery," in C. V. Davis (ed.), *Handbook of Applied Hydraulics,* McGraw-Hill, New York, 1969.
4. R. Camerer, *Vorlesungen über Wasserkraft—Maschinen,* Verlag von W. Engelmann, Leipzig, 1924.
5. M. Nechleba, *Hydraulic Turbines—Their Design and Equipment,* Artia-Prague, Czechoslovakia, 1957.

6. E. Mühlemann, "Zur Aufwertung des Wirkungsgrades," *Schweizerische Bauzeitung,* 1948, p. 331.
7. S. P. Hutton, "Component Losses in Kaplan Turbines and the Prediction of Efficiency from Model Tests," *Proc. Inst. Mech. Engrs.,* vol. 168, 1954, p. 743.
8. R. W. Kermeen, "Water Tunnel Tests of NACA 4412 and Walchner Profile 7 Hydrofoils in Noncavitating and Cavitating Flows," Calif. Inst. Technol. Hydrodyn. Lab., Report 47-5, 1956.
9. R. W. Kermeen, "Water Tunnel Tests of NACA 66, -012 Hydrofoil in Noncavitating and Cavitating Flows," Calif. Inst. Technol. Hydrodyn. Lab., Rep. 47-7, 1956, 12 pp.
10. J. W. Daily, "Cavitation Characteristics and Infinite-Aspect Ratio Characteristics of a Hydrofoil Section," *Trans. ASME,* vol. 71, 1949, pp. 269–284.
11. R. E. A. Arndt, "Semiempirical Analysis of Cavitation in the Wake of a Sharp Edged Disk," *J. Fluids Engrg.,* vol. 98, 1976, pp. 360–362.
12. J. W. Holl, R. E. A. Arndt, and M. L. Billet, "Limited Cavitation and the Related Scale Effects Problem," *Proc. 2nd Int'l Symp. Fluid Mech. and Fluidics,* JSME Tokyo, September 1972, pp. 303–314.
13. R. E. A. Arndt, "Cavitation in Fluid Machinery and Hydraulic Structures," *Ann. Rev. Fluid Mech.,* vol. 13, 1981, pp. 273–328.
14. R. E. A. Arndt, "Recent Advances in Cavitation Research," *Adv. in Hydroscience,* vol. 12, V. T. Chow (ed.), 1981, pp. 1–78.
15. R. E. A. Arndt, J. W. Holl, J. C. Bohn, and W. T. Bechtel, "The Influence of Surface Irregularities on Cavitation Performance," *J. Ship Res.,* vol. 23, 1979, pp. 157–180.
16. D. R. Stinebring, R. E. A. Arndt, and J. W. Holl, "Scaling of Cavitation Damage," *J. Hydronaut.,* vol. 11, no. 3, 1977, pp. 67–73.
17. W. M. Deeprose, N. W. King, P. J. McNulty, and I. S. Pearsall, "Cavitation Noise, Flow Noise and Erosion," in *Cavitation,* Inst. Mech. Eng., Herriot-Watt University, Edinburgh, 1974, pp. 373–381.
18. Howard H. Mayo, Jr., "Low-Head Hydroelectric Unit Fundamentals," 1979 Engineering Foundation Conference, *Hydropower: A National Resource,* Dept. of the Army Institute for Water Resources, Corps of Engineers, March 11–16, 1979.
19. N. N. Kovalev, *Hydroturbines, Design and Construction,* Israel Program for Scientific Translations (translated from Russian), 1965.
20. J. W. Daily, "Hydraulic Machinery," in H. Rouse (ed.), *Engineering Hydraulics,* New York, 1950.
21. Bureau of Reclamation. "Selecting Hydraulic Reaction Turbines," *Engineering Monograph No. 20,* 1966.
22. Tudor Engineering Company, "Reconnaissance Evaluation of Small, Low-Head Hydroelectric Installations," Final Report, Contract No. 9-07-83-V0705, July, 1980.

7

Generation and Electrical Equipment

Wm. L. Hughes

R. G. Ramakumar

Dan Lingelbach

This chapter introduces the reader to the electrical component parts of small hydroelectric systems. A discussion of the technical and economic considerations, advantages and limitations of specific hardware is given. Expectations and system performance which are reasonable and unreasonable are also discussed.

7.1 GENERAL CONSIDERATIONS

Hydroelectric systems ranging in size from a few kilowatts to a few megawatts have been in existence for many decades and were the obvious outgrowth of waterwheel power systems which have been used for centuries. Most early hydroelectric systems were of the direct current (dc) variety to match early commercial electric systems. The dc motor and dc generator were the first machines developed to a practical level for common usage. Today's commonly used alternating current (ac) technology was considered too complex and too dangerous.

It was relatively easy to control dc electric generator systems. Accurate speed-control mechanisms were not necessary. Voltage regulation as a function of connected load could be handled by means of a simple field rheostat with most connected loads primarily being lighting.

As ac technology evolved, electricity utilization became much more diverse. Lighting loads were joined by electric-motor loads, heating loads, and finally electronic equipment. Each type of load put more stringent requirements on the power source. Frequency-regulation requirements, in particular, became very critical, and continuity of service became the expected norm. Intermittent outages, instead of being expected, became intolerable if they occurred frequently. A freezer full of food could be lost, electric motors could be destroyed by low voltages, and industrial plant production schedules could be disrupted.

Solutions to these regulation and continuity problems for electrical service were provided by the electricity supplier in a number of ways. In contrast to dc systems, ac systems became the standard because it was easier (and less costly) to build them to large capacity. Alternating current could be transmitted longer distances at less loss than direct current because of the application of transformers. Loads represented by motors, switches, lighting, and electronic devices were easier and cheaper to build when operated with alternating current, particularly when the voltage was well regulated. As electric systems became larger, it became feasible to interconnect them with high-voltage power grids making them more reliable and less subject to power outages. That most conventional electric power plants operated at a maximum of perhaps 85 percent of the time is completely unknown to most customers, because today they may have electricity available 99 percent of the time.

The evolving energy crisis requires the exploitation of sources of energy previously ignored. In many cases, small hydroelectric systems in a large grid may not be practical because connection cost is too high for the amount of power available. In those cases, interconnection may be technically feasible, but the cost of that connection (on a per kilowatt basis) may be a significant fraction of the total system costs. The small isolated hydroelectric system will be required to start and run electric motors thus providing the same (per unit) surge current capability formerly required of the large interconnected systems. It is possible to engineer the small systems so that they are able to provide those surges, but only at a significant inconvenience and added cost to the user.

An additional critical problem with an isolated electric power system is continuity of service. Hydroelectric systems may not always have water flow available continuously, and electrical systems must occasionally be shut down for maintenance or repair. Fortunately these times may be short for well-designed modern systems, but they still occur occasionally. In the early 1980s, no satisfactory electrical energy storage systems of low or even moderate costs exist. Thus, any load requirements for isolated power systems must account for the possibility of discontinuous service. If such discontinuous service is intolerable, then increased system costs to provide the required built-in redundancy are necessary.

Hydroelectric systems have some special problems associated with their electrical components not common to other types of power systems. In some cases, if the load diversification is small, these problems are compounded. For example, a large power system may have load diversification such that the largest single load served is but a small fraction of the total system output. When that largest load is switched on or off (even when it is a large motor drawing 6 times its normal load current on starting), the total system is relatively undisturbed and voltage-regulation devices need make only small corrections. Further, the percentage adjustment of the speed regulator on the ac alternator (or dc generator) prime mover is also a small fraction. The effect of sudden load changes on the rest of the system is thus minimized because of the massive capacity in the total system.

Small hydroelectric systems, on the other hand, may have individual load components which represent large fractions of their total capacity. When those large load components are switched on or off, the effect on the overall system is significant. The prime mover (water turbine) load can be instantaneously increased or decreased radically causing large overspeed or underspeed transients, large torque impulse loads on shaft couplings, and severe voltage and current transients which may damage other connected loads.

The techniques used to deal with these problems can vary, depending on the nature of the water turbine used. In large systems, fast-acting vanes surround the turbine and can quickly regulate the turbine speed and prevent runaways. If the turbine is interconnected with other turbines, the problems are alleviated because all interconnected devices automatically share in the problems created by load changes.

In smaller hydroelectric systems which may be isolated and not interconnected, radical load changes must be dealt with in other ways. In system sizes of several megawatts, the problem may be handled conventionally, with fast-acting alternator field regulators and fast-acting turbine vanes. In small systems, these problems can be handled with a fast-acting field regulator to help with voltage changes, but wide frequency swings resulting from wide turbine speed variations are an additional problem. Normally, turbine vanes are not always practical in small systems. Sometimes dummy (resistive) loads are switched in and out to keep turbine speeds approximately constant. Such techniques work but waste energy.

Alternatively, one can specify what are often referred to as *variable-speed constant-frequency systems* (VSCF) in which turbine speed (and thus electrical generator or alternator speed) is permitted to fluctuate widely. However, the electrical output is conditioned to produce only constant-frequency (usually 50 or 60 Hz) and constant-voltage electricity. The technique for accomplishing this is discussed later.

7.2 REVIEW OF BASIC ELECTRICITY AND THE NATURE OF ALTERNATORS AND GENERATORS

Because of the different ways in which small electrical sources such as small hydroelectric systems respond to conventional electrical loads, it is useful to review the basics of electric power. The purpose is first, to quickly and quantitatively review basics for readers whose technical background in electricity may not have been used recently, and second, to give readers without a technical background some enumeration of the factors involved.

Basic Concepts (dc)

The basic parameters of electric circuits are voltage, current, and energy. It is assumed that the reader has a good grasp of the physical concept of energy and power; however, it is useful to briefly review unit relationships as given in Table 7.1.

TABLE 7.1 Unit Relationships

Energy	
SI units:	Formula
One joule equals one newton-meter	1 J = 1 N·m
One kilogram-meter equals 9.80 kilojoules	1 kg·m = 9.80 kJ
One kilowatt-second equals one kilojoule	1 kW·s = 1 kJ
One kilowatt hour equals 3,600 kilojoules	1 kWh = 3,600 kJ
USCS units:	
One horsepower second equals 550 foot-pounds	1 hp·s = 550 ft·lb
One horsepower hour equals 1,980,000 foot-pounds	1 hp·h = 1,980,000 ft·lb
Combination units:	
One horsepower hour equals 0.746 kilowatt hours	1 hp·h = 0.746 kWh
Power	
SI units:	Formula
One watt equals one newton-meter per second	1 W = 1 (N·m)/s
or one joule per second	= 1 J/s
One kilowatt equals one kilojoule per second	1 kW = 1 kJ/s
One kilogram-meter per second equals 9.8 kilowatts	1 (kg·m)/s leq 9.8 kW
USCS units:	
550 foot-pounds per second equals one horsepower	550 (ft·lb)/s = 1 hp
Combination:	
One horsepower equals 746	1 hp = 746 W
watts or 0.746 kilowatts	= 0.746 kW
One kilowatt equals 1.34 horsepower	1 kW = 1.34 hp

Simple electrical relationships can easily be described by the electric circuit given in Fig. 7.1.

The unit of *resistance* (opposition to current flow) is the ohm (symbol Ω). If 1 Ω has a dc voltage of 1 volt (symbol V) connected across it, 1 ampere (symbol A) of current will flow in the resistor and 1 watt (symbol W; or 1 joule per second, J/s) of power will be dissipated as heat in that resistor. The unit of current is the ampere which is the flow of 1 coulomb (symbol C) of electric charge past a point per second. One coulomb of charge equals the charge on 6.24×10^{18} electrons. If the voltage is increased to 2 V, 2 A will flow, and the power dissipated will be 4 W (proportional to the square of the current in the resistor). If the voltage is reduced to 0.5 V, the current will be 0.5 A, and the power will be 0.25 W. If the voltage is returned to 1 V, but the resistor is increased to 2 Ω, the current will be 0.5 A (inversely proportional to the resistor), and the power will be 0.5 W, and so forth. The basic dc relationships are shown in Eq. (7.1) through Ohm's law.

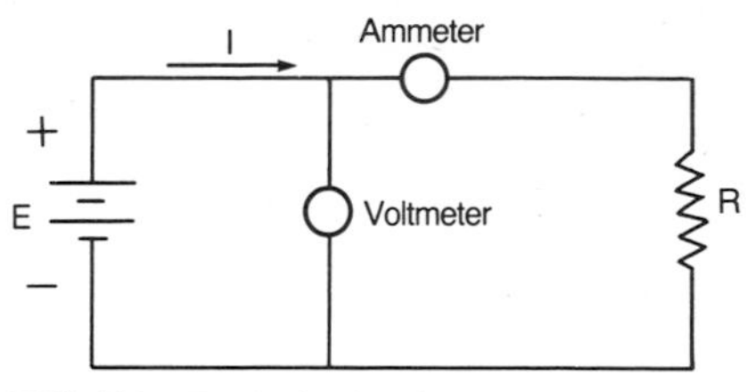

FIG. 7.1 Basic dc circuit.

$$E = IR \tag{7.1}$$

where E = voltage, I = current, and R = resistance.

The power P relationships are given by

$$P = I^2R = \frac{E^2}{R} = EI \tag{7.2}$$

Basic Concepts (ac)

If we make a slight change in our circuit by replacing the battery with a time-varying voltage which changes according to a sinusoid as shown in Fig. 7.2 we have a basic ac circuit.

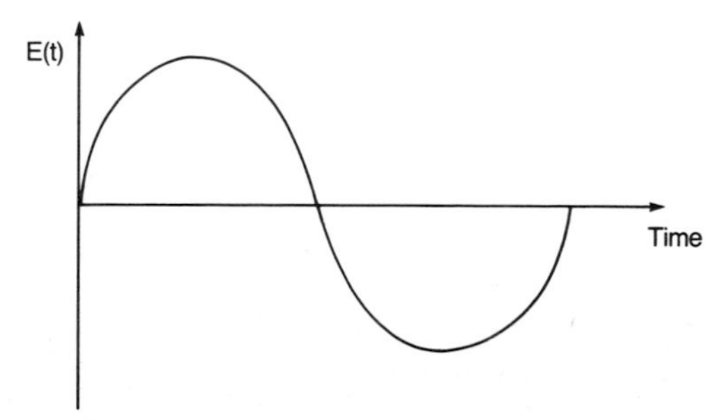

FIG. 7.2 Basic ac sinusoid.

We can say mathematically

$$e(t) = E_o \sin \omega t \tag{7.3}$$

where

$e(t)$ = the voltage as a function of time
E_o = the peak value of the sinusoid
ω = the angular frequency in radians per second (rad/s) or $2\pi f$, where f is the frequency in hertz (Hz), formerly called cycles per second (cps).

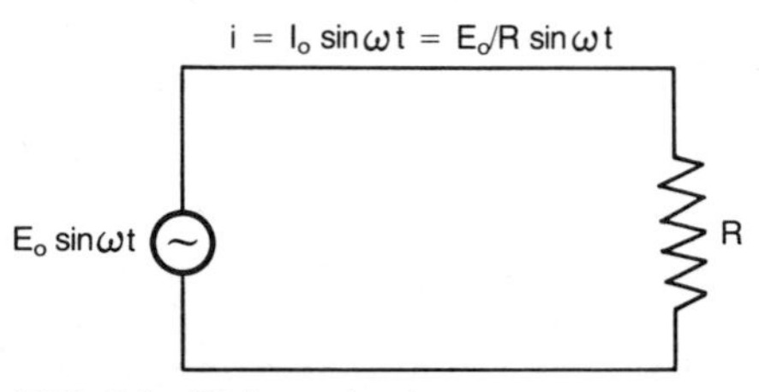

FIG. 7.3 Basic ac circuit.

If this voltage is inserted in the circuit shown in Fig. 7-3, the current that flows $[i(t)]$ results from Ohm's law.

$$i(t) = \frac{e(t)}{R} = \frac{E_o}{R} \sin \omega t \tag{7.4}$$

The power dissipated in this circuit at any given instant of time, $p(t)$, is given by

$$p(t) = e(t)i(t) \qquad \text{(J/s or W)} \tag{7.5}$$

In the period of 1 s, the energy dissipated is clearly

$$\begin{aligned} P &= \int_0^1 e(t)i(t)\,dt \\ &= \int_0^1 (E_o \sin \omega t)\left(\frac{E_o}{R} \sin \omega t\right) dt \\ &= \frac{E_o^2}{R} \int_0^1 \sin^2 \omega t\,dt \\ &= \frac{E_o^2}{R} \int_0^1 \left(\frac{1}{2} - \frac{1}{2} \cos 2\omega t\right) dt \\ &= \frac{E_o^2}{2R} \qquad \text{(J/s or W)} \end{aligned} \tag{7.6}$$

If we define the root mean square ac voltage E_{rms} as $E_o/\sqrt{2}$ (for a sinusoid), then as in dc circuits, power is given by

$$P = \frac{E_{\text{rms}}^2}{R} \tag{7.7}$$

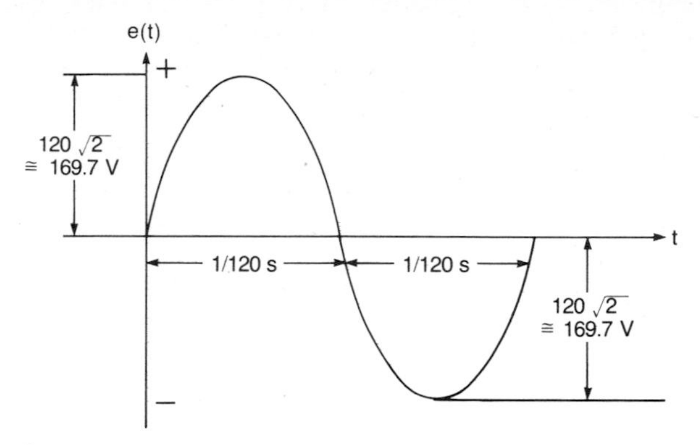

FIG. 7.4 Typical residential voltage sinusoid.

Whenever we indicate the rms voltage (of a sinusoid) we mean the peak voltage divided by $\sqrt{2}$. Thus, the ordinary 120-V, 60-Hz outlet in a house is a voltage that varies with time as shown in Fig. 7.4.

The rms value of a sinusoidally varying current is defined the same way. As Fig. 7.5 shows, 1 A rms at 60 Hz is a time varying current.

The set of ac relationships can be summarized

$$I_{rms} = \frac{E_{rms}}{R} \tag{7.8}$$

$$P = I_{rms}^2 R = \frac{E_{rms}^2}{R} \tag{7.9}$$

Notice that we did not write power as $P = E_{rms} I_{rms}$. If the electric circuit had only resistance, then it would have been perfectly correct to do so. However, most ac circuits are not limited to resistance loads. We can now introduce the concepts of *inductance* and *capacitance* which are found in ac circuits.

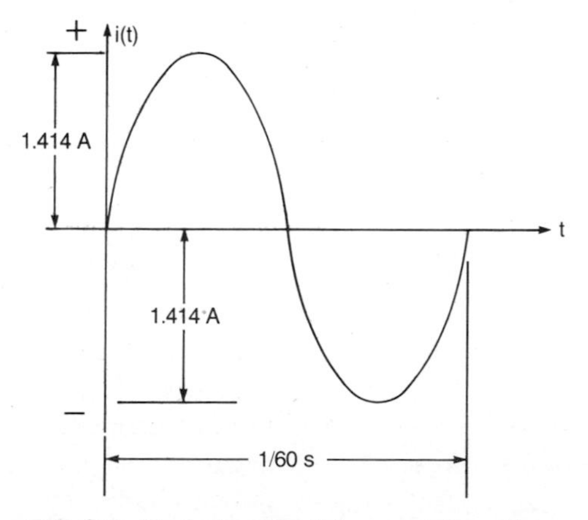

FIG. 7.5 Typical residential current sinusoid.

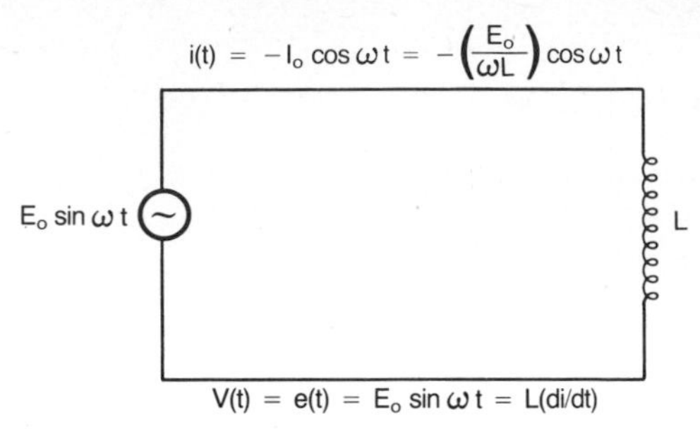

FIG. 7.6 Simple inductance circuit.

Inductors

Basically, inductors and capacitors are devices which momentarily store energy rather than dissipate it. An *inductor* in its simplest form is a coil of wire. The magnetic field created by the electric current through the coil is concentrated in a relatively small region inside and surrounding the coil, and the energy in the magnetic field is proportional to the square of the magnetic field intensity. It therefore follows that the energy stored in an inductor is proportional to the square of the instantaneous current through it, or

$$\text{Energy} = \frac{1}{2} Li^2 \qquad \text{J} \tag{7.10}$$

where the constant of proportionality L is called the inductance, with the unit of henry (symbol H), and i is instantaneous current in amperes. The instantaneous voltage across an inductor is proportional to the time derivative of the current passing through (A/s) it or simply

$$v(t) = L\frac{di}{dt} \tag{7.11}$$

Thus, if we have a circuit as shown in Fig. 7.6, by simple integration we obtain

$$i(t) = -\frac{E_o}{\omega L} \cos \omega t \tag{7.12}$$

If we plot the voltage across an inductor and the current through it for 1 cycle, Fig. 7.7 emerges.

We note that the current in the inductor is also sinusoidal but lags the voltage by 90°. During the first and third quarter cycles, $p(t) = e(t)i(t)$ is negative and during the second and fourth quarter cycles $p(t)$ is positive. Thus, an inductor in an ac circuit

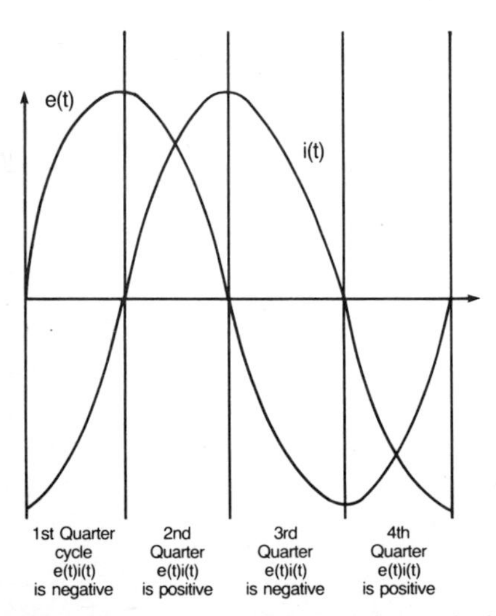

FIG. 7.7 Voltage and current for a single cycle.

takes energy from the circuit for half the cycle and returns it to the circuit for half the cycle. The net energy dissipated (in an inductor without resistance) is therefore zero.

We can now express the voltage and current in rms values as

$$I_{\text{rms}} = \frac{V_{\text{rms}}}{\omega L} \tag{7.13}$$

where $\omega L = 2\pi f L$ is called the *inductive reactance* and is normally symbolized by X_L. The current through an inductance lags the voltage by 90°.

Shorthand notation developed by Steinmetz years ago is customarily used. We symbolize inductive reactance by jX_L where $j = \sqrt{-1}$ in the notation of complex variables. Thus, we can summarize the inductive voltage, current, and reactance relationships

$$V_{\text{rms}} = I_{\text{rms}}\,(jX_L)$$

or (7.14)

$$I_{\text{rms}} = \frac{V_{\text{rms}}}{jX_L} = -j\,\frac{V_{\text{rms}}}{X_L}$$

Capacitors

The *capacitor* is simply a pair of unconnected parallel conducting plates closely spaced. When a dc voltage is applied across the plates as shown in Fig. 7.8, the top plate becomes positively charged and the bottom plate becomes negatively charged. The charge on each plate Q, in coulombs, is related to the voltage as follows

$$Q = CE \tag{7.15}$$

where the proportionality constant C is called the capacitance with units in farads.

The energy stored in the capacitor (actually in the electric field between the plates) is given by

$$\text{Energy} = \frac{1}{2}\,CE^2 = \frac{1}{2}\,\frac{Q^2}{C} \quad \text{J} \tag{7.16}$$

Next, if an ac voltage is applied to the capacitor as shown in Fig. 7.9, the plates

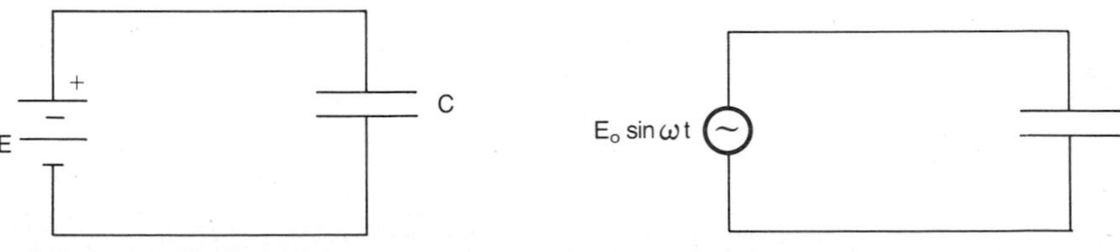

FIG. 7.8 Simple dc capacitor circuit.

FIG. 7.9 Simple ac capacitor circuit.

are continually charging and discharging such that the charge can be related to voltage as follows

$$q(t) = CE_0 \sin \omega t \tag{7.17}$$

Clearly the current through the capacitor is given by

$$i(t) = \frac{dq(t)}{dt} = \omega CE_0 \cos \omega t \tag{7.18}$$

Again, the current is out of phase with the applied voltage, but now it leads the voltage by 90° instead of lagging it (as was the case with the inductance).

In terms of rms currents and voltages, we can write

$$I_{\text{rms}} = j\omega CE_{\text{rms}} = \frac{jE_{\text{rms}}}{X_c} = \frac{E_{\text{rms}}}{-jX_c} \tag{7.19}$$

where X_c, the *capacitive reactance,* is given by

$$X_c = \frac{1}{\omega C} = \frac{1}{2\pi fC} \tag{7.20}$$

Kirchhoff's Laws

There are basic relationships between circuit elements allowing us to compute currents and voltages for a variety of interconnections. These relationships, first defined by Kirchhoff, are stated as follows:

1. Kirchhoff's voltage law (KVL). The sum of the voltage drops (or rises) around any closed loop equals zero.
2. Kirchhoff's current law (KCL). The sum of the currents entering (or leaving) any nodal (interconnection) point in an electric circuit equals zero.

We shall first examine a simple electric circuit using these laws. Consider the circuit shown in Fig. 7.10. We would like to determine the current through each resistor.

From KVL (around the first loop) we write

$$10 - 6I_1 - 8I_2 = 0$$

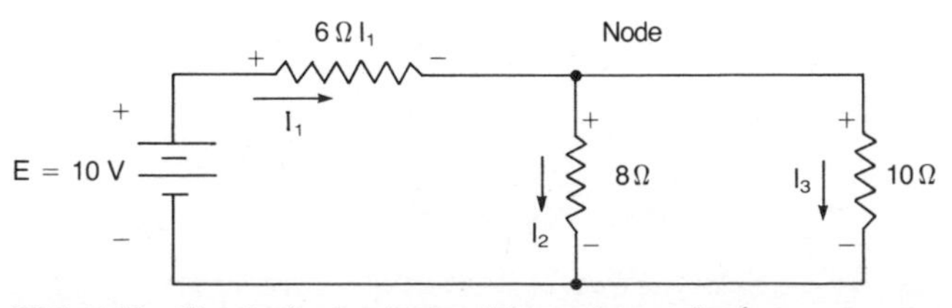

FIG. 7.10 Simple dc circuit containing resistance loads.

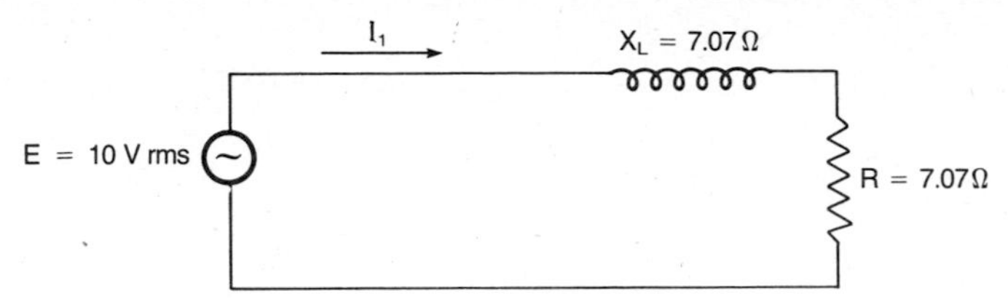

FIG. 7.11 Simple ac circuit containing resistance and inductance loads.

from KCL (at the node)

$$-I_1 + I_2 + I_3 = 0$$

and from KVL around the second loop

$$-8I_2 + 10I_3 = 0$$

solving all three equations simultaneously yields

$$I_1 = 0.957 \text{ A}$$
$$I_2 = 0.532 \text{ A}$$
$$I_3 = 0.425 \text{ A}$$

We can use the same technique for solving linear ac circuits. Consider the circuit shown in Fig. 7.11.

Using KVL

$$I\,(7.07 + j7.07) = 10$$

or

$$I = \frac{10}{7.07 + j7.07} = \frac{10(7.07 - j7.07)}{(7.07) + j7.07)\,(7.07 - j7.07)}$$
$$= 0.707 - j0.707 \text{ A}$$

This indicates that the total current is 1 A, since the two components of the current are 90° apart so $\sqrt{0.707^2 + 0.707^2} = 1$, and this current lags the supply voltage 45°.

Let us now compute the power dissipated in the circuit. We know that the rms current is 1 A through the 7.07-Ω resistor, indicating that

$$P = I^2R = 1^2 \times 7.07 = 7.07 \text{ W}$$

To compute the power using voltage and current, we have to go back to fundamentals and integrate the current times the voltage over one second. The average power becomes:

$$P = 10 \times 1 \times \cos \alpha$$

where α is the angle between the current and voltage, in this case 45°. Since cos 45° is equal to 0.707, the power is 7.07 W which is the correct value.

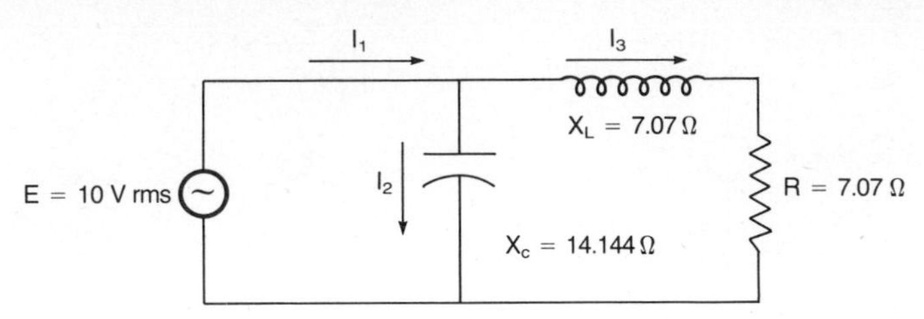

FIG. 7.12 Simple ac circuit containing resistance, inductance, and capacitance.

The cos α term is called the *power factor* in ac circuits, and the power delivered by an ac source is

$$P = E_{\text{rms}} I_{\text{rms}} \text{PF} \tag{7.21}$$

where PF = power factor.

Finally, adding a capacitor to our circuit across the source of Fig. 7.11, we obtain Fig. 7.12.

We note from KVL and Ohm's law that the current through the capacitor is

$$I_2 = j \frac{10}{14.144} = j0.707$$

From Fig. 7.11

$$I_3 = 0.707 - j0.707$$

and from KCL,

$$-I_1 + I_2 + I_3 = 0$$

or

$$I_1 = 0.707 + j0.707 - j0.707 = 0.707 + j0.0$$

The angle between voltage and current = 0, and the power factor = cos α = cos 0 = 1. The power = $E_{\text{rms}} I_{\text{rms}}$PF = 7.07 which is the correct value. We can say that for these values, we have corrected the power factor to unity. Note that the total power dissipated is unchanged after power factor correction.

We can now generalize our discussion of ac circuits by simply defining the concept of *impedance*.

1. The impedance of a resistor equals its resistance R.
2. The impedance of an inductive reactance equals jX_L.
3. The impedance of a capacitive reactance equals $-jX_c$.
4. When impedances are connected in series, the equivalent impedance is the sum of the individual impedances. (See Fig. 7.13.)

$$Z_T = Z_1 + Z_2 + \cdots \tag{7.22}$$

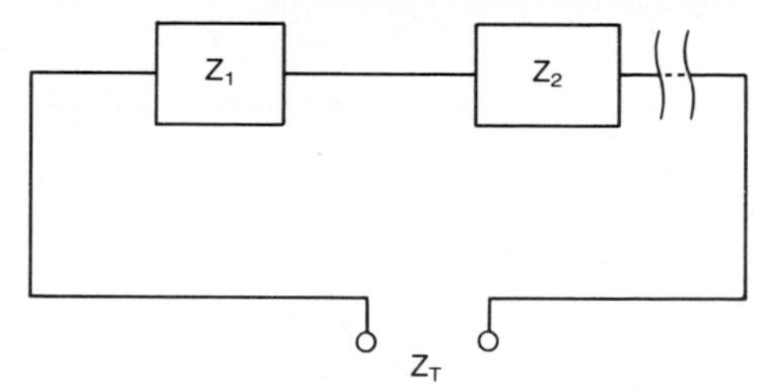

FIG. 7.13 Series impedance circuit.

5. When impedances are connected in parallel, the equivalent impedance is the reciprocal of the sum of the reciprocals. (See Fig. 7.14.)

$$Z_T = \frac{1}{\dfrac{1}{Z_1} + \dfrac{1}{Z_2} + \cdots} \tag{7.23}$$

In the case of only two impedances in parallel, we have

$$Z_T = \frac{Z_1 Z_2}{Z_1 + Z_2}$$

Thus, the impedance of the series circuit shown in Fig. 7.15 is

$$Z_T = R + j(X_L - X_c)$$

When $X_L = X_c$, the situation is called *resonance,* and they cancel each other out, leaving a total impedance of simply R. (A unity power factor condition.)

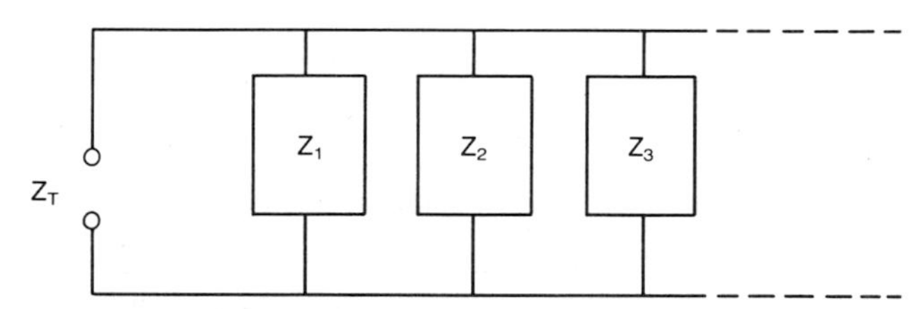

FIG. 7.14 Parallel impedance circuit.

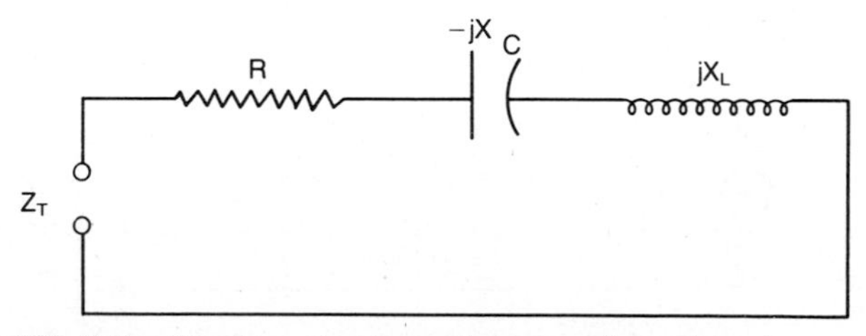

FIG. 7.15 Simple ac circuit containing impedances in series.

Polyphase Circuits

Quite often it is desirable to generate alternating current not as a single ac voltage, but rather as three individual ac voltages. These three voltages are generally of the same amplitude and frequency but displaced in phase from each other by 120°. They may be connected to each other in Y ("wye" or star) arrangement or in Δ (delta) fashion as shown in Figs. 7.16 and 7.17 respectively. Note that, in the case of Δ connection, KVL is not violated as is shown below.

$$E + j0 - 0.5E + j0.866E - 0.5E - j0.866E = 0$$

The reasons for building polyphase ac systems are as follows:

1. Generators and transformers for any given power size are smaller, cheaper, and more efficient.
2. Power line losses and motor losses, again for any given size, are less.

Each type of connection (Y and Δ) has certain advantages and disadvantages in any given situation and both are widely used. A variety of schemes have been devised for switching from one to the other as the need arises.

Typically, ac systems of 1 or 2 kW will be single phase for simplicity (though not always). Larger systems will generally be polyphase for better efficiency and lower cost.

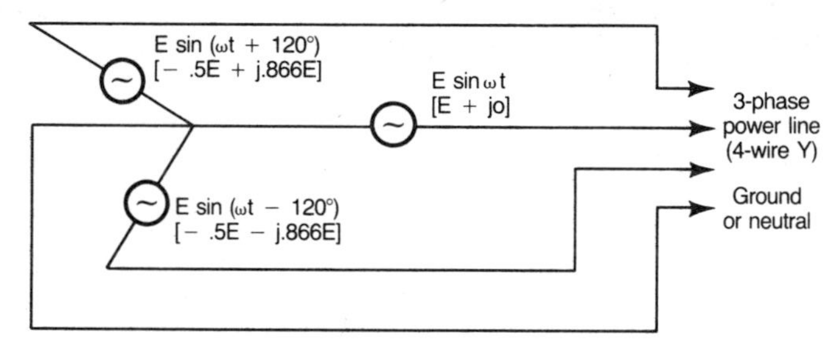

FIG. 7.16 Typical Y or star connection.

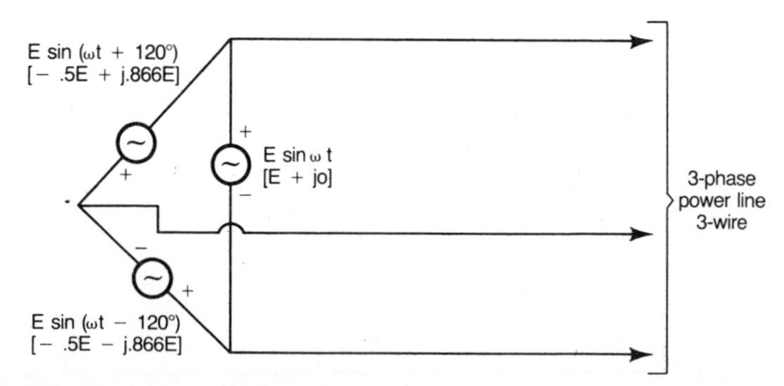

FIG. 7.17 Typical delta connection.

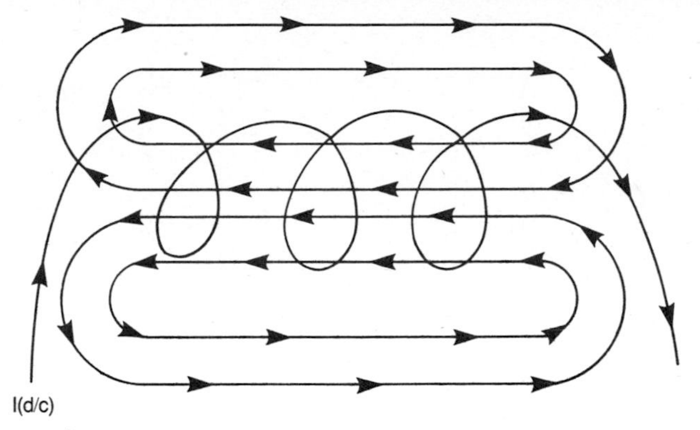

FIG. 7.18 Magnetic field around a coil.

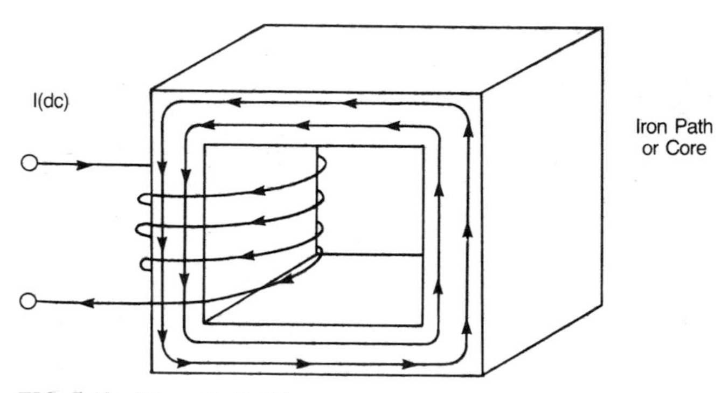

FIG. 7.19 Magnetic field in a core.

Magnetic Circuits

A magnetic field is generally associated with inductances. If the inductance coil was simply wound in air, the magnetic field would be dispersed around the coil roughly as shown in Fig. 7.18. However, if the coil is wound around an iron path with a permeability (ability to conduct field flux lines) much higher than air, the magnetic field will principally be confined inside the iron path as shown in Fig. 7.19. Further, if an ac voltage is placed on one coil and another coil is placed on the iron path, as shown in Fig. 7.20, we have a simple transformer.

The voltage E_1 in side 1 will induce a voltage E_2 in side 2 causing a current to flow in the resistor. The ratio of these two voltages will be

$$\frac{E_2}{E_1} = \frac{N_2}{N_1} \tag{7.24}$$

where N_1 and N_2 are the respective numbers of turns. However, the currents I_1 and I_2 will be inversely proportional to the numbers of turns.

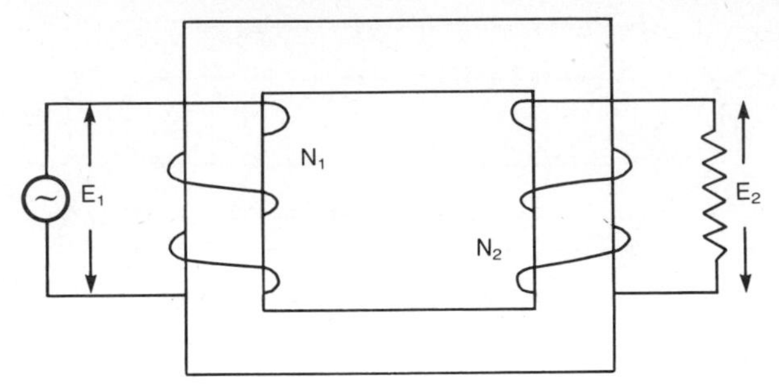

FIG. 7.20 Simple transformer.

$$\frac{I_2}{I_1} = \frac{N_1}{N_2} \tag{7.25}$$

From these relationships emerges

$$E_1 I_1 = E_1 I_2 \tag{7.26}$$

This voltage-transforming phenomenon can occur in such a device for alternating current only. This is the principal reason why power is generated and transmitted in the form of alternating current: because we can freely use transformers.

Magnetic fields are used, not only in transformers, but also in electric generators and alternators. Consider the system shown in Fig. 7.21.

A coil is placed in a magnetic field and rotated. The magnetic field is created by the battery and field winding and is conducted through the rotating coil by the iron path. As the coil rotates, it "cuts" the field lines in such a way that a sinusoidal voltage is produced as an output. The device is a simple *single-phase alternator*. We could put a *commutating system* on the rotor which would switch the current direction every half cycle, thus creating a unidirectional (or dc) current. The device would then be called a *generator*, or sometimes a *dynamo*.

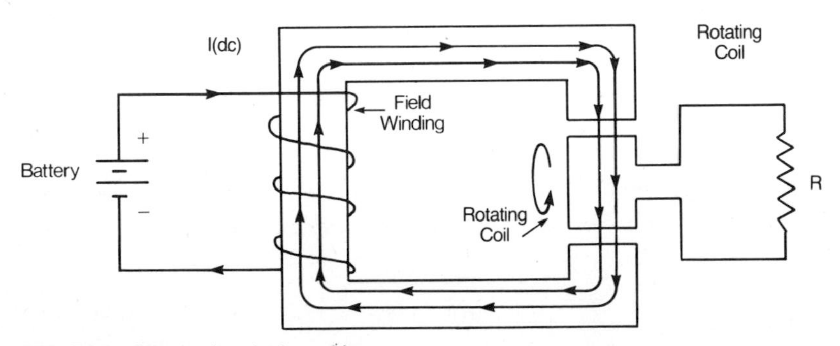

FIG. 7.21 Simple electrical machine.

Clearly, we can give only a simple discussion of transformers and alternators. Some general comments about transformers and alternators or generators follow:

1. The iron path generally consists of thin laminations of iron. If it were solid, internal heat losses would be excessive because of stray induced currents called *eddy currents.*
2. Alternators, generators, and transformers tend to be highly efficient and reliable devices.
3. Hydroelectric systems will have an alternator (or generator) and a transformer to step up voltage.
4. For larger systems, these devices are generally built to produce or transform polyphase rather than single-phase power because of the greater efficiencies possible. Further, single-phase electric motors require special starting mechanisms, while polyphase motors automatically start when power is first applied.

There are other applications of magnetic fields such as the *magnetic circuit breaker.* This device (which is generally a spring-loaded switching mechanism) will open circuit a power line when overload currents occur. Thus, if an electrical short circuit occurs, the excessive current creates a magnetic field which trips the circuit-breaker mechanism. It performs the same function as a fuse, but does not need to be replaced after use.

Solid State Devices, Rectifiers, and Inverters

Sometimes, it is desirable to use direct current for certain functions when only ac power is available. A rectifier system is one possible solution as seen in Fig. 7.22. The circuit shown is a *bridge rectifier,* although there are other types of rectifiers. The rectifiers (diodes) are simple solid-state devices which can pass current only in one direction, as indicated by the direction of the arrow. We note that current can pass through the resistor only in one direction, and is thus a direct current. The capacitor helps smooth out the fluctuations which can cause serious problems in many types of electrical loads.

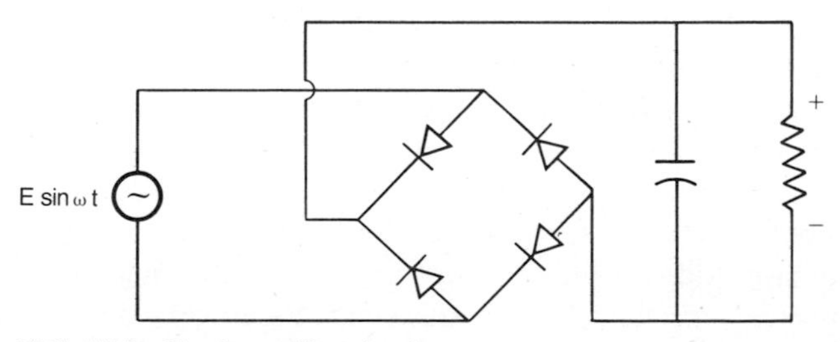

FIG. 7.22 Simple rectifier circuit.

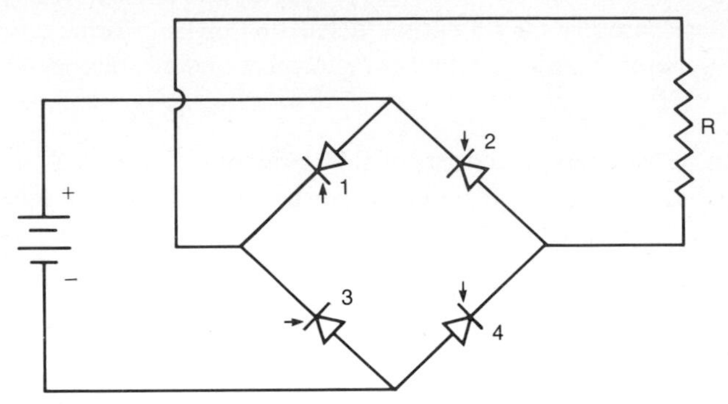

FIG. 7.23 Simple SCR circuit.

On the other hand, we may have direct current available and would like to have alternating current. It is possible to build such an *inverter* device which may be either with transistors or *silicon controlled rectifiers* (SCRs). The SCR is simply a rectifier which can be turned on at any chosen time, and turned off by special commutating techniques beyond the scope of this discussion. However, let us for the moment assume we are using such devices (transistors or SCRs) and can turn them on and off by some suitable technique. Then if we replace the rectifiers in the previous circuit by transistors or SCRs, omit the capacitor, and replace the alternator by a battery, we obtain the circuit shown in Fig. 7.23.

Note that when devices 1 and 4 are on (and devices 2 and 3 are off), the voltage across the resistor is positive at the top and negative at the bottom. Next (one-half cycle later), we will turn off SCRs 1 and 4 and turn on SCRs 2 and 3 causing the voltage across the resistor to be positive at the bottom and negative at the top.

Thus, we can alternate the voltage across the resistor, resulting in a square wave and not a sine wave as we would like. One obvious problem is that, if SCR 3 happens to be turned on before SCR 1 is shut off, a *short circuit* (or shoot through as it is sometimes called) occurs which can do considerable damage. The same can be said of SCR 2 and SCR 4. Improving waveform quality and inverter reliability has been a long, hard struggle in the electrical engineering field. Satisfactory devices with some reliability (much more complex than discussed here) are becoming available. Their absence has caused considerable difficulties in small hydroelectric systems in the past, but those difficulties can now be addressed as new and better inverter devices become available.

Three-phase outputs can be obtained from dc sources by employing three single-phase inverters operated in such a way as to produce the proper phase differences between the three voltages. Alternatively, the dc voltage can be fed to a three-phase bridge inverter, employing six SCRs which are "gated" in a proper sequence to produce a set of three-phase voltages.

With both single-phase and three-phase inverters, techniques are available to reduce the harmonic (higher frequency components) content of the output. In addition, commutation and snubber circuits are necessary to obtain proper operation without exceeding the various limitations of the solid-state devices. All these factors combine to make inverters operating at power levels under 25 kW quite expensive on a per kilowatt basis.

This discussion of basic principles should provide enough understanding that the reader can follow the systems and problems discussed subsequently.

7.3 SELECTION OF ELECTRICAL GENERATION EQUIPMENT

Quality of Electrical Energy

Electrical energy can be utilized in various ways, such as for lighting, heating, mechanical power using electrical motors, and electronic devices. Each type of electrical load or device is affected by what is called the quality of the electrical energy. The quality of electrical energy generally refers to the constancy of the voltage and frequency for alternating current. Most electrical devices are designed for constant voltage and frequency.

Frequency

The frequency of the alternating current is determined by the speed and the number of poles of the generator given by the equation

$$f = \frac{n}{60}\left(\frac{p}{2}\right) \tag{7.27}$$

where n is the speed in r/min, f is the frequency in Hz (or cycles per second), and p is the number of poles.

Depending on geographic location, the frequency may be different but is usually 50 or 60 Hz. To obtain constant frequency from an alternator driven by a water turbine, it must run at constant speed and drive the alternator through a fixed-gear ratio. The speed of the water turbine is controlled by a governor which opens or closes a valve or gate to hold the speed constant as the load changes.

Alternating current can be obtained from a direct-current generator system by means of an inverter as previously described. The inverter is the electronic device that essentially connects and disconnects the load to the dc supply reversing the polarity each time and at a rate to give the desired frequency.

Voltage

The voltage is controlled at the generator or alternator by changing the field excitation through the action of a feedback-type voltage regulator. The voltage can be held constant anywhere in the system by means of the voltage regulator. Usually, only voltage at the generator or alternator is controlled.

When the electrical load changes or the water flow to the turbine changes, frequency as well as voltage will change. Because of the nature of the regulating device, a change in both must occur before any correction or regulation can take place. The electromechanical system is such that as the electrical load increases, both the frequency (speed) and voltage decrease slightly. Conversely, if the electrical load decreases, both speed and frequency increase slightly.

Power

The electric output power of a hydroelectric system is given by

$$P_{\mathrm{kW}} = \frac{\rho g Q H \eta}{1{,}000} \tag{2.3}$$

where

P_{kW} = electrical output, kW
η = efficiency of turbine-generator
Q = water flow, $\mathrm{m^3/s}$
H = net head, m
ρ = density of water, 1,000 $\mathrm{kg/m^3}$
g = acceleration due to gravity, 9.8 $\mathrm{m/s^2}$

The efficiency of small hydraulic turbines ranges from 80 to 90 percent, with a maximum value approaching 95 percent for large units. Typically, a mean value of 85 percent is chosen for preliminary calculations. Efficiencies of electrical generators range from 85 to 95 percent, with a maximum value approaching 98 percent for large machines. A mean value of 90 percent may be chosen for preliminary calculations.

The power that the water turbine and correspondingly, the power that the electrical generator can deliver, is determined by the characteristics of the machine. The generator is rated in terms of voltage, current, frequency, power factor, and temperature rise. If the voltage, current, power factor, and frequency are rated values, then the manufacturer guarantees that the temperature will not rise above the value specified for a standard ambient temperature. The temperature is a critical parameter since it determines the life of the generator.

For short time periods, the generator can deliver more power than its rated value (assuming the turbine provides that power plus the generator losses). However, if operated above rated power for some time, the machine will overheat and a decrease in life will result. When the generator is interconnected to a large electrical system with other generators, there is a maximum power it can deliver, provided the turbine has the capability. For a single generator supplying a given load, with the terminal voltage specified, there is a maximum power the generator can deliver at constant frequency. Under any condition, the generator cannot deliver any more power than it obtains from the turbine minus the operator losses.

Parameters Influencing the Selection of Electrical Components

The basic parameters to be considered in the selection of a suitable electrical generator are

1. Type of electrical output desired: constant-frequency utility-grade alternating current, variable-frequency alternating current, or direct current
2. Hydraulic turbine operations mode: constant-speed operation, nearly constant-speed (± 5 percent) operation, or variable-speed operation
3. Type of electrical load: interconnection with an existing utility grid, storage in batteries, or an isolated system supplying a variety of household or industrial loads

The technology of controlling the speed of a hydraulic turbine within acceptable tolerances is well developed and mature. Both mechanical and electrical hydraulic governors are in use to control the flow of water through the turbine by adjusting the position of the wicket gates (in the case of reaction turbines) or the spear rod (needle) stroke in the case of impulse turbines. No special advantages are seen in operating the turbine in a variable-speed mode except (1) in the case of small systems, use of an induction generator tied to a utility grid may reduce the overall cost of the system by simplifying the control hardware necessary since line frequency will maintain the turbine rotational speed within narrow (± 2 to 5 percent) limits and (2) in the case of a small isolated system with dedicated battery storage and a highly variable water flow rate. In the second case, the rotational speed and the output frequency may be allowed to vary since the output will be rectified for energy storage in a battery bank.

Constant-Speed Systems Using Synchronous Generators

Synchronous machine technology for generating constant-frequency alternating current from a constant-speed prime mover is well-known. The generator runs at constant speed, called the *synchronous speed,* and it is related to the number of poles p in the machine and the output frequency f (in Hz) as shown in Eq. (7.27).

Synchronous machines operating in parallel with utility grids require constant speed, tolerating only minor fluctuations (± 0.5 to 1 percent) for short durations (fractions of a second). There is no special problem to satisfy this requirement with currently available hardware. Synchronization of the unit with power lines is also routinely done with available equipment.

In small synchronous alternators found in micro hydroelectric systems, the rotating exciter supplying the required dc excitation is usually replaced by a static self-regulating self-excitation scheme. The required dc excitation is provided from the stator through a three-phase bridge rectifier and filtering arrangement. A set of current transformers included in this circuitry boosts the excitation current to the required value as the generator is loaded. A capacitor bank connected across the stator terminals assists the voltage buildup starting from a small, residual value to the rated voltage, even if the unit has remained idle for a long period of time.

Both horizontal-axis and vertical-axis synchronous generators are in use in micro hydroelectric systems, depending on the turbine type and the location of the unit. A flywheel is invariably used to smooth out speed variations due to sudden changes in load, and a robust damper winding located in the rotating field (to prevent hunting) assists in satisfactory parallel operation of the unit with utility grids.

Nearly Constant-Speed Systems Using Induction Generators

When the stator of an induction motor is connected to a utility grid, if the rotor is driven at speeds above synchronous, the machine becomes a generator and will deliver constant line-frequency power to the grid. The per-unit slip s of an induction machine is defined as follows

$$s = \frac{\text{synchronous speed} - \text{actual rotor speed}}{\text{synchronous speed}} \tag{7.28}$$

As a motor, the full-load speed is slightly below synchronous and the per-unit slip usually lies between 0 and +0.05. Rated output conditions as a generator are achieved at speeds slightly above (about 105 percent or less) synchronous. Therefore, micro hydroelectric systems using induction generators operate at nearly constant speeds.

Induction generators are simpler than synchronous alternators. They are easier to operate, control, and maintain, they have no synchronization problems, and they are cheaper to build. However, they draw their excitation from the grid and consequently impose a reactive voltampere (VA) burden. This can be corrected by adding static capacitors or synchronous condensers. However, such procedures add to the cost and complexity of the unit. Another point of interest is that the efficiencies of induction generators are slightly (2 to 4 percent) lower than the efficiencies of synchronous alternators over the entire operating range.

The electrical output of a micro hydroelectric system using an induction generator is uniquely determined by the operating speed. An increase in speed results in an increase in output as long as pull-out (maximum torque) conditions are not reached. When this occurs, the electrical output starts to decrease, speed continues to increase, and the system may "run away." Therefore, overspeed protection is vital. The absence of synchronization and excitation equipment makes the adoption of induction generators simple and convenient. They are preferred in situations where long (several months), continuous, and unattended operation in parallel with a local utility network is required. They cannot be used reliably in stand-alone systems (i.e., without being connected to the utility grid).

Systems Using Permanent-Magnet Generators

In remote locations (with no utility grid nearby) it may be necessary to store energy in battery banks for peak time use. With dedicated battery banks for storage, the output of the generator must be rectified to direct current if not in that form. Under these circumstances, the frequency of the alternator output is not important. Coupled with the simplicity that may result from allowing the hydraulic turbine to operate over a range of speeds, a situation is created in which the use of permanent-magnet alternators becomes attractive. Highly varying water flow rates may also necessitate a variable-speed operation.

In a permanent-magnet alternator, the stator is often wound for polyphase (two- or three-phase) output, and the rotor consists of permanent magnets of alternating polarity mechanically embedded around the rotor periphery. The output frequency f (in Hz) is strictly proportional to the rotational speed and the number of poles in the rotor according to Eq. (7.27).

Under open-circuit (no-load) conditions, output voltage is also proportional to the rotational speed. However, when supplying a load, armature reaction effects and internal impedance drops contribute to a departure from this proportionality. If the hydroelectric system is operating at a constant speed, the electrical output under load will be at a constant frequency, but the voltage will be variable. If operating in a variable-speed mode, the electrical output will be variable both in voltage and in frequency. Either of these outputs can be rectified to obtain direct current for charging a battery bank. The performance of the permanent-magnet alternator can be significantly improved by connecting suitable capacitors across the machine terminals. The improvement is primarily due to the leading currents drawn by the capacitors and the resulting self-excitation of the alternator. The dc output of the battery bank can be converted to a constant-voltage constant-frequency ac supply using solid-state inverters.

Cost and Configuration

Assuming direct mounting (without any mechanical interfaces, such as a gearbox) of the generator or alternator to the turbine runner, slow-speed units will require slow-speed generators. For electrical generators, as a first order of approximation, the following relationship applies

$$P_g \propto D_g^2 L_g n_g \qquad (7.29)$$

where D_g and L_g are the diameter and the active core length of the generator rotor, P_g is the power output, and n_g is the rotational speed.

It is noted that slow-speed generators require a large $D_g^2 L_g$ for a given P_g. They are typically shaped like a disk (large diameter and small axial length), heavy, and have high moment of inertia. High-speed (3,600 r/min is the highest speed for 60-Hz output, corresponding to a 2-pole generator) generators are typically shaped in the form of long cylinders (small diameter and large axial length) and weigh less on a per kilowatt basis. Salient-pole structures are most common in generators employed in hydroelectric systems. The damper winding is made up of damper bars embedded in pole shoes with short-circuiting rings at each end of the rotor.

From the electrical viewpoint, slow-speed machines are considered to be "stiff," meaning that they tend to have lower internal reactances; therefore small angular changes in the rotor position can cause large power swings when operating in parallel with a power grid. They also have high short-circuit currents, another consequence of their low reactance, which requires good protection against faults occurring at or near the machine terminals. High-speed generators typically have higher reactances and lower short-circuit currents, and they are flexibly coupled to the grid into which they feed their power output.

In terms of cost, slow-speed generators can be expected to be more expensive than high-speed machines, primarily because of the additional material needed for fabrication. Efficiencies of high-speed generators are only slightly better than slow-speed machines, not enough to solely influence the selection process.

7.4 LOAD CHARACTERISTICS

Depending on how the demand centers are affected by voltage and frequency, the loads can generally be divided into three categories: heating, lighting, and motor and transformers.

Heating

The least critical load is a heating load. Heat output is affected primarily by the voltage, but also by the temperature at which it is operating. Generally, the load (heat developed or power) varies as the square of the applied voltage. Undervoltage is not a problem for heating loads, but devices such as broilers, toasters, space heaters, fryers, grills, and irons generally have maximum specified voltage ratings. Too high a voltage will burn out the heating elements. Any voltage greater than 10 percent over the rated voltage should be avoided.

Some heaters, such as hair dryers, have electric motors, and the motor has to be considered separately as to the effect of voltage and frequency on it.

Lighting

Lighting devices usually exist in three forms: incandescent, fluorescent, and metal- (mercury- or sodium-) vapor lamps.

Incandescent Lamps: These lamps respond like heating elements where undervoltage reduces the light output and extends the life significantly, but overvoltage increases the light output and reduces the life. Any overvoltage greater than 5 percent should be avoided since a 10 percent overvoltage reduces the life by 70 percent for an average lamp. Conversely, a 5 percent reduction in voltage increases the life by 80 percent and reduces the light output by 5 percent.

Fluorescent Lamps: These lamps are affected by both under- and overvoltage. If the voltage decreases more than 10 percent, the lamp may not start and repeated attempts to start shorten the life of the lamp and starter. If the lamp is operating and the voltage drops more than 25 percent, the lamp may be extinguished. Fluorescent lamps use a transformer device and operation in an overvoltage and/or underfrequency can be a problem. (See following section Motors and Transformers.) Lamp efficiency decreases with an increase in voltage as contrasted to incandescent. The normal recommended operating range for the lamp is $\pm$ 10 percent for voltage variations.

Mercury- or Sodium-Vapor Lamps: These lamps use a choke or ballast to stabilize the arc and to limit the current during starting. Since the choke or ballast is an iron-core device, the voltage and frequency variation limitations of transformers apply to them. The operating circuits are normally designed to only allow a 15 percent reduction in supply voltage before arc instability and extinction occur.

Motors and Transformers

Motors and transformers are both iron-core devices and are therefore affected by the variations that affect such devices. The basic governing relationships for iron-core devices are

$$E = 4.44NfAB_m \tag{7.30}$$

where E is the voltage, N is the number of turns across which E is applied, f is the frequency of the voltage, B_m is the maximum value of the core magnetic flux density, and A is a constant. The core losses can be defined as

$$\text{Core losses} = k_e f^2 B_m^2 + k_n f B_m^x \tag{7.31}$$

where k_e and k_n are constants for eddy current and hysteresis losses respectively, f is the frequency of the voltage, B_m is the core magnetic flux density [see Eq. (7.30)], and x is a constant approximately equal to 2 (1.5 to 2.5) for a given material. The heating (and consequent temperature rise) of a given device is determined by its losses (the difference between input and output). If the losses increase, the temperature increases and vice versa.

The core losses vary approximately as the square of the voltage for a transformer operating at fixed frequency. Therefore, overvoltage can become a problem. The usual allowable overvoltage is +5 percent at constant frequency and rated load and +10 percent at no load. At constant voltage, if the frequency decreases, the hysteresis losses increase because of an increase in B_m [see Eq. (7.30) or (7.31)]. Therefore, with a frequency decrease, the allowable voltage increases must be reduced proportionately. Undervoltage is not a significant problem with transformers. At constant output, when the voltage decreases, the current has to increase, but the core losses vary as the square of the voltage, while the copper losses vary as the square of the current.

The effect of voltage and frequency variations on induction motors is more involved than for most other electrical devices because of the relationship existing between torque and speed for both motor and load.

The basic relationship is that both *starting torque* and *running torque* vary directly with the square of the voltage. In the normal operating (running) range, the slip of the induction motors varies inversely as the square of the voltage since there is an inverse relationship between torque and slip. *Slip* is defined as the difference between synchronous speed and actual speed, divided by synchronous speed. Also, the relationships given in Eqs. (7.30) and (7.31) apply to the induction motor.

As an example, at constant frequency and a 10 percent *increase* in voltage, the following changes result: torque increases 21 percent, slip decreases 17 percent, speed at full load increases 1 percent, efficiency change depends on the load usually increasing about 1 percent at full load to decreasing about 2 percent at half load, power factor decreases 3 to 6 percent depending on load, starting current increases 10 percent, full-load current decreases 7 percent, temperature rise decreases 3 to 4°C, and maximum overload increases 21 percent.

At constant frequency and with a voltage *decrease* of 10 percent, the following changes result: torque decreases 19 percent, slip increases 23 percent, full-load speed decreases 1.5 percent, efficiency decreases 2 percent at full load and increases 2 percent at half load, power factor increases 1 to 5 percent depending on load, starting current decreases 10 percent, full-load current increases 11 percent, temperature rise increases 6 to 7°C, and maximum overload decreases 19 percent.

At constant voltage and with a change in frequency, speed is affected since synchronous speed varies directly with frequency. The reactance elements also vary directly with frequency; therefore, the following changes result from a 5 percent *increase* in frequency. Torque decreases 10 percent, speed increases 5 percent, and starting current decreases 5 percent. The other parameters have slight changes with increases for efficiency and power factor and decreases for temperature and maximum overload.

For a 5 percent *decrease* in frequency, torque increases 11 percent, speed decreases 5 percent, efficiency and power factor slightly decrease, starting current increases 5 percent and a slight increase in temperature results. Combined voltage and frequency change should not exceed 10 percent.

For three-phase motors, a small voltage imbalance among the phases can cause significant current imbalance ranging from 6 to 10 times the percentage voltage imbalance.

Control devices, such as relays, solenoids, and contactors, are magnetic core devices and are affected similarly as transformers, except for low voltages. At low voltage they may not operate depending on the temperature of the operating coil.

7.5 DETERMINATION OF DEMAND

In selecting system size, both *power demand* and *energy demand* need to be estimated. The first is usually called *demand* and represents the instantaneous power required by the various electrical devices connected to the system simultaneously, given in watts (W) or kilowatts (kW). The second parameter, electrical energy, often called *production,* involves not only the demand, but how long the device is connected to the system. Electrical energy production is given in watthours (Wh) or kilowatthours (kWh). The physical size of the turbine and generator selected is determined primarily by the power demand while the flow of water and head required is determined by the energy demand.

The demand at any particular time is determined by how many and which devices are connected to the system. The energy used is a function of the living pattern and life style of the user.

Table 7.2 contains typical values of demand (both power, column 1, and energy, column 3) of specific household appliances. However, these devices have a range of sizes and corresponding demand, making the figures only representative. The table also contains average energy uses associated with these appliances which are particularly sensitive to life styles. This energy is best determined by estimating how many hours per month the device is used. The demand (watts) of most electrical appliances is indicated on the device along with rated voltage and frequency.

The size of turbine-generator required is determined by the peak demand (i.e., the largest demand occurring or expected to occur during the year). Peak demand can be estimated by determining what devices would be connected to the system during a typical day for each season of the year. If air conditioning is used, then a summer day would probably have the peak demand. If electric heating is used, then a typical winter day may have the peak demand. Devices that are controlled by thermostats cycle off and on and may all be on during

TABLE 7.2 Typical Household Appliance Loads

Appliance	Power, W	Avg. use/ mo, h	Total energy consump., kWh/mo
Household equipment:			
Blender	600	3	2
Car block heater	450	300	135
Clock	2	720	1
Clothes dryer	4,600	19	87
Coffee maker	600–900	12	7–11
Electric blanket	200	80	16
Fan (kitchen)	250	30	8
Freezer (chest, 15 ft^3)	350	240	84
Hair dryer (hand-held)	400	5	2
Hi-Fi (tube type)	115	120	14
Hi-Fi (solid state)	30	120	4
Iron	1,100	12	13
Light (60-W)	60	120	7
Light (100-W)	100	90	9
Lights (3 extra, 75-W)	225	120	27
Light (fluorescent, 4 ft)	50	240	12
Mixer	124	6	1
Radio (tube type)	80	120	10
Radio (solid state)	50	120	6
Refrig. (standard, 14 ft^3)	300	200	60
Refrig. (frost free, 14 ft^3)	360	500	180
Sewing machine	100	10	1
Toaster	1,150	4	5
TV (black & white)	255	120	31
TV (color)	350	120	42
Washing machine	700	12	8
Water heater (40-gal)	4,500	87	392
Vacuum cleaner	750	10	8
Shop Equipment:			
Water pump (½ hp)	460	44	20
Shop drill (½″, ⅙ hp)	250	2	0.5
Skill saw (1 hp)	1,000	6	6
Table saw (1 hp)	1,000	4	4
Lathe (½ hp)	460	2	1

the same time. If loads are electric motors, as found in refrigerators and air conditioners, then the starting load of the motors must be considered. The principal effect of motor starting is a momentary drop in voltage at the loads. These drops in voltage cause *voltage flicker,* affecting the output of lamps and picture size in television sets.

The amount of voltage change during motor starting can be calculated, but requires more technical data than is usually given on the nameplate. Representative values for devices of a given size range can be obtained from technical references. Accurate values have been determined through tests and can usually be found in a typical textbook about transformers and machines. Any motor load having a size comparable with the size of the generator will result in about a 50 percent voltage drop for a fraction of a minute during startup. This is well beyond the range of recommended voltage drops and would probably also result in control devices not operating properly. It could also result in damage to generator, motor, and other equipment connected to the line.

Calculating the voltage drop requires determination of the starting impedance of the motor Z_m and the impedance of the power system to which the motor is connected, referred to as Thévenin's impedance Z_{th}. The system voltage before the motor is connected is called Thévenin's voltage E_{th}. The starting voltage E_s across the motor is given by

$$E_s = \frac{E_{th} Z_m}{Z_m + Z_{th}} \tag{7.32}$$

The change in the voltage ΔE is given by Eq. (7.33).

$$\Delta E = E_{th} - \frac{E_{th} Z_m}{Z_m + Z_{th}} = \frac{E_{th} Z_{th}}{Z_m + Z_{th}} \tag{7.33}$$

For ΔE to be small, say less than 5 percent, then Z_{th} must be less than $0.0426 Z_m$ or the motor impedance Z_m at start must be greater than 19 times the system impedance.

This same procedure is used to determine the impedance of the total load except the complex mathematical form of the impedance needs to be considered. In the case of motor starting, it was assumed the angle on each of the impedances Z_{th} and Z_m were approximately equal during start. During running conditions and with other loads, these angles are likely to be different. The equation looks essentially the same. However, for a 5 percent voltage drop, $|Z_m + Z_{th}| = 20\,|Z_{th}|$ where the bars $|Z_{th}|$ indicate magnitude of the impedance.

To determine the peak demand, the demands must be summed for that time period when most devices will be connected to the system. Devices such as freezers, refrigerators, and air conditioners or electric furnaces, which are thermostatically controlled should, for purposes of peak demand, all be considered on at the same time. To these loads, lights, electric irons, electric hot plates and other light loads can be added.

To determine the total energy required for a given month or year, it is necessary to estimate how long (hours per month or year) each device will be used. Table 7.2 gave some representative values which are very useful for freezers, refrigerators, furnaces, water heaters, water pumps, etc. For other devices, personal habits and choices will determine how long each is used. The basic question to be addressed is: "Given a hydroelectric system, how many households can be supplied from it?" Simple summing of the demand would give a result larger than the actual demand. The point is that all devices in all households would not be on simultaneously. That is, a certain amount of diversity exists in the times that the loads occur. The amount of reduction in demand that can be expected from diversity is a function of the kinds and number of loads found on the system.

For example, if the loads were completely random, then the probability of a given load could be computed from knowing the demand of each load and its duty cycle (percentage of time the load is on). For example, if four loads of 100, 200, 300, and 400 W were connected to the system and were random with duty cycles of on-time of 50, 40, 30, and 20 percent respectively, then the probability of all being on at the same time is $0.5 \times 0.4 \times 0.3 \times 0.2 = 0.012$ or 1.2 percent. This means that for 4 days of the year, the demand would be as large as 1 kW. If the generator can only stand a 1-kW load for 1 hour, then the probability that all four loads are on simultaneously for this example must be less than $1/8760 = 0.014$ percent. Not many loads can be considered completely random in a household. The amount of demand reduction from diversity can only be accurately determined by measurement over an extended period of time.

7.6 OPERATING REQUIREMENTS

If the hydroelectric unit is to be operated in parallel with an existing grid system, then the size and type of switch-gear and protective devices selected for the unit must be capable of handling and interrupting safely the maximum short-circuit current available from the grid system as well. This information can usually be obtained directly from the electric utility or calculated from its data. In addition, equipment to synchronize the hydroelectric unit with the grid system must be installed with appropriate metering to determine energy bought or sold. The utility can usually recommend the size and type of equipment to install and explain how to operate such an interconnection.

At isolated hydroelectric systems, certain other switch-gear and protective equipment will be required. The generator connected to the turbine must be capable of withstanding the runaway speed of the turbine. Runaway speeds can be as high as 290 percent for variable-pitch blade units and 200 percent for fixed-pitch blade units. An automatic clutch can be used between the turbine shaft and generator shaft to uncouple the generator if its overspeed capacity is not sufficient. Principal protective equipment required is

1. Overspeed shutdown
2. Overcurrent disconnection or interruption, such as remotely operated circuit breakers or fused disconnects
3. Overvoltage relaying to disconnect the electrical load, which is in addition to the voltage control or voltage regulator of the generator

4. Loss of voltage relaying to disconnect the electrical load in case of system failure

7.7 SAFETY CONSIDERATIONS

Safe application of generators and motors involves both electrical and mechanical considerations. Energized electrical parts and rotating or moving mechanical parts must be enclosed, covered, or guarded by some means. The degree of enclosure depends on the location of the installation and application.

In general, the more accessible the equipment is to untrained personnel, the more it should be enclosed. Machines which are accessible to the public should be limited to the following: drip-proof machines, semiguarded machines, and totally enclosed fan-cooled machines. The totally enclosed fan-cooled machine is the only one allowed in hazardous locations.

Hazardous locations are usually those where volatile flammable liquids or liquefied flammable gases or air mixtures of gases, vapors, or dusts exist. These are easily ignited by sparks, open flames, or hot surfaces. Other types of enclosures are allowed, and in the United States, the details may be obtained by consulting the latest edition of the **National Electrical Code®** (NFPA 70), issued every 3 years by the National Fire Protection Association.

Open-type machines can be utilized in power plants, generating stations, or other locations, where only qualified electrical personnel are allowed.

The degree of enclosure should be determined after considering the following general questions:

1. Will the equipment be installed in:
   a. residences
   b. locations open to the public
   c. locations only open to employees
   d. locations accessible only to qualified personnel?
2. Will the equipment be attended by an operator when it is in use?
3. Is the location, size, appearance, and working arrangement such as to discourage inappropriate use?
4. Will the equipment be serviced or used in such a manner that its condition is not visibly obvious to qualified servicing personnel?

Proper Selection of Equipment

Motors and generators should be properly selected with respect to their usual and unusual service conditions, both of which involve environmental conditions to which the machine is subjected as well as operating conditions.

Service conditions may involve some degree of hazard. This additional hazard may result from over-heating due to improper ventilation, mechanical stress due to coupling misalignment or improper belt arrangements, abnormal deterioration of the electrical insulation, or corrosion due to toxic fumes or excessive moisture, fire, and explosion. One important requirement is not to use open or non-explosion-proof generators or motors in the same room with charging or discharging batteries.

For any service condition beyond those specified as usual, the user should consult the equipment manufacturer indicating the conditions to be expected. The usual conditions regarding voltage and frequency were discussed under load characteristics. The other usual service conditions are as follows:

1. Exposure to an ambient temperature in the range of 10 to 40°C
2. Exposure to an altitude which does not exceed 3,300 ft (1,000 m)
3. Installation on a rigid mounting surface
4. Installation in areas which do not interfere with ventilation of the machine

Unusual Service Conditions

Many conditions must be considered but the following are the most common:

1. Exposure to:
   a. Combustible, explosive, abrasive, or conducting dusts
   b. Lint or dust where normal ventilating ducts could be plugged up
   c. Chemical fumes or explosive gases
   d. Steam, salt spray, or oil vapor
   e. Damp locations conducive to growth of fungus, radiant heat, or vermin infestation
   f. Abnormal shock, vibration, or mechanical load from external sources
   g. Abnormal axial or side thrust shaft loads
2. Operation where the voltage and frequency variations discussed in Sec. 7.4 Load Characteristics are exceeded
3. Operation at speeds above the highest rated or below rated speed in an inclined position
4. Operation where it is subjected to torsional impact loads, repetitive abnormal overloads, and excessive starting or reversing conditions

Electrical Considerations

Motors that are used on equipment which is accessible to personnel should not have automatic reset thermal protectors. An injury could result with the unexpected starting of the equipment. These motors should have a manual reset thermal protector.

To obtain the most from the capability of the system, the system power factor should be approximately unity. When power factor correcting capacitors are used on the load side of the motor controller, the total kvar capacity of the capacitor must not exceed the inductive kvar taken by the motor at no load. Corrective kvar in excess of this value may cause self-excitation or over-excitation, resulting in high transient voltages, currents, and torques that can be a safety hazard to personnel and equipment.

The condition given for motors, motor circuits, and controllers in the **National Electrical Code®**, Article 430, should be followed in determining line wire sizes and in selecting protective equipment, starters, and controllers for motors. The details are too numerous to cover in this discussion, but some possible *general* requirements can be stated. These are (*a*) all energized wires and parts are to be insulated or enclosed in approved enclosures, (*b*) all motor frames and enclosures are to be grounded by approved means, (*c*) overcurrent or overload protective devices must open the circuit to all energized conductors (the grounded conductor is not disconnected by the overcurrent or overload devices except and unless all the other conductors are opened simultaneously), and (*d*) the disconnecting means must be in sight of the motor controller and can be in the same enclosure.

7.8 BIBLIOGRAPHY

Alward, R., et al. *Micro-Hydro Power, Reviewing an Old Concept,* report prepared for the U.S. No. DOE/ET/01752, 1979.

Cost of Controls for "Small Hydroelectric Plants" on River Systems, DOE/ET/28310-1, February 1979.

Creager, William P., and Joel D. Justin. *Hydro-Electric Handbook,* Wiley, New York, 1927.

Energy for Rural Development—Renewable Resources and Alternative Technologies for Developing Countries: National Academy of Sciences Report, Washington, D.C., 1976, pp. 137–164.

Idaho Falls Hydroelectric Project: *Design Criteria,* Ido-1699-1, December 1978.

"IEEE Recommended Practice for Protection and Coordination of Industrial and Commercial Power Systems," IEEE Std. 242-1975, New York, 1975.

Jyoti Small Hydro-Electric Generating Units, Bulletin No. HEGE-1, Jyoti Limited, Baroda, India.

National Electrical Code®, NFPA 70, National Fire Protection Association, Boston, 1984.

"Small Low-Head Hydroelectric Power," *Proceedings from the Midwest Regional Conference May 23–25, 1978,* IDO-10076, August 1978.

Solar Photovoltaic Applications Seminar: Design, Installation and Operation of Small Stand-Alone Photovoltaic Power Systems, DOE/DS/32522-TI, July 1980.

Wachter, G. F.: Sec. 9 "Hydroelectric Power Generation" in D. G. Fink, Ed., *Standard Handbook for Electrical Engineers,* 11th ed., McGraw-Hill, New York, 1978.

8

Environmental Impact

Willis E. Jacobsen

Comparing hydroelectric projects with other sources of electrical generation, hydro plants produce power that is compatible with the environment. In contrast with thermal electric generating systems, hydropower projects produce no air or thermal pollution. Also hydropower impacts are reasonably well understood and, therefore, predictable to a moderate degree of accuracy. Hydroelectric generation is an economic use of water that does not consume the resource. Additionally, small and mini hydropower projects tend to create reduced environmental impacts relative to large-scale plants.

Design trade-offs establish project scale and the resultant nature and extent of environmental impacts during construction and operation. Essential design variables include dam height (e.g., from foundation to spillway crest) or design head (between tailwater and reservoir surfaces), reservoir capacity, and power-generation capacity. These variables are interrelated: The dam height customarily controls both resultant reservoir volume and potential power generation capacity which would be installed at a given site.

Environmental analysis must be undertaken early during project planning. Findings from such investigations help establish project feasibility.

The decision to proceed with a particular project should be based on more than conventional evaluations of economic benefits: External (or societal) costs and benefits of the power project under consideration must also be assessed. In many proposed developmental areas of the world, relative influences of specific externalities may differ. For example, development of hydropower in a high-elevation, unpopulated area would generally be expected to produce lower environmental stress and resultant environmental costs than a plant built in the lower river valleys, in more heavily developed regions. Also downstream dams may more readily act as barriers to the upstream migration of species of anadromous fish.

When carefully analyzed, design or operating decisions are frequently found to produce mixed impacts, neither totally unfavorable nor wholly beneficial. Such in-depth evaluations are often essential in discerning and tracking the many possible and complex cause-effect relationships. A recommended analytical approach is to weigh desirable and undesirable influences and then base project design and operating decisions on net effects of the interacting combinations of results. One complicating factor is that costs typically arise immediately while benefits may accrue over many years: Also, costs are usually easier to measure than benefits. Another point which must not be overlooked is that different segments of nearby groups may represent many different interests, concerns, and purposes; therefore, they might be influenced differently by a given project and its mode of operation.

Prevalent viewpoints should be identified and assessed, to equitably satisfy the many interests. For example, a direct and immediate effect of small hydropower system development could be the availability of power for pumping irrigation water, which improves the productivity of nearby arid, but fertile, lands. Concurrently, another segment of the population may find its livelihood stymied because it had been dependent on the stream, now dammed, to transport barges of produce to downstream markets. Should commerce on the stream warrant it (and should stream discharges be adequate), construction of a navigation lock at the dam site would be a remedial option for economic consideration.

In addition to primary or direct environmental effects, indirect ramifications of hydropower development (second- or even third-order effects) must not be overlooked in project evaluations. As a case in point, a new, reliable supply of hydroelectric energy may attract a number of power-consuming industries to the area, drawing numerous workers and their families to the expanding area. In turn, new construction to accommodate the expanding population may require considerable clearing and grading of previously undeveloped land. Clearing of formerly vegetated slopes could create increased dislodgement of sediment during periods of heavy precipitation. If developmental planning was insufficient and if runoff interception and control regulations were inadequate, sediment loads of increased intensity might wash into the stream and reservoir. An accumulation of sediment in the reservoir not only could prove environmentally adverse but also could severely shorten the economic life of power production from the pondage. Thus, it is appropriate or even necessary to evaluate extended chains of possible events.

Relative environmental sensitivities to hydropower development and operation are highly site specific; one aim of the guidelines for environmental assessment is to list and briefly describe key considerations which potentially pertain to a diversity of prospective sites. These analytical guidelines will be supplemented with a checklist of useful environmental information to support feasibility determinations of candidate developmental sites.

General environmental influences and possible measures to reduce undesirable impacts, if identifiable, are conveyed in the next subsections. Included are (1) influences of stream impoundments, particularly in regard to impacts on fish and wildlife; (2) environmental aspects of spillways and outlets; (3) impacts associated with generation of electrical energy at mini and small-scale hydropower projects, including electrical transmission; and (4) potential damages related to project construction and maintenance.

The importance of project permits as well as typical requirements for applying for licenses or permits are discussed. Next, factors for estimating project costs associated with environmental control activities are summarized. Finally, a suggested checklist of project information for environmental assessments is outlined, and a synopsis of techniques possible to employ in making environmental assessments is presented.

8.1 STREAM IMPOUNDMENTS

Several prevalent environmental effects are associated with creation of an impoundment behind a power dam. The dam may act as a stream barrier to upstream or downstream movement of objects (including migrating fish or river traffic). Also the slackwater pool which forms upstream of the dam will inundate land formerly lying along the banks of the flowing stream. Depending on such factors as ratio of the volume of usable water impounded (active storage) to the average streamflow,* or the pattern of withdrawing water from the pool, downstream discharges and water quality (physical, chemical, and biological) may be altered—beneficially or adversely. Furthermore, variations in water level, in the impoundment or downstream, may affect aquatic and/or terrestrial organisms. Flow excursions due to regulating the releases from an impoundment often can be expected to attenuate with distance downstream, however. Attenuation would be more rapid if the steam discharge were supplemented by inflowing groundwater or tributary flows.

Flow regulation controlled downstream from a large impoundment, espe-

*The ratio is termed the *storage fraction,* a parameter which represents the time to fill the usable reservoir volume at average inflow and zero outflow conditions. Simple diversion structures (e.g., leading to an out-of-stream channel, penstock, thence to a power plant) may have nearly zero storage capacity. Run-of-river projects incorporate pondages which are capable of storing only 1 or 2 days of average river flow; conversely, some storage projects can impound weeks, months, or even years of river flow—but large storage capacities would not typify small-scale projects. Environmental impacts customarily increase for systems having larger reservoir volumes.

cially should outflows be primarily a result of baseload power generation, could produce beneficial effects. The benefits would be derived from the sustained, uniform (nonpeaking) flow in comparison with the pulsations of discharge which naturally occur in an intermittent stream. Seasonal fluctuations in an intermittent stream may vary from periods of zero discharge to rampaging floodwaters following high, sustained precipitation. Downstream benefits attributable to flow regulation include maintenance of improved quantities and qualities of water for industrial or drinking supplies, improved navigation, and reduced flood damage. Other potential benefits of flow regulation include a greater annual energy production, both at the planned site and at downstream power projects. With improved flow regulation is the possibility of installing increased dependable generating capacity at downstream power stations.

Reservoir-Related Ecological Impacts

Direct relationships frequently exist between sizes of hydropower projects and magnitudes of environmental impacts.[1] In illustration, the area inundated by a reservoir is directly dependent on the proposed project head. Reservoir lands may prove environmentally sensitive to the permanent change from a terrestrial to an aquatic ecological system. Additionally, most reservoirs, in preparation for flooding, are cleared and graded, with associated environmental influences. The design head of the proposed dam, plus the type of structure (e.g., concrete gravity, earthfill, rockfill, arch), dictates the localized disturbance zone associated with field activities such as excavations for dam and power house and the construction of tunnels, chutes, roadways, storage and maintenance yards, crushing and mixing plants, approach channels to powerhouses, channels to spillways and locks, and downstream aprons for stilling basins and tailraces.

Assuming, for example, a damming and inundating of a trapezoidally shaped valley of constant cross section (Fig. 8.1), constant bottom width b, and constant channel slope upstream-to-downstream, 1 on S, the reservoir plan area A_r as a function of the head d_0 behind the dam, may be obtained. Sketches of the cross section of the trapezoidal basin and its side elevation along the

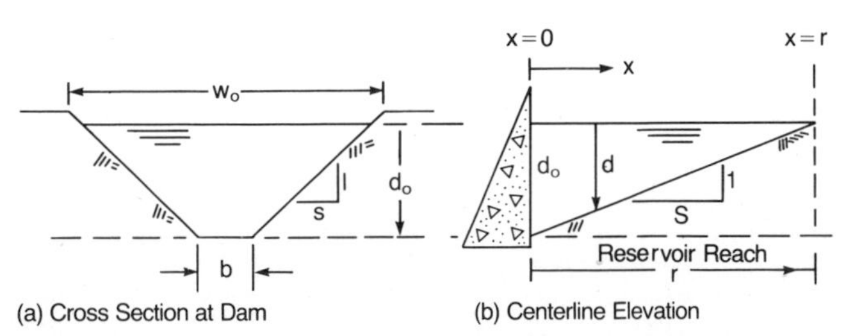

FIG. 8.1 Cross section and side elevation of trapezoidal basin.

stream centerline are depicted in Fig. 8.1*a* and *b*, respectively. The independent variables are shown in the figure.

A simplifying assumption is that the pool behind the low-head stream barrier is horizontal, i.e., there is no backwater curve.

The reservoir surface area A_r is an isosceles trapezoid that is widest at the dam and uniformly narrows upstream. It can be expressed by

$$A_r = (w_0 + w_r)\frac{r}{2} \tag{8.1}$$

where w_0 is the width of the reservoir immediately behind the dam, w_r is the width at the head of the reservoir, and r is the reach (total length) of the reservoir. Based on a channel bottom width b, a horizontal projection c of either sloping side, a head d_0, and a channel side slope 1 on s, the reservoir width behind the dam is

$$w_0 = b + 2c = b + 2d_0 s \tag{8.2}$$

and the width at the upstream end is nearly the base width, or

$$w_r = b \tag{8.3}$$

With a bottom slope along the stream of 1 on S, the reach of the reservoir becomes

$$r = d_0 S \tag{8.4}$$

Setting the channel base width to

$$b = kd_0 \tag{8.5}$$

and substituting these expressions in Eq. (8.1) gives

$$A_r = (kd_0 + 2d_0 s + kd_0)\frac{d_0 S}{2}$$

or (8.6)

$$A_r = (k + s)Sd_0^2$$

The area shown to be inundated by the reservoir and thus subject to environmental disturbance varies as the square of the head. The variation in the dimensionless designation of reservoir area is shown in Fig. 8.2. The dimensionless surface area increases directly as the sum of the channel bottom width ratio and the channel side slope. The actual surface area increases directly as the reservoir bottom slope term S and as the head, squared.

The total storage volume in the example reservoir, of trapezoidal cross section may be obtained from integrating the cross-sectional area along the centerline of the reservoir

$$V_r = \int_0^r A\,dx \tag{8.7}$$

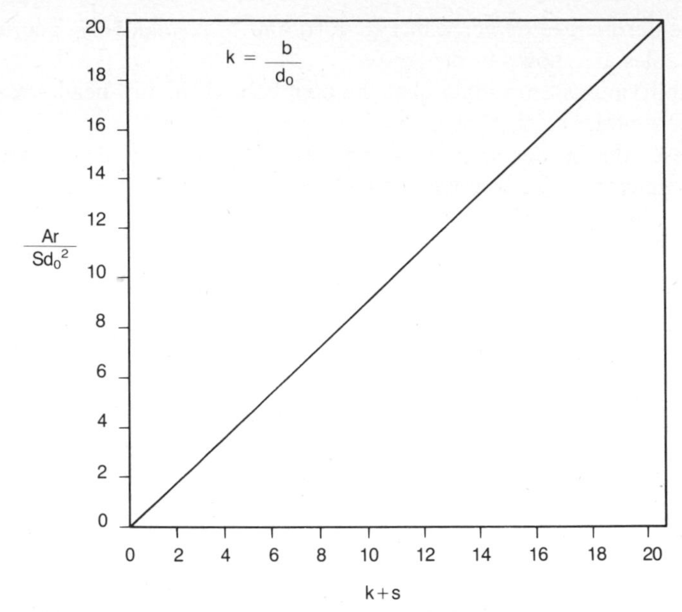

FIG. 8.2 Disturbed reservoir area, dimensionless.

where the reservoir cross section A varies in the x-direction upstream of the dam. As shown in the sketch (Fig. 8.1b), water depth is assumed to decrease linearly from the dam ($x = 0$) to the upstream end of the reservoir ($x = r$). At a general upstream distance x from the dam, the reservoir depth is d. From geometric similarities, the depth may be expressed as

$$d = d_0 - \frac{X}{S} \tag{8.8}$$

The trapezoidal area of the channel cross section at any upstream distance x is

$$A = (b + ds)d \tag{8.9}$$

Substituting Eq. (8.9) into Eq. (8.7) gives for the volume of the reservoir:

$$V_r = \int_0^r (b + ds)d\, dx \tag{8.10}$$

Then, substituting expressions (8.5) and (8.8) for b and d, respectively, gives

$$V_r = \int_0^r \left[kd_0 + \left(d_0 - \frac{X}{S} \right) s \right] \left(d_0 - \frac{X}{S} \right) dx \tag{8.11}$$

which yields a reservoir volume

$$V_r = \left(\frac{k}{2} + \frac{s}{3}\right) Sd_0^3 \tag{8.12}$$

The total volume stored in the reservoir thus varies as the cube of the head (as would be the case for any similarly proportioned reservoir cross-sectional shapes that were to be compared). Impacts of controlling the storage and the release of water from the reservoir, i.e., through regulation of natural flow, depend on the volume available for storage.

Consideration of another reservoir geometry would help to convey effects of constants in disturbed reservoir area expressions. Were the base of the trapezoidal canyon assumed to possess a uniform flare upstream of the dam, the base could increase from $b_0 = k_1 d_0$ at the dam to $b_r = k_2 d_0$ ($k_2 > k_1$). At distance x upstream, the generalized bottom width of the channel would be represented by

$$b = b_0 + (b_r - b_0)\frac{x}{r} \tag{8.13}$$

As previously, the bottom of the channel is assumed to slope uniformly upward in the upstream direction such that, again, the depth at any upstream section is

$$d = d_0 - \frac{x}{S} \tag{8.8}$$

The surface area of the reservoir is given by

$$A_r = \int_0^r w\,dx \tag{8.14}$$

where the reservoir surface width is

$$w = b + 2ds \tag{8.15}$$

Substituting expressions (8.13) and (8.8) and the defined base widths gives

$$w = (k_1 + 2s)\,d_0 + \left(\frac{k_2 - k_1 - 2s}{S}\right)x \tag{8.16}$$

Solving the integral (8.14) based on the reservoir width term and substituting the value, $d_0 S$, for the limit r gives

$$A_r = \left(\frac{k_1 + k_2 + s}{2}\right) Sd_0^2 \tag{8.17}$$

The similarity between this and the first example can be readily observed. For the special case involving a constant base width of the flooded channel $k = k_1 = k_2$ and equation (8.17) becomes identically equal to equation (8.6), applicable to the initially assumed channel geometry.

Environmental Effects of Stream Impoundments

A summary of environmental influences associated with impoundments, or resulting from the utilization of impoundments in small hydroelectric power system operations, is presented in Table 8.1. This and the ensuing summaries of impacts include the following: (1) basic causes of environmental influences, (2) the possible resultant impacts, and (3) potential remedial measures that could be employed for reducing or changing specific impacts.

Major concerns in the formation of an impoundment on a stream include the barrier effects, the existence of the slackwater pool, reservoir leakages, reservoir warming, altered flow velocities in the pool and downstream, land inundation, variable pool levels, and algae stimulation. Again to be emphasized is that most of these parameters are strongly dependent on the size (head) of the project because dam height controls disturbed land areas and stored water volumes. Additionally, project size determines time of disturbance. Time is likely to be a more critical factor in construction, e.g., to complete the project and restore the disturbed ground cover. Long-term irreversible effects or periodic effects generated by operations must clearly be differentiated in environmental evaluations from short-term (e.g., construction) effects.[2]

The alteration in ecology from a land-stream biome to a quiescent water biome is difficult to assess as adverse or beneficial unless site-specific critical investigations are able to establish net influences. Location of the proposed facility and prior development of a planned small- or mini-scale hydropower facility are important factors; if the facility is built in an existent concrete-lined canal or concrete pressure culvert (to replace a pressure-dissipating drop structure) net positive benefits would always result.

In Table 8.1, another possible undesirable influence of cause 2, related to the slackwater pool, would be the impact of synergism, e.g., the combined effects of multiple stresses such as a simultaneous increase in temperature (cause 4) with a decrease in dissolved oxygen because of aerobic decomposition along the floor of the reservoir. Synergistic effects are, in general, poorly understood at present.

A possible remedy for slackwater pools is to build a stepped series of low-head dams in the same reach of the stream as would otherwise be utilized by a single high dam having considerable storage capacity. However, there are other adverse influences associated with a string of lower dams, including: (1) They are more costly, (2) they provide less total storage capacity, (3) they may drown out and produce little or no power during peak streamflows, (4) the downstream projects may suffer degraded operations because of unregulated flows, and (5) a closely spaced series of dams may produce a higher concentration of nitrogen in solution.

Increased temperature in reservoirs and downstream is a serious influence. Salmon and trout are particularly affected by even slight temperature rises. For example, the optimum range for salmonids is 5.8 to 12.8 °C (42 to 55 °F).[3]

TABLE 8.1 Environmental Influences of Impoundments

Effect	Remedy
Cause 1. Stream barrier, flow blockage	
Disrupts or prevents fish migration	Provide fish-passing system, e.g., a ladder
Blocks passage of water traffic	Include navigation lock in multipurpose project
Blocks local movement or migration of land animals which use narrow, shallow, or intermittent stream as crossing	Construct causeways with culverts
Cause 2. Slackwater pool forms behind dam	
Creates or improves stream navigation	————————
Stagnant reservoir subject to thermal buildup, chemical buildup (salts, H_2S), uncontrolled marine-plant growth, oxygen depletion (less reaeration than in naturally flowing stream); all subject marine life to stresses, diseases, or disablement; shallow or marshy shorelines may provide conditions for propagation of vectors	Construct multiple lower head, in-series projects (possibly run-of-river operation), rather than a single higher head project; during decreased power demand, establish supplemental downstream discharges via reservoir outlet works; implement vector-control measures
Sensitive stream-type organisms are decimated in altered ecosystem; new varieties of lake-type aquatic organisms adapt to newly formed conditions	————————
Cause 3. Reservoir floor or walls contain pervious formations	
Loss of stored water through leakage creates net reduction in downstream flow, dependent on inflows of returning groundwater	Detect anomalies in basin in careful, extensive geological exploration; install flow barriers or seals (grout curtains, cutoff walls, sheet-piling, reservoir liners)
Cause 4. Thermal increases in reservoir	
Damage to marine life plus possible increased water losses via evaporation (augmented by large surface areas of pool)	————————
Cause 5. Decreased flow velocities in pool and downstream channel	
Reservoir sediment buildup lowers active storage volume of pool and is likely to shorten economic life of project; dependent on upstream conditions that produce stream sediment loadding/turbidity (e.g., fine clays and silts in upland soils, exposed soil surfaces, steep slopes, high precipitation over watershed, excessive channel scour); downstream hydropower operations would degrade if dependent on regulated outflows from ample upstream storage	Periodically dredge accumulated sediment from reservoir; construct settlement pond upstream of operational reservoir; any intercepted sediment in upstream pool may also require removal

TABLE 8.1 Environmental Influences of Impoundments (*Continued*)

Effect	Remedy
Unnaturally clarified tailwater downstream—based on interception of sediment in reservoir—can create abnormally high erosion of streambed and persist for considerable distances downstream, dependent on rate of restoration to equilibrium sediment transport	———————
Sediment deposition/buildup in a reservoir can bury and suffocate benthic communities (bottom marine plant and animal life) and alter character of bottom plants	———————
Transporting velocity of small downstream migrants in pool may become inadequate, causing fish losses as result of disease, predation, increased exposure to higher temperature water, decreased dissolved oxygen, elevated nitrogen, or other deleterious chemicals in solution	Release fish-carrying or flushing flows from reservoir although such flows represent generation loss at hydropower plant
Decreased flows admitted to downstream channel during lower power demand or reservoir filling (assuming low discharges are not counterbalanced by supplemental groundwater or tributary inflows) may create shortages for downstream uses and/or degrade quality, e.g., create increased temperatures, decreased dissolved oxygen levels, and cause marine life mortalities, species substitution, and more difficult fish migration	Release fish-carrying or flushing flows from reservoir although such flows represent generation loss at hydropower plant
Cause 6. Inundation of reservoir area	
Displaces former terrestrial wildlife species inhabiting area	———————
Pool may flood scenic areas, built-up areas, archeological sites, or historic landmarks	Relocate structures and rights-of-way; build dikes or cofferdams to preserve sites (at high cost); complete archeological studies
Sensitive marine species requiring specialized aquatic ecosystem along flowing stream may not survive stress and shock of lacustrine ecology, e.g., for hatching of eggs and development of young; some species need specific depth of flowing water over suitable bottom composed of gravel beds at former stream confluence	Attempt to reestablish aquatic species in nearly equivalent environments
Inadequate clearing of vegetation from flooding zone may lead to death and decay of plants and other matter, unesthetic floating debris, blockage of screened intake passages, water pollution (e.g., formation of phenols), hazards to navigation, and degraded recreation	Properly clear reservoir
Improvements in communities or villages adjoining stream to be impounded may be periodically or permanently flooded	Relocate or rebuild improvements on adjoining higher ground

TABLE 8.1 Environmental Influences of Impoundments (*Continued*)

Effect	Remedy
Cause 7. Fluctuations in pool level	
Power pool excursions (i.e., daily drawdown for power generation or variable levels established seasonally for flood control) may degrade pool's value as recreational water body (marooning docks and beaches, forming mud flats and dangerously shallow areas for boating) or diminish its value as a lake or water supply; such changes in level may shock, isolate, or cause the extinction of sensitive, less hardy types of marine life	Pool level variations are produced by powerhouse and dam operations, dependent on electrical production, inflows, and nonpower releases; control pool drawdown and filling judiciously by policies governing project operations, e.g., a major trade-off is in optimizing hydropower production at near-maximum pool vs. increasing risk of downstream flooding
Cause 8. Stimulated algal growth	
Excessive algae can alter trophic state of stream or reservoir system, choke body of water, pollute water, deprive it of oxygen, create fish kills, lead to high and costly maintenance of intake screens	Use algal management-control techniques (which themselves may present environmental dangers)

The lethal limit for adult fish is 28°C (82°F), and that for juveniles is about 6°C (11°F) lower. Indirectly, the increasing temperatures delay migration of adult fish, foster the development of fungus and disease organisms, affect food supplies, and favor some species which compete with, or prey on, cold-water fish like salmon or trout.

In the preservation and protection of sensitive ecosystems, seemingly harmless concentrations of pollutants or slight changes in ambient conditions may be hazardous to some species that have special sensitivities or are subjected to prolonged exposure. During the preliminary site exploration the existence of any species or subspecies of fishes, reptiles and amphibians, mammals, or birds which are either considered threatened or endangered must be determined. (Those threatened are likely to become endangered unless current trends are reversed; those endangered are in imminent danger of extinction in the wild.)

8.2 SPILLWAYS AND RESERVOIR OUTLETS

Major factors associated with spillway operations (Table 8.2) include the violent dissipation of energy in the stilling basin and the entrainment of supersaturated atmospheric nitrogen in flows downstream of the spillway. Excess nitrogen in the bloodstream of fish can cause sickness and mortality from the "bends." Problems related to the operation of outlets (controlled outlet conduits for releasing water from behind the dam and inlet passages to the hydraulic turbines) include blockages by trashrack grid or screen structures,

TABLE 8.2 Environmental Influences of Spillways and Reservoir Outlets

Effect	Remedy
SPILLWAYS	
Cause 1. Energy dissipation in stilling basin	
High velocities and turbulence in hydraulic jump, roller bucket, etc., may injure species of fish carried over spillway; injuries to fish may result from encountering baffle piers on stilling basin apron	Minimize releases of flow over crest of gated spillway, especially during peaks of downstream fish migration; low-head projects do not usually injure fish, since fall velocities decrease with the square root of the head
Cause 2. Turbulent mixing in stilling basin/plunge pool entrains air, producing supersaturated nitrogen levels in flow continuation	
Fish mortality from excessive nitrogen in bloodstream causes "bends"; problems augmented if dams are built in series with quiescent pools in staircase profile along stream; deeply plunging spillway flow with minimal free flow between dams prevents water degassing	Include a reach of natural flow between tailwater of given project and pool of next downstream project for stream deaeration; structurally alter flow pattern of downstream face of spillway to reduce plunge depth
OUTLETS	
Cause 3. Flow through in trashracks	
Larger species of marine life may impinge on and be entrapped on trashracks	Increase bar or grid spacing to enlarge intake flow area to reduce carrying velocity in plane of trashrack
Cause 4. Water releases from variable depths of pool	
Operation of multilevel outlets may produce variable-temperature water in downstream channel; stratification into warmer surface layers and colder deep layers increases in quiescent pools; variable oxygen content and dissolved chemical content of released water may occur; can either improve or degrade conditions in downstream channel for water supplies or sustenance of aquatic organisms	Selectively mix flows from outlets to control resultant water quality or temperature; monitor and control flows to help maintain uniformity of quantity and quality of reservoir releases
Cause 5. Fish passage through intakes and turbines	
Marine organisms can undergo shock or destruction in passing through submerged intake channels and hydraulic turbines as a result of sudden pressure changes, contact with moving turbomachines, and long exposures to changed ambient conditions (pressures); aquatic biota concentrations at intake depths may increase inflicted damage; injured or stunned fish can be subject to predation	Use lateral diversion or bypass channels near intakes; incorporate outlet passage geometry to minimize low pressure or cavitation zones; operate turbines at maximum efficiency point to reduce undesirable pressure gradients in flow passages. At peak downstream fish migrations, shut down multiple generating units and use gated spillways and open outlets to augment downstream flows

variable-quality water releases through multilevel outlet works, and passage of marine life through hydraulic machinery.

Optimization of screen-grid openings, streamlining, and the approach flows requires complex trade-offs. For example, if grid spacing on trashracks is too large, floating objects in the flow may damage the rotating hydroturbines, or large marine life may pass through the turbomachinery. Grid spacing that is too small will create excessive head losses, will readily block by debris and require excessive maintenance, or will trap small- to moderate-sized fish that impinge on the surface.

Some attempts have been made to reduce pressure gradients through water turbines by admitting air in the flow passages. This procedure slightly reduces the operating efficiency of the system and may create dynamic oscillations. In most turbine installations of less than 15-m (50-ft) head, fish mortalities through the units should be low or nonexistent, even without any special operating procedures.

8.3 POWERHOUSES AND ELECTRICAL SYSTEMS

Environmental factors associated with hydroelectric powerhouses, switchyards, and transmission systems, as indicated in Table 8.3, include: The blockage of upstream-moving marine species (similar to the dam as a stream barrier), baseloading of the power plant (generally beneficial), downstream surges of flow due to power-peaking operations, dangers to trespassers, electromagnetic radiation from circuit elements, and preemption of the wildlife habitats by structures.

In the overall hydropower system, these factors tend to be least severe, and many of them can be suitably remedied.

8.4 CONSTRUCTION OR MAINTENANCE

As shown in Table 8.4, major concerns related to project construction or maintenance include: stream pollution by sediment from construction activities, improper clearing of the reservoir area, underwater blasting, construction (or maintenance), chemical spillage, chemical treatment of the reservoir, reservoir filling (initial or following drainage for maintenance of facilities), and reservoir maintenance dredging.

Construction damage is generally temporary and fast recovery can take place, especially if the project is small or if field activities are accomplished in stages, where only small parcels are disturbed at a time and then quickly revegetated. Small streams, seriously disturbed by construction, have recovered when sediment was controlled and natural processes were permitted to replen-

TABLE 8.3 Environmental Influences of Powerhouses and Electrical Systems

Effect	Remedy
Cause 1. Power plant structure blocks upstream-moving aquatic species	
Sudden losses of anadromous species	Provide means for bypassing upstream migrants around barrier, e.g., fish ladders
Cause 2. Power plant baseloaded (storage at-site or upstream)	
Regulated uniform flow produced downstream—potentially improves water quality and quantity available seasonally; concomitant improvements possible in boating, fishing, navigation, water supplies, and power generation at downstream hydropower plants (especially if plants have minimum at-site power storage available)	———
Cause 3. Power plant utilized for peaking	
Flow excursions in downstream channel or tailwater-level fluctuations can damage, kill, or produce substitutions in types of aquatic organisms; flow fluctuations also degrade or endanger downstream recreation; surges of high flow tend to scour or erode downstream channel, with worst damage nearer powerhouse	Implement judicious policies and procedures for powerhouse operations; provide re-regulating dam and pool downstream of main project
Cause 4. Trespassing	
Danger of slipping and falling into water, electrical shock, or falls from high towers, catwalks, or ladders	Install secure fencing with conspicuous warning signs; provide perimeter protection by remote intrusion alarms
Cause 5. Electromagnetic radiation from electrical conductors	
Implicated as potential hazard to humans and livestock in vicinity of transformers, transmission lines, etc.	———
Cause 6. Preemption of wildlife habitat	
Powerhouse structures separate from dam, switchyards, penstocks, etc., can create loss of wildlife habitat, block wildlife movement, or create barriers to low-flying birds	———

TABLE 8.4 Environmental Influences of Construction or Maintenance

Effect	Remedy
Cause 1. In-stream sediment addition from clearing, excavation, and grading	
Increase in local and downstream turbidity is major impact of construction associated with dam and powerhouse foundations, reservoirs, approach and downstream channels and stilling basins, locks, roadways, diversion channels, tunnels, switchyards, transmission corridors, borrow areas for fill or aggregates (plus aggregate washing); sediment buildup in stream can result in disablement of marine life from gill blockage and suffocation or can subject species to severe stress and disease; higher sediment loading can also decrease levels of dissolved oxygen in water	Remove sediment from surface-runoff channels at construction sites by interception or filtration; physical facilities include collection and drainage ditches, porous (e.g., straw) filters, gravity settling basins including ponds used in recirculating system for treating washwater; reestablish vegetation on exposed land, especially sloping land
Cause 2. Inadequate or excessive clearing of reservoir area	
Decaying logs, brush, and debris may create pollution, e.g., offensive tastes and odors from phenols released in drinking-water supplies; elevated oxygen demand severely depletes dissolved oxygen required to support native marine life; snags and floating debris create navigational hazards, degrade reservoir esthetics and recreational value, and plug turbine intakes; clearing vegetation above future shorelines degrades esthetics and exposes rim area to future erosion	Thoroughly clear reservoir of all growth and debris to highest pool elevation
Cause 3. Underwater blasting	
Detonations from blasting for in-stream cofferdams or dam foundation excavations may produce underwater shock waves severely injuring or killing fish in sphere of influence	Use smaller, localized, and more directed charges
Cause 4. Chemical spills infiltrate soil or enter surface waters	
Spilled chemicals (oil, grease, paints, acids, cleaning solvents, concrete curing compounds, soil additives, fertilizers, pesticides, etc.) may contaminate ground or surface waters, injuring or decimating marine life or wildlife; toxic substances are hazardous in drinking-water supplies; fish may concentrate poisons which, in turn, enter the food chain	Use care in chemical handling; properly dispose of excess substances and empty containers; for mixing and maintenance areas, build sloping slabs with gutters or berms and collection drains and sumps
Cause 5. Addition of chemical agents to operational reservoir	
Excessive additives, e.g., coagulants for turbidity control or algicides, may poison marine life	Use great care in applying chemicals to use lowest concentrations that achieve intended control

TABLE 8.4 Environmental Influences of Construction or Maintenance (*Continued*)

Cause 6. Reservoir filling	
River closure and creation of impoundment may severely decrease downstream flows, destroying marine life and causing inadequate dilution of lower river pollution	Carefully control and sustain discharges from power plants and low-elevation outlets during reservoir-filling operations
Cause 7. Reservoir dredging	
Maintenance dredging of sediment deposits can be extremely costly if required frequently; during dredging, drastic increases can occur in turbidity and sediment load carried downstream (sometimes several kilometers distance); problems associated with in-stream construction activities can be recreated by dredging; if bottom sludges contain entrapped toxic precipitates (e.g., heavy metals or chlorinated hydrocarbons), such materials may be reintroduced into suspension and ingested or concentrated in marine plant and animal species and biologically magnified in food webs	Use dredging equipment and practices that minimize reintroduction of finely suspended particles in water column; transport dredged material to dump in drainage or containment ponds on land; stockpile must filter and retain particulates and remove clarified liquids for either return to stream or to supplemental treatment, if required

ish stream life. Construction causes some loss of terrestrial habitat and food supply because of the removal of vegetation for surface earthwork.

Although not listed in Table 8.4, safety of the structure during construction and operation must be an overbearing consideration, commencing from the earliest exploration and design. The most catastrophic impacts could result from flooding accompanying sudden dam failure. The project head, the stored volume of water behind the dam, and the character of the downstream development all determine the damage impact.

Sediment remains one of the more serious aspects of construction activities, whether arising from instream or out-of-stream disturbances or even from aggregate-washing operations. Suspended and settleable solid particles produce deleterious effects by coating, burying, suffocating, and abrading living organisms. Aquatic vegetation may be adversely affected by a reduction in photosynthesis caused by suspended solids. The chemical character of the silts and other fine materials may be toxic or the materials may act as transport media for adsorbed toxic materials such as heavy metals or organics. Settleable solids also impede navigation, reduce reservoir capacities, and increase flooding in downstream reaches of streams.

In addition to burying bottom organisms, including eggs and larvae, high concentrations of suspended sediment can injure aquatic biota, including fish, by abrasion. Many aquatic invertebrates and fish respond to high concentrations of suspended solids, lower population densities, and changes in community composition. Aquatic communities may also be critically impacted or eliminated by habitat destruction, e.g., filling of the gravel and rock voids on the bottom.

8.5 REQUIRED PERMITS

Careful preparation of permit applications as well as objective review and issuance of permits for project construction and operation help to ensure that water resources are developed in a logical, coordinated, and an environmentally acceptable manner. Integrated and equitable development of a stream basin helps sustain the rights of citizens in the region—all sharing the public ownership of the water resource.

Furthermore, the permitting process provides a means of reducing interference or friction among diverse potential developers of a common water resource: Permits not only help assure a fair allocation of the resource among potential users but also allow proper scheduling and tracking of stream development. Thereby, institutional and environmental interests may be reasonably factored into the plans, and long-range planning and zoning goals may be adhered to and achieved. While a fully controlled improvement of a natural resource can be an elusive target, even rudimentary requirements for permits can enhance the mutual benefits and satisfaction of all parties.

The level of information to be provided in the permit application should be related to the scale of the proposed project; i.e., a small-scale hydro project should be required to submit only minimal information to the regulator, namely, fundamental plans and project characteristics. Pertinent types of environmental and operational information include:

- Location (vicinity and site maps), present and future topography, and structural layouts
- Dam type, height/head, length, volume
- Cost estimates
- Land ownership; competing water uses or users
- Planned turbine-generator unit types, numbers, and capacities
- Findings of preliminary site investigations, including soil types, engineering properties of the soils and foundation structures, hydrogeology, topography, hydrology (including drainage area, precipitation, runoff, stage-discharge relationships, hydrographs, flow-duration curves, flood frequencies, and floodplain delineation), existing land development and demography, and required relocations
- Maximum reservoir area and stage-volumes
- Design discharges and flow control—spillway bays (gated and ungated) and their discharge capacities; elevations and discharge capacities of outlet works
- Appurtenant structures, e.g., navigation lock, fish ladders

- Preliminary construction/diversion plan, including planned control or multipurpose-project pool levels
- Environmental assessments for both baseline and projected impacts, e.g., stream water quality, including turbidity, sediment load
- Assessment of institutional implications in the locale
- Marketing plans for generated hydropower, including expected rates and revenues
- Planned intertie arrangements

8.6 COST IMPLICATIONS OF ENVIRONMENTAL CONTROL TECHNOLOGY

Costs of project environmental control may be considered as required increments over the conventional front-end investments for acquiring land and land rights, constructing dam and powerhouse and miscellaneous structures and improvements, and installing turbogenerators, accessories, and switching and transmission equipment. Most of these conventional capital costs are mandatory for an operating plant whether or not environmental measures are implemented. These basic project costs are highly variable, being governed by site characteristics and design parameters (e.g., type, height, and length of planned dam; size and operating head of the power-generating units; state of the regional infrastructure—remoteness or ruggedness of the terrain around the selected site, existence of access roadways, and distance to the planned transmission line, if existent).

Costs of typical new (nonretrofit) hydropower projects fall in the range of \$1,000 to \$1,500 per installed kilowatt. Project capital expenditures, currently rising rapidly, have been cited[4] for large dams as now averaging about \$120 million per cubic kilometer of reservoir capacity (about \$150 per acre-foot). Unit costs for smaller dam and power plant projects may be assumed to be greater, likely nearer \$150 million to \$300 million per cubic kilometer (\$185 to \$370 per acre-foot).

Environmental control costs are typically a small fraction of basic project costs. Estimated bands of the more significant and frequently encountered environmental costs are shown in Table 8.5. In many project cost estimates, budgeted amounts for environmental control are difficult or impossible to distinguish from conventional pricing for labor, materials, and equipment. For example, costs of new hydroelectric projects generally include standard entries having environmental implications such as clearing and grubbing; water diversion and care; fish-handling facilities; drilling, grouting, and water-pressure testing of the foundation; preliminary and final cleanup; application of topsoil and mulches; revegetation of disturbed slopes and landscaping; and perimeter fencing. Also, cost estimates for such construction projects customarily include labor for activities such as inspections, safety monitoring, and spill prevention.

TABLE 8.5 Estimated Costs of Environmental Control

Capital outlay	Percent of conventional capital cost of project
Sediment treatment facilities during construction	1–3
Fish-passing facilities (ladders, skimmers, traps, transport vehicles or systems)	10–25
Treatment plant for washwater reclamation or recycling	1–2
Temporary erosion control, before final landscaping	<1
Baseline environmental investigations	<1
Reregulating dam and reservoir (assuming no installed hydropower generation units)	15–25

In addition to the capital outlays required for project environmental control (Table 8.5), certain operating costs may be ascribed to environmental preservation. These annual costs persist for the life of the project. Depending on the type and degree of river-basin management (coordinated operating policies and agreements and regulatory requirements), a significant operating cost could be power lost due to water releases from the reservoir to maintain downstream flow. Nonpower releases may be required to aid migration of downstream-bound fish (to achieve adequate carrying velocities), to sustain minimum discharges or water quality, to provide subsequent flood storage in the reservoir, or to benefit downstream navigation and recreation. Also, should the project require a navigation-lock facility, each lock-operating cycle represents loss of stored water, and therefore, power generation losses if no turbine-generators are installed in the lock filling-emptying system. The same applies to fish ladders, which generally operate continuously.

Another annual operating cost is associated with scheduled environmental monitoring and analysis throughout the project lifetime. However, annual costs for tracking the project environment would not be expected to exceed about 10 percent of the original cost of acquiring baseline environmental information that would have been incurred at the outset of the project.

On the other hand, many possible environmental benefits accrue from sound system operation and management practices that require minimal or no monetary disbursements.

8.7 CHECKLIST OF ENVIRONMENTAL INFORMATION

Fundamental types of information customarily acquired and analyzed in environmental impact assessments are summarized in the checklist, Table 8.6. These data represent, for small-scale projects and sites, about the minimum likely to be required. Much of the information could probably be found in publicly documented sources (maps, reports, brochures, government records) as well as through contacts (local universities, agencies, societies, associations,

TABLE 8.6 Suggested Environmental Analysis Information

1. Large-scale topographic maps [e.g., 1:25,000 with 2-ft (0.6 m) contour interval], plus reservoir hydrographic soundings
2. Detailed aerial photographs of candidate site and watershed
3. Multiyear stream discharges (e.g., preferably at least a 10-year record of average daily flows or average weekly flows, supplemented by minima and maxima)*
4. Multiyear precipitation (e.g., average daily or monthly, supplemented by peak storm intensities)
5. Rated design discharge of hydraulic turbines and planned locations of intake and discharge structures
6. Planned variation in power pool elevation—filling and drawdown
7. Competing uses of water in the stream basin (e.g., identification of intakes, outfalls, and flow rates; also, quality of returning flows)
8. Existence of environmentally sensitive wetlands in the planned reservoir area or downstream, the latter potentially subject to periodic draining or to filling with sediment
9. Existence of alluvial floodplains along the river valley which would no longer be seasonally flooded as a result of flow regulation by the project
10. Maps of present and future land use in the development area, and ultimate developmental plans
11. Important historical, cultural, archeological, geological, or soil resources that would be lost by construction or flooding; location of nearby protected areas (wilderness, preserve, sanctuary, refuge, dedicated area)
12. Special sensitivities of the stream waters and of the general area, considering the ecological life zones in the watershed—includes fish spawning, feeding, rearing, migration routes; aquatic and land species that are commercially or recreationally important, rare, or endangered; species that are potentially affected by project construction and/or operation
13. Identification of waterborne diseases or parasites endemic to the area and which could be spread or accentuated by project development
14. Local aquatic plant species that could interfere with hydroelectric production or other planned water needs
15. Characterization and possible quantification of ambient environmental stresses (natural or induced) in the area
16. Temporal quality of the stream, patricularly temperature and suspended sediment load, but also available parameters, such as dissolved solids, salt, nutrients, dissolved oxygen levels, toxic materials, and disease organisms
17. Identification of unique project design features (e.g., interbasin diversions, unusual construction techniques, extraordinary reservoir volumes) which might necessitate special consideration in evaluations of environmental impacts, including means for either their possible promotion or their mitigation and avoidance

*This information is not only important for environmental appraisals, such data are mandatory for facility sizing and analyses of projected economics.

local property owners). Other information may originate from site reconnaissance, inspections, and some on-site monitoring, testing and sampling.

Based on these types of data, environmental impact assessments, frequently involving experienced judgments, must be made. Such assessments hinge on weighing the total costs of measures to prevent and mitigate environmental damage, plus societal costs of any irreversible impacts, against stream-basin-wide environmental benefits, plus the benefits of hydroelectric production.

8.8 ENVIRONMENTAL ASSESSMENT TECHNIQUES

A diversity of methods for assessing the overall impact of a hydroelectric project on the environment are available to the project planner.[5] Choice of a particular evaluation technique or a combination of techniques hinges on factors such as data availability, analytical staff experience and preferences, available time and funds, access to laboratory and field equipment, and availability of computers and computer programs. Five assessment methods, described in Ref. 5, are summarized.

Field Surveys

Field surveys are a commonly used technique to provide information and analyses for project environmental assessments, as well as for the preparation of engineering structural designs and specifications. Since much of the data are essential anyway for design, construction, and operational planning, and since environmental assessment requires little additional or exclusive data, this is perhaps the most frequently used approach.

Data from preliminary surveys range from (1) findings obtained in brief site inspections or reconnaissance (by one or two technical people with multidisciplinary backgrounds and experience) to (2) in-depth field investigations, testing, and monitoring (involving teams of experts in collecting and analyzing data). Categories of site-exploratory data typically required for planning and assessments include: physical and demographic, engineering, geological, hydrological, environmental, and institutional. Example items within these categories are conveyed in the data for permit applications (Sec. 8.5) and in the summary of suggested environmental information (Table 8.6, Sec. 8.7).

Case Studies

The case study approach generalizes or extends environmental causes and effects from known defined cases to unknown and uninvestigated cases. It is based on the existence of similarity between projects and their operating conditions. However, near-equivalency of cases frequently is difficult to establish and may remain open to considerable question.

Geography may be a basis for assumed similarity; i.e., Project A is said to correspond with Project B because both are situated on streams in the same region, or possibly both are situated in the same river basin. Another criterion for similarity between existent and future cases might be corresponding size of a dam of a given type, e.g., gravity overflow dams, each about 15 m (49 ft) high from foundation to spillway crest. Still another criterion might be the mode of operation, both, for example, may be comparable run-of-river projects.

The "run-of-river" designation relates to small-volume reservoirs that can be drawn down in about a day or less at the average rate of inflow of the stream.

The objective in each case is to extrapolate from known cases to unknown cases. Generally, considerable information would have been acquired for the predecessor case (from investigations and reconnaissance) and reported in previously conducted technology assessments with which the new project is to be compared. For improved assurance that the cases fall in comparable categories, the analyst may employ multiple similarity criteria—the more areas of similarity between projects, the better the match. Furthermore, there is more confidence in results obtained for well-matched projects.

Modeling

While flow parameters can be investigated in physical scale models of aquatic systems, the predominant simulations of the broader aspects of hydraulics (water quality, operations, and ecosystem stresses) are investigated by means of *mathematical models.* Such models incorporate representative functions, should such exist. Useful model results hinge on accurate tested models and the availability of ample data obtained from existing sources and field investigations, as previously described. Verification of the model with known check conditions is an essential part of employing this technique.

Should a new mathematical model require development to assess a given project, or should uncertainty exist as to the applicability of an available model, considerable time and expense may be involved in model validation, possibly requiring extensive field test data in the process.

Description

From the dual standpoints of least cost and time required for environmental assessments, this technique—involving preparation of narrative descriptions of cause and effect by individuals knowledgeable in this type of data acquisition and appraisal—is certainly much utilized and advantageous. Since the technique is judgmental, the nonquantitative results hinge on the thoroughness and experience of the technical analyst(s). This approach, with care in selecting experienced staff, may be entirely warranted for many low-cost small-scale hydropower projects.

Delphi

Similar to the preceding "Description" technique, the Delphi method possesses advantages of allowing rapid evaluation when limited site information is available. Delphi techniques comprise intensive discussion by a team of experts formed into a working group. Then the group votes to determine the best estimate of their combined professional judgment regarding the issues being addressed. The result sought is a consensus of specialists. Success depends heavily on a dynamic moderator and a good discussion format.

8.9 REFERENCES

1. G. F. Cada, Oak Ridge National Laboratory, and F. Zadroga, Universidad Nacional, Heredina, Costa Rica, "Environmental Issues and Site Selection Criteria for Small Hydro Power Projects in Developing Countries" (draft), Final Report to National Rural Electric Cooperative Association, Washington, D.C., September 1980.
2. R. K. Linsley and J. B. Franzini, *Water-Resources Engineering,* McGraw-Hill, New York, 1979.
3. R. T. Oglesby, C. A. Carlson, and J. A. McCann (eds.), *River Ecology and Man,* Academic Press, New York, 1972.
4. R. P. Ambroggi, "Water," *Scientific American,* September 1980, pp. 101–116.
5. D. R. Zoellner, National Rural Electric Cooperative Association, International Programs Division, "Environmental Assessment of Small Scale Hydropower." Paper presented at Conference on Small Hydro-electric Power Plants, Quito, Ecuador, August 19–21, 1980.

9

The Systems of Regulation of Hydroelectric Power in the United States

Peter W. Brown

This chapter will describe the legal, regulatory, and institutional context of the development of hydropower in the United States. It will also discuss important policy trends which have, and which the author thinks will have, a significant bearing on hydroelectric development.

9.1 HYDROPOWER—A DIVERSE AND DISPERSED RESOURCE

The potential hydropower resources in the United States have varying characteristics. For purposes of regulatory and institutional analysis the important characteristics of the resource are the ownership of the sites, likely developers of the sites and potential environmental impacts of hydroelectric power development at the sites. This chapter examines the regulatory problems associated with the expansion of existing small-scale hydroelectric capacity, rehabilitation of existing sites, development of small-scale projects at new sites, and installation of hydropower capacity in conduits and other constructed water diversion structures.

Developers of these sites are diverse and include the U.S. Army Corps of Engineers (Corps), Bureau of Reclamation (Burec), the Tennessee Valley Authority (TVA), public and investor owned utilities, state power authorities, municipalities, entrepreneurs, and manufacturing or industrial concerns with access to a site.

Ownership is similarly diverse. In the west, federal land and site ownership is important. In the east many owners acquired sites at little or no cost when the sites were abandoned by local electric utilities. Environmental impacts of hydropower development will be the most serious at large new sites, including pumped storage projects, and the least serious at conduit or other constructed diversion sites. While site-specific environmental data are always important and broad assumptions concerning environmental impacts are dangerous, it is reasonable to suppose that there will be some, but not significant, environmental impacts resulting from expansion of existing sites and development of small, new sites.

According to available inventories and studies there are, with the exception of Alaska, few feasible, large-scale sites in the United States. The regions with the largest potential capacity at existing sites are the northeastern United States and the Pacific northwest. There is also considerable potential in the southeastern and Rocky Mountain states.[1]

9.2 THE LEGAL, REGULATORY, AND INSTITUTIONAL CONTEXT FOR HYDROPOWER DEVELOPMENT

Although the small-scale hydropower resource in the United States is diverse and dispersed, common features of development can be identified: (1) the problem of resource allocation, (2) the problem of external costs (environmental regulation), (3) the problem of the market for the output, and (4) the problem of the external benefits (public subsidies).[2] The legal, regulatory, and institutional system in which hydropower is developed will be examined in the light of these common features.

Every proposed development must initially decide to allocate resources (land and water, capital, etc.) to the development of a hydropower project. Given the extensive system of environmental regulation in this country, every project must examine its environmental impacts (external costs) and will be required to internalize all or a portion of those costs. Each development, to assure its economic feasibility, will be required to determine the market for its output. Because hydropower projects are perceived as increasing this country's energy independence, improving its national security, providing recreational opportunities and flood-control benefits, there have been public expenditures (subsidies) in support of development in general and of specific projects. The availability of these subsidies is obviously important to development of the resource.

Allocation of Resources of Hydropower Development

In the U.S. legal system property interests are defined by rules established, for the most part, by the states. The exception to this general proposition involves the definition and determination of property interests of the federal government and the Indian tribes. In the case of the federal government there is a developing body of law concerning federal water rights.[3] There have also been very recent attempts to define the property interests of the federal government in several western states.[4] Similarly, there is a developing body of law concerning Indian land and water rights, and recent court decisions have significantly extended those rights.[5]

For purposes of hydroelectric development, there are in place state systems of property law which may be used to determine ownership of a site and the necessary water rights. With respect to water rights, however, two different legal systems have been developed.

Riparianism, the first system, is utilized principally by states east of the Mississippi. Under riparianism *riparian owners,* persons who own land bordering a stream, pond, or river, have the right to use the reasonable flow of the waters flowing past the property. A reasonable use is hydroelectric power generation. Out-of-stream diversions are not permitted under the riparian doctrine unless a substantial portion of the water diverted is returned to the stream.

Prior appropriation, the second system, is utilized principally in states west of the Mississippi. Under this system, a state agency issues a permit or a certificate to a person who wishes to use or consume *(appropriate)* a quantity of water for some beneficial use. The categories of beneficial use vary from state to state but generally include domestic uses, agricultural uses, and mining uses. The appropriation certificate or permit in many states requires that water be diverted from the stream and *consumed* by the appropriator. The rights to the water are subject to the rights of persons who have perfected their water rights earlier in time. Failure to *use* the quantity of water appropriated for the purpose prescribed in the permit will result in a forfeiture of the water right.[6]

In the east, the system of water law does not appear to pose obstacles for development of hydropower. A developer may acquire the requisite property interest, including water rights, by purchase and utilize those water rights for hydropower production.

The system of water law in the west would not, at first blush, appear to pose an obstacle to development. Although the water used for hydropower production is an *in-stream use,* in almost all states an appropriation certificate will be necessary. In these states, water rights for hydropower production may be subject to a number of prior uses but they nevertheless can be obtained through a complex administrative process.

Under either system of water law, hydropower projects which seek to store large quantities of water for release through turbines at times of system peak load, will encounter difficulty. Under the riparian system, *store and release* may be an unreasonable use of the flow of water. Under the appropriation system, store-and-release projects may conflict with existing water allocation schemes.

As noted earlier, considerable new law is being made in the area of federal and Indian property law and, most particularly, water rights. This emerging body of law will affect water law systems of the western states to a considerable extent. The federal government, either through its power to regulate interstate commerce or in its capacity as a proprietor of large amounts of property in the west, is seeking to regulate minimum stream flows notwithstanding existing appropriation certificates permitting a *consumptive use*. Indian water rights, which have consistently been recognized by the courts, may grant to Indian tribes, because of their sovereign status and treaties with the United States government, rights to water which will diminish substantially or render worthless water rights recognized by state appropriation systems. To a great extent the quantity of water owned by Indian tribes is unknown at this time. The uncertainties created by the assertion of federal water interests and recognition of Indian water rights are obvious problems for hydropower development.

Given this body of property law, one would assume that decisions to use a particular site and water rights to develop a hydropower project would be made by persons who were willing to "purchase" the necessary property interests. If the price asked for the property were too high, given the risk and expected return to the hydropower project, presumably the hydropower project would not be built and there would be higher valued uses to which the property would be put. Ordinarily, markets tend to operate in this manner and society, for the most part, accepts the choice made by the market to allocate a resource for one purpose over another.

With respect to hydropower development, the decision to allocate the resource is not as simple as the market choice and, in fact, the market is not permitted to make the choice. Interference with the market decision in the case of hydropower development is a product of a long and contentious history of federal hydropower regulation and development, federal land ownership (especially in the west), and the public benefits *(public goods)* aspects of hydropower development.[7] In recent years, with the development of a complex system of environmental regulation at the state and federal level, individuals and groups have also attempted, through litigation or legislation, to influence the choice to develop hydropower resources.

The Federal Power Act (FPA)[8] passed in 1920 after a bitter struggle between conservationists and private development advocates, in major respects determines *whether* a hydropower project will be built and *who* will build it. Under the provisions of this act, the Federal Energy Regulatory Commission (FERC) is given the authority to determine whether any particular site will be developed by *nonfederal developers* and which of a group of potential developers will be chosen. FERC and its predecessor, the Federal Power Commission, has developed a complex system of permitting and licensing for nonfederal development of virtually all sites in the country. At the present time FERC has jurisdiction to issue a preliminary permit, a minor project license for a project of 1.5 MW or less, a major project license at an existing site, a major project license at a new site, an exemption for certain conduit hydropower facilities, an exemption for hydropower facilities at existing sites of 5 MW or less, and a new license upon expiration of a preexisting license for a project.[9] With the exception of the exemptions for conduit hydro projects and existing sites of 5 MW or less, state and municipal entities, assuming their plans for development are equally well adapted to the comprehensive development of the water resource, are granted a preference for a preliminary permit or license.[10] Any developer selected must develop a project which is best adapted to the comprehensive development of the water resource.[11] This requirement imposes responsibility on the developer to provide for recreational uses of the site, mitigate environmental impacts, assure dam safety, provide fish passageways or other fish protection or enhancement facilities where necessary, and maximize capacity and output in the context of the physical and environmental characteristics of the site.

The process pursuant to which FERC issues a permit or license is open to intervention by interested groups and individuals and must be coordinated with other federal and state agencies. In the event that the developer selected for the license does not own the requisite property interest to construct and operate the project, that licensee-developer is empowered under FPA to exercise eminent domain powers.[12]

In the event that federal lands or other property are utilized in developing or operating the project, FERC is empowered to issue permission to the licensee-developer to use those federal lands. The powers and requirements of the FPA and FERC also extend to federally owned structures and impoundments capable of generating electricity. FERC may issue a permit or a license to a nonfederal developer to study or develop a site owned by the federal government and under the management of a federal agency.[13] The principal federal agencies which maintain sites on behalf of the federal government are the Corps, Burec, and the TVA. However, if a site owned by the federal government is designated by Congress for study as a potential hydropower project, FERC will not issue a license.[14] Moreover, FERC, on its own motion, may refrain from issuing a permit or license on a federally owned site and recommend that Congress authorize a federal development agency to study the hydropower potential of the site.[15]

The extraordinary powers contained in FPA and conferred on FERC, obviously create a system whereby FERC, through its permitting and licensing process, substitutes for market choices favoring development. As noted earlier, because of the pervasive jurisdiction of FERC to regulate interstate commerce, virtually every site in the United States in which there is interest by nonfederal developers is subject to FERC's allocation rules and decisions.

Recently, through powers granted to FERC under Title II of the Public Utility Regulatory Policies Act (PURPA)[16] and Section 408 of the Energy

Security Act[17] to exempt conduit hydro projects of 15 MW or less and projects at existing sites of 5 MW or less, FERC may choose to exempt certain sites or classes of sites from the licensing process. The effect of such exemptions will be to remit the choice of whether to develop a site or sites subject to an exemption to the market, i.e., to nonfederal developers who have acquired the necessary ownership of the sites under preexisting state property law and have made the investment decision.

It is submitted that the permitting and licensing system created by the FPA and administered by FERC is premised on the view that hydropower sites are "public" resources[18] and have so many of the attributes of public goods[19] to warrant allocation of the resource by the government rather than the market. The consequences which flow from this view are a complex licensing and permitting system which is accessible by a variety of interested parties. There is no question that the system is a burden on developers and invites conflict. However, there is also no question that FERC takes its responsibilities seriously and invariably seeks to assure that the hydropower project confers additional benefits on the public in the form of recreational opportunities, environmental enhancement, and public safety.

As noted earlier, FERC is authorized under FPA to designate hydropower sites for study and recommend to Congress that it authorize the study. This authorization is part of another complex system which operates to allocate hydropower resources and, accordingly, substitute for market choice. This additional complex system is the federal water projects development process which has evolved over the entire history of the United States. In the past, Congress has authorized hundreds of water projects, some of which have included hydroelectric projects. In the 1930s, Congress also established TVA,[20] a major function of which was and is to construct and operate various water projects including hydroelectric projects in the Tennessee Valley. Congress, acting directly or through TVA, in all instances which involve federal construction and operation of water projects, has made the choice to devote federal capital to construction of these projects. Again, as with the premises of FPA, the natural resources associated with water projects have historically been viewed as *public resources* or have important attributes of *public goods* so as to warrant federal (i.e., governmental) development.

As noted above in the discussion of the resource-allocation system of FPA and the FERC licensing process, a designation of a natural resource as a *public resource* immediately implies that the *public* has a right to determine how that resource should be used and that state property law systems which have heretofore allocated the resource will be superseded by use of the eminent domain powers of the United States.

Theoretically, the public is represented by the Congress which enacts legislation which authorizes or fails to authorize a particular water project. Theoretically the *public,* in the case of TVA, is represented by the Board of Directors of TVA who are appointed by the President with the advice and consent of the Senate. However, as surrogates for the public, Congress and the TVA Board have been viewed as inadequate in some quarters. Other critics have observed that regardless of whether Congress and the TVA Board truly represent the public, the public interest is composed of such divergent interests as to require greater public participation in the process. What has emerged from this debate and other conflicts surrounding federal water project development is a complex authorization process.

Water projects directly authorized by Congress typically proceed through three phases of administrative and legislative review.[21] Each phase will involve a different federal budget cycle and enactment of separate legislation. The first phase of any project involves a request by a federal development agency, typically the Corps or Burec, to study a site. The request for authorization to study will be referred to the appropriate House and Senate committees. An authorization for funding the study will be included in the legislation. In all probability the authorization for study will become part of an "omnibus" water projects bill which will contain various authorizations for study, design, or construction for a number of water projects. If the legislation is passed and funds are appropriated for the study, the development agency proceeds with the study.

The study is conducted pursuant to the *Principles, Standards,* and *Procedures* promulgated by the Water Resources Council (WRC) of the United States.[22] These regulations have been revised recently and most notably to require careful consideration of nonstructural alternatives to the project. The major result of the study is a *benefit-cost analysis* which, if greater than 1, prompts preparation of a draft *environmental impact statement* (EIS).[23] The draft EIS examines in detail the environmental impacts of the project and is circulated for review by interested state and federal agencies and the public. Upon receiving comments on the draft EIS, the study agency prepares a final EIS and submits to Congress a request for authorization to undertake advanced design and engineering.

The Congressional process is repeated for the authorization for advanced design. If design is authorized and funds are appropriated, the development agency proceeds to design the project pursuant to the WRC *Principles, Standards,* and *Procedures* and its own design requirements. Depending on the length of time and changes in circumstances between preparation of the final EIS at the study phase and development of the project design, the developing agency may prepare another draft EIS or a supplemental draft EIS and circulate it for comment. Upon receiving comments, a final EIS will be prepared and submitted to Congress along with the request to authorize construction.

The Congressional process is repeated for authorization of construction. If construction is authorized and funds are appropriated, the development agency, in all probability, will proceed to prepare a supplemental or second draft EIS. The draft EIS will be circulated for comment and upon receiving comments a final EIS will be prepared. Construction will commence and, assuming sufficient subsequent appropriations from Congress during the construction period and no litigation resulting in court orders staying construction, the project will be completed.

Study, design, and construction of TVA projects proceed under the WRC *Principles, Standards,* and *Procedures* and will likely involve the preparation and circulation of a draft and final EIS. The major difference in the development process for TVA projects is, of course, the absence of the requirement of congressional legislation to authorize various phases of the project. Moreover, because TVA has access to its own funds supported by revenues generated by supplying electrical services to its service territory, TVA does not need an appropriation from Congress to construct a project. There is, of course, internal review of a project within TVA and by the Board of TVA.

The foregoing discussion is a simplified and somewhat stylized description of a complex, costly, and extremely lengthy process of direct congressional authorization of federal projects. The TVA process, although obviously less time consuming, is still complex. With respect to the direct congressional authorization process, public participation occurs during the commenting period for at least two draft EISs and, in the form of lobbying, with the Congress during the legislative process. Ultimately, as will be discussed more fully later in connection with the discussion on environmental regulation, the public and public interest groups may force participation through litigation over the adequacy of the impact statement or some other provision of environmental law or regulation. While this litigation may concern immediately the adequacy of an EIS or a question concerning a particular provision of one of the several environmental regulatory statutes applicable to federal water projects, it may also be the most effective way for a public interest group to insist that it have a say in the decision to build the project or to prevent its construction.

It is submitted that, although the detailed procedures and steps to be followed for development of nonfederal and federal projects are very different and the institutions, FERC, Congress, the Corps, Burec, and TVA, are all different, there are a number of common traits of each process which are particularly useful for policy analysis:

1. Each process allocates a resource and is premised on the view that the resource allocated is a *public resource* or has attributes of a *public good.*
2. The process which results in the decision is open to public participation.
3. The process of deciding to develop a project invites conflict. If one accepts the premise that the natural resources of a particular site are *public resources,* one can hardly deny individuals and groups the right to insist that their views on how "their" resources should be used are the correct views on a particular project.
4. The processes ultimately disregard state property law systems which have allocated the resources in the first place. Presumably, these state systems

have developed over a period of time and are designed to meet the special circumstances of the particular site.

Treatment of External Costs—Environmental Regulation and Hydropower Development

In the last several years, major pieces of legislation enacted at the federal and state levels have attempted to identify and mitigate the environmental impacts of a variety of human activities. At the risk of oversimplifying the main purposes and effects of this legislation, it is submitted that this legislation, the institutions and agencies which the legislation creates, and regulations emanating from this legislation perform three principal functions or are based on three principal premises:

1. That economic markets tend to ignore the external costs of a particular economic activity with the effect that third parties who are not participants in the market transaction bear these costs. Microeconomic theory views the failure of the market to internalize external costs as resulting in a subsidy by the affected members of society.[24]
2. That there are certain values inherent in the natural environment which are absolute and any attempt to mitigate the harmful effects of a project or activity on these values will be unsuccessful.
3. That the persons who will be affected by a particular activity but who are not parties to the market transaction giving rise to the activity or project should have the right to participate in the decision-making process and, under certain circumstances, compel changes in the activity or project or be able to prevent the activity or project from going forward altogether.

With respect to the attempt to regulate the external costs of a particular project or activity, the National Environmental Policy Act of 1969, the Clean Water Act of 1972 and its amendments of 1977, the Fish and Wildlife Coordination Act, the Anadromous Fish Conservation Act, the Historic Preservation Act, the Federal Land Policies Management Act of 1976,[25] and state legislation, which in many states has replicated this federal legislation, are designed to identify and regulate the externalities of particular projects and activities.

The Wilderness Act, the Wild and Scenic Rivers Act, and the Endangered Species Act[26] are examples of legislation which establish as absolute values the esthetic and natural environments of wilderness areas, certain rivers, and the continued existence of certain animal and plant species. Projects or activities which intrude on the natural and esthetic environments of wilderness areas or wild and scenic rivers or endanger the continued existence of certain animal and plant species will be prohibited altogether.

The Federal Administrative Procedure Act,[27] FERC regulations, the statutes described, and judicial decisions[28] have provided opportunities for participation by individuals and groups in the decision-making process of federal and state administrative agencies, federal development agencies, and other nonfederal private and public entities. Direct participation in the political process of the Congress of the United States and the state legislatures is, of course, also available to individuals and groups.

Again, at the risk of some oversimplification, our system of regulation has, for the most part, consigned the responsibility of regulating and internalizing external costs to state and federal administrative agencies. For example, the Council on Environmental Quality regulates the environmental impact statement and environmental assessment process by which externalities are to be identified and mitigation alternatives examined.[29] The Environmental Protection Agency of the United States government (or, in the case of a delegation of authority, a state water quality agency which has the responsibility of administering the Clean Water Act), assures that adverse environmental impacts on water quality and water supply are mitigated by project developers. The Fish and Wildlife Services, the Heritage Conservation and Recreation Service, and the Bureau of Land Management all have the responsibilities of assuring that adverse environmental impacts on fish species, historic places and sites, natural areas, and federal lands held in federal trust are mitigated or eliminated. Value judgments are made initially by state and federal legislatures and are represented by the Wild and Scenic Rivers Act and the Endangered Species Act. Of course, administrative agencies responsible for administering environmental legislation make more subtle value judgments with respect to individual and classes of projects.

The Congress and state legislatures are institutions which must deal with public participation in their processes. Administrative agencies are also required to provide opportunities for participation under various administrative procedure acts enacted at the federal and state levels and under the Constitution of the United States. Various statutes conferring jurisdiction on the federal courts and similar bodies of law in the states involve courts in the decision-making process.

The system of laws and regulations discussed, and the institutions created to develop and administer the system, all have a substantial bearing on the development of hydropower projects. With respect to a particular hydropower project, the impact of the system of environmental regulation and its institutions will have a greater or lesser effect on the development depending upon two characteristics of the project. (1) If the project has been identified as imposing large external costs (significant adverse environmental impacts), the system of environmental regulation will subject that project to intense scrutiny and elaborate and lengthy administrative processes. (2) If the project is one with respect to which there is intense local opposition or public objection, the project will also be subjected to intense scrutiny and lengthy administrative processes by the institutions charged with responsibilities of environmental regulation. Lesser environmental impacts and strong local or public support will tend to reduce administrative scrutiny and procedural delay. Although there are some important procedural differences in how the system of environmental regulation bears on nonfederal development and federal development, the system asks identical questions of each group of developers.

As discussed in the preceding subsection, FERC has jurisdiction over virtually every hydroelectric project to be developed by nonfederal developers. In addition to authorizing FERC to determine whether a hydro project is to be built and who builds it, the FPA requires FERC to engage in extensive environmental review of license applications and to coordinate carefully its environmental review with other federal and state environmental protection agencies.

Since issuance of a federal license or an exemption is a federal action which may have an impact on the human environment, FERC at the very least is required to perform an environmental assessment in conjunction with the issuance of any license or exemption. This requirement is imposed by the National Environmental Policy Act of 1969 (NEPA). As a general rule, FERC will undertake an environmental assessment with respect to minor project licenses, major project licenses at existing sites, conduit hydro exemptions, and exemptions of sites of 5 MW or less. Upon preparing the assessment, FERC, unless unusual circumstances are presented in the assessment itself, will prepare a Finding of No Significant Impact (FONSI).

With respect to major project licenses at new sites and licenses and exemptions of projects with unusual environmental circumstances, FERC will prepare a draft environmental impact statement (EIS) and circulate the draft for comment. Upon receiving comments FERC will prepare a final EIS prior to the issuance of the license.[30] As noted in the previous subsection, development of hydropower projects by federal development agencies also involves NEPA and the preparation of an EIS. Most certainly during the study phase of a hydropower project, the Corps or Burec will prepare a draft EIS, submit the draft for comment, and prepare a final EIS. The final EIS will be submitted to Congress by the Corps or Burec along with the request for authorization of advanced design or engineering.[31] As noted earlier also, supplemental or additional EISs may be prepared during the subsequent phases of the process. TVA, the other major federal hydropower development agency, most likely will prepare a draft EIS and a final EIS in conjunction with any of its hydropower projects.

The EIS process under NEPA is designed to compel federal development agencies and federal regulatory agencies to identify external costs and environmental impacts of proposed projects. As such the effects of the statute are procedural, and the statute and the regulations cannot compel mitigation activities for individual projects. The EIS process is one that is open and accessible to state and federal agencies and the interested public.

Since hydropower development may affect water quality, a number of provisions of the Clean Water Act of 1972 and its amendments of 1977 are invoked.[32] FERC will require each license applicant to obtain a Section 401 water quality certificate from the authorized state water quality agency or, in the absence of an authorized state water quality agency, from EPA. Federal development agencies, if the project is located in a state to which authority to issue 401 water quality certificates has been delegated, must obtain the water quality certificate from the state water quality agency.[33] For the most part water quality certificates prescribe minimum stream flows which must be maintained in the stream on which the impoundment structure is located.

Section 404 of the Clean Water Act[34] confers jurisdiction on the Corps to regulate the dredging and discharge of fill materials into waters of the United States. FERC requires that every license applicant planning a project involving dredge and fill activities obtain a Section 404 permit from the Corps. Federal development agencies in their construction of hydropower projects are bound by the standards implicit in Section 404 and the standards and requirements imposed by the Corps as conditions of its permits. Obviously the Corps does not require a Section 404 permit from itself for a hydropower project that it is constructing. Similarly, Burec is not required to obtain a 404 permit from the Corps. Pursuant to a memorandum of understanding between the Corps and Burec, Burec, upon consultation with the Corps, will agree to abide by the terms and conditions of the Corps with respect to dredge and disposition of fill material. TVA is also not required to obtain a 404 permit from the Corps but agrees with the Corps to abide by the requirements imposed by the Corps on dredge and fill material.

The two most important pieces of federal legislation designed specifically to protect plant and animal species are the Fish and Wildlife Coordination Act[35] and the Endangered Species Act.[36] Not surprisingly, both these pieces of legislation apply to hydropower development in the United States. Under the Fish and Wildlife Coordination Act, the Fish and Wildlife Service (FWS) of the Department of the Interior must be given the opportunity to comment on any FERC license or federal development project with the exception of a TVA project. Both FERC and the federal development agency must give careful consideration to the comments of FWS. It should be noted at this juncture that through memoranda of understanding between the Corps and FWS and between Burec and FWS, FWS has agreed to undertake certain environmental studies involving protection of animal, fish, and plant species which may be affected by a hydropower development project. Under these arrangements, funds are made available to FWS out of the funds appropriated to the Corps or Burec by Congress. On the other hand, in conjunction with the FERC licensing process, FWS is not funded to perform environmental studies. The comments of FWS and interest of FWS under the Fish and Wildlife Coordination Act have concerned maintenance of base streamflows and construction and maintenance of fish passageways or other fish protection or enhancement devices.

The Endangered Species Act, as noted earlier, is a statute which bars development if it is determined that the development will adversely affect an endangered plant or animal species. As a result of the Tellico Dam litigation,[37] Congress in 1979 amended the Endangered Species Act to authorize exemptions from the Act in certain narrowly prescribed circumstances. Notwithstanding, the narrow exemption provision of the amendements to the Endangered Species Act, the Endangered Species Act remains a bar to hydropower development in those instances where endangered and threatened species will be adversely affected.

FERC and federal development agencies must also examine a project from the standpoint of its impacts on historic and scenic places and areas. In this regard, several pieces of legislation have recently been passed by Congress designed to protect historic, scenic, and natural areas and places. The Historic Preservation Act[38] requires consultation with the Heritage Conservation Recreation Service of the Department of the Interior and the Advisory Council on Historic Preservation. Under this legislation a National Register is maintained of historic sites. Federal development agencies and FERC are required to take into account the effect of any project on a historic site included in or eligible for inclusion in the National Register.

Similar responsibilities are imposed on the federal development agencies and FERC under the National Wilderness Preservation System, the National Trail System, and the National Wildlife Refuge System legislation. As the titles of these statutes imply, they attempt to protect respectively wilderness areas of the United States, hiking trails systems within the United States, and wildlife refuges within the United States.

FERC and federal development agencies must confer with the Secretaries of the Departments of Agriculture or the Interior with respect to projects that may affect wilderness areas. Under the Wilderness Act, FERC no longer has the power to license a project in a wilderness area. Rather, the authority to authorize development of hydropower projects in wilderness areas has been transferred to the President.

Under the National Trail System[39] legislation, FERC and the federal development agencies must confer with the Secretaries of the Interior and Agriculture concerning projects that may affect trails established by those secretaries on lands within their respective jurisdictions.

FERC and the federal development agencies will be required to confer with FWS on any projects which may adversely affect a wildlife refuge maintained under the National Wildlife Refuge System.

Finally, the Wild and Scenic Rivers Act establishes a system whereby rivers or segments of rivers may be designated as either wild or scenic or recreational. Rivers or segments of rivers may be nominated for designation by federal and state agencies and private individuals and groups. Designation occurs by resolution of Congress. Designation of a river or a section of a river as wild or scenic prohibits the issuance of a FERC license for a hydropower project on that river.[40] Similarly, federal development agencies will be barred from developing sites on wild and scenic rivers. The status of development on recreational rivers (rivers which have already been impounded) is in doubt at this writing. FERC has taken the position that nomination of a river or a segment of a river as a recreational river will prohibit FERC from issuing a license on that river.

Because of the interest of local agencies, organizations, and groups and of state agencies in hydropower development and in environmental regulation, both FERC and the federal development agencies are required to follow a process of coordinating the licensing or development of a project with state and local agencies. Under the FERC licensing process, as a precondition for filing a license application, FERC will require a license applicant to consult with and seek approvals from interested state agencies.[41] These agencies will typically be state water quality agencies, state departments of natural resources with jurisdiction over fish and wildlife protection, state agencies concerned with dam safety, and state agencies concerned with historic preservation.

Federal development agencies are required to coordinate any federal hydropower project with state and local entities pursuant to the processes established by Part II A-95, of the OMB Circular A-95.[42] Under A-95 review the federal plans for development will be reviewed by state and local agencies. Comments received by the state and local agencies must be considered by the federal development agency and must accompany any final EIS prepared and submitted by the federal development agency.

The foregoing discussion illustrates the various concerns of state and federal legislatures as to the effects on the environment of various activities including hydropower development. There can be no question that the system of review described involves significant transaction costs and many different agencies. The environmental regulatory system described compels external costs to be internalized. These costs cannot be specified in the abstract. Most certainly, the layered review of hydropower projects by state and local agencies at the first level and the federal coordinating agencies at the second level duplicates effort and incurs "unnecessary" transaction costs. Moreover, as noted throughout the discussion of environmental regulation, the environmental regulatory system welcomes participation and, accordingly, invites conflict.

While it would be naive to suggest that the system of environmental regulation is designed and used exclusively to assure that any activity bears its external costs, it is helpful for purposes of policy analysis to view environmental regulation as properly fulfilling that function. If individuals and groups are consistently using the system of environmental regulation to challenge the initial decision to allocate natural resources to a particular activity, one may conclude that the *process* of making the decision in the first place is a poor one. If individuals and groups are using environmental statutes and environmental regulation to participate in the allocation decisions, then there needs to be some reform of the FERC licensing and federal development process whereby there can be effective participation in the decision to proceed or not to proceed with development and some finality to that decision.

When one focuses on environmental regulation as serving the principal purpose of requiring projects to bear external costs, it is easier to identify and describe some of the problems of the environmental regulatory system as it applies to hydropower projects. As discussed earlier, there is a significant risk, under the present environmental regulatory system, that the transaction costs of administering and complying with that system exceed the value of the external benefits associated with the project. These high transaction costs occur as a result of the plethora of agencies involved in the decision-making process of both federal and nonfederal developers, the potential conflict among agencies over jurisdiction, and the cost of obtaining the requisite information to comply with the complex environmental regulatory system.

The Marketing Conditions for the Output of Hydropower Projects

A few observations are in order here concerning the marketing of electric power from hydropower projects for two reasons.

First, the principal purpose of any hydropower development is to market electric power. To the extent that there are constraints on the marketing of power, there will be constraints on development. To the extent market constraints are eased, there will be an increase in hydropower development. As will be discussed briefly later, market constraints for small-scale nonfederal developers have eased considerably in recent years.

Second, the way power is marketed will influence substantially which type of developer, federal or nonfederal, develops the project. Under present law, power developed by a federal development agency is marketed by regional federal power marketing agencies which are required by law to transport power and to sell at low cost to certain preference customers, i.e., municipal, state, and rural electric cooperative entities.

The industry which generates, transmits, and distributes electrical power to persons and entities in the United States has been subject to pervasive economic regulation by state and federal regulatory agencies. Traditional economic theory, until recently, viewed the industry as a *natural monopoly*.[43] Under this theory the industry was perceived as an industry with continuing declining average costs over the long run. Under these circumstances, competition could not exist and one firm would come to dominate the market.

The legislative response to this phenomenon in the early part of this century, was to recognize the inevitable, i.e., the emergence of a dominant firm in the electric utility market, and to license that firm as a state franchised monopoly and to subject that firm to pervasive economic regulation. Under the system of economic regulation the firm would have the obligation to serve all customers in its franchise territory, could not abandon service without permission of the regulating agency, would have its prices reviewed by the regulatory agency, and would be limited to *reasonable* profits derived from its business. To administer this system of pervasive economic regulation, the states established public utility commissions. *Public utility commissions* are administrative agencies which administer the system of rate of return and price regulation of electric utilities doing business in their states.[44]

In 1935, Part II was appended to the Federal Power Act.[45] Part II authorized and required the Federal Power Commission (FPC) to regulate the interstate transmission and sale of electric power. What this legislation failed to make clear was the relationship of the FPC to state regulatory commissions when a utility subject to state regulation also was engaged in interstate sales and transmission of electrical power. For a period of 29 years the FPC and state agencies were uncertain about their relationship. In the *Colton* case,[46] the relationship was clarified; it was determined that the FPC had jurisdiction over investor-owned utilities selling electric power at wholesale, and state regulatory commissions had jurisdiction over electric utilities selling electric power at retail. Given that the bulk of revenues derived from the sale of electric power are derived from retail sales,[47] the most significant regulatory effort is by the state regulatory commissions. Moreover, under Part II of the FPA, states continue to regulate the siting and construction of nonnuclear facilities exclusive of the FERC and determine the extent of the regulated utilities' franchise territory.[48]

There are exceptions to the system of pervasive economic regulation by administrative agencies of electric utilities. While the investor-owned utilities have close to 80 percent of the retail sales of electric power in the nationwide electric power market, there are public entities and rural electric cooperatives which also sell electric power at retail in substantial quantities. Public utilities are state-authorized entities, political subdivisions or municipalities which are engaged in the electric utility business. Much like their private counterparts, they market power to a service territory defined by state law or regulation. Many states have chosen not to subject public utilities to regulation by the utility commission of the state but rather have chosen to leave the determination of rates to the political process.

In 1935, by executive order, the President of the United States established the Rural Electrification Administration (REA). Shortly after the establishment of REA, Congress enacted legislation confirming the establishment of REA and expanding its power.[49] REA is an agency within the Department of Agriculture which provides financing and other technical assistance to nonprofit cooperatives engaged in the distribution of electric power to their membership. As with public entities, many states have chosen not to subject cooperatives organized within their states to the regulation by the state regulatory commission.

The market relationship between investor-owned utilities and public and cooperative utilities is one of unequal market power. Investor-owned utilities control 80 percent of the generating capacity in the United States and 70 percent of the miles of high-voltage transmission lines within the United States.

There are certain sectors of the United States, such as the state of Washington, where the public utilities are dominant and have access to or control a substantial portion of the hydroelectric generation capacity in their regions. The relationship between the investor-owned utilities and the public and rural cooperatives is regulated by FERC, the sales from investor-owned utilities to public and cooperative utilities being wholesale sales of electricity in interstate commerce.

The public utilities and electric co-operatives have been critical of FERC regulation of wholesale rates. One criticism is that FERC permits the investor-owned utilities to practice price discrimination against public utilities and rural cooperatives.[50] This price discrimination takes the form of permitting the investor-owned utility to charge high wholesale rates to wholesale customers while, at the same time, state regulatory commissions which regulate the retail rates of that same investor-owned utility, will permit lower retail rates to the same classes of customers. Another criticism directed at FERC is that it does not subject requests for wholesale rate increases by investor-owned utilities to careful regulatory scrutiny. A third criticism is directed more at the system of federal regulation of the electric utility industry than at FERC. Public utilities and cooperatives may have access to electric power from remote sources. However, to obtain access to that power, the power must be shipped over transmission lines owned by an intervening investor-owned utility. Pursuant to a number of legal theories, public and cooperative utilities have sought to impose an obligation on the intervening investor-owned utility to "wheel" the power to the purchasing public or cooperative utility.[51] The position of the investor-owned utility industry and FERC has been that the Federal Power Act does not authorize FERC to order "wheeling" under these circumstances.

Viewed from the perspective of an economist, the electric utility industry of the United States is a highly concentrated industry in which considerable market power resides with the investor-owned utility portion of that industry. The regulatory system imposed on the industry by the administrative agencies charged with the regulation is complex and time consuming. Rate cases before state agencies often take 6 to 9 months to decide and involve voluminous exhibits and extensive testimony by experts. In recent years the economic literature has questioned the description of the electric utility industry as a *natural monopoly*.[52] This literature has pointed out that the industry, especially the generating portion of it, no longer appears to be a declining average cost industry. This literature further suggests that the generating portion of the industry be deregulated and that the transmission portion of the industry, which continues to bear natural monopoly characteristics, be regulated as a common carrier. As noted, the electric power market is not an open market which is accessible by persons wishing to purchase and sell electric power.

In an attempt to overcome the structural deficiencies of the electric power industry, Congress has established a number of institutions which directly benefit public and cooperative utilities and has enacted legislation which provides preferences for public and cooperative utilities. The institutions are the five federal power marketing agencies which are under the administration of the Department of Energy (DOE). These power marketing agencies provide transmission services for electric power generated at federally owned stations. The principal source of federally generated power is, of course, hydroelectric power from federal hydropower projects. The rates for the transmission service and for the power itself are set by DOE and are subject to review by FERC.

Under the Flood Control Act of 1944[53] and the Bonneville Project Act,[54] public utilities and electric cooperatives are granted a preference to the power marketed by the federal marketing agencies. In the event that cooperative and public utilities are unable to utilize the power generated at federal projects, the excess is sold to industrial and investor-owned utilities customers. Given the weaker market position of public utilities and cooperatives in most parts of the United States, the existence of the marketing agencies and the "public power" preference under the Flood Control Act of 1944 and the Bonneville Power Act are of vital interest to public and cooperative utilities. These institutions and policies suggest that public utilities and cooperatives will strongly support federal development of hydropower.

Until recently, access to the electric utility market was, for the most part, prohibited by virtue of the existence of monopoly enterprises and pervasive state and federal regulatory systems. Accordingly, hydropower development was confined by this market structure to private, public, and cooperative utilities, federal development agencies or, in rare instances, a manufacturing plant which had access to a nearby site. For the investor-owned utility, access to the market was easy since it controlled the market.

For the public utilities and cooperative utilities, if a site was located in their service territory, obviously there was a ready market for the power. If the site was not located in the service territory of the public or cooperative utility, then the public or cooperative utility confronted the difficulties of persuading the private utility to wheel power from the site to the public or cooperative utility's service territory. With respect to those sites owned and operated by manufacturing establishments, the manufacturing establishment would market the bulk of the power to itself with the excess being sold to the electric power system in its territory at "dump" (very cheap) power rates.[55] There was, however, little room for the private entrepreneur who sought to develop generating capability and sell the output to a customer or customers at fair rates.

Title II of PURPA, enacted in 1978,[56] contains the potential for significant change in the market structure of the electric utility industry in the United States. By that legislation, development of electric generation capacity by persons not engaged in the electric utility business is encouraged and developers of this capacity are guaranteed access to a market at fair rates. The access to the market provided by PURPA, however, is limited to certain forms of generation. Electrical capacity developed by cogeneration technology is guaranteed access to electric power markets. Generating capacity 80 MW or less in size and utilizing renewable energy sources will similarly be provided access to the market at fair rates.[57] Under FERC regulations promulgated pursuant to the requirements of Title II of PURPA, hydroelectric generating stations at existing and new sites qualify for the protections of Title II of PURPA.[58]

Since the enactment of Title II of PURPA, FERC has promulgated extensive regulations defining those entities and persons who qualify for the protection of the Act, establishing formulas and standards for the determination of the rates for the exchange of power between the small power producer and the electric utilities and exempting the small power producers from traditional forms of state and federal electric utility regulations.[59] Implementation of the FERC regulations is presently underway within the various states. It is too early to tell precisely what the effect of the legislation and the FERC regulations will be on the development of hydroelectric power at sites of 80 MW or less. However, there can be no question that the passage of the legislation and the promulgation of the regulations by FERC address significant market imperfection and make accessible a market which otherwise would not be accessible to *private* developers of electric generation.

The market structure of the electric utility in the United States and the recent development under Title II of PURPA are instructive to future hydropower development in the United States. Given the market structure of the industry, the federal power marketing agencies were a rational response to the problem of marketing federally developed power. The public power preference under the Flood Control Act of 1944 and the Bonneville Power Act can also be seen as the attempt by Congress to redress an imbalance in market relationships. The existence of the marketing agencies and the public power preference has stimulated support for federal hydropower development among public and cooperative utilities. It is likely that this support will continue even after the enactment of Title II of PURPA. PURPA provides no special benefits to public and cooperative utilities but rather makes electric power markets accessible to nonutility public and private developers.

Public Expenditures Favoring Hydropower Development

In the past, large expenditures of federal funds have been made to support hydropower development in the United States. These expenditures are best illustrated by federal development in the Columbia River Basin in the 1930s, the Boulder Dam, and development by TVA in the Tennessee Valley. More recently, provisions in the Internal Revenue Code[60] and programs authorized by Title IV of PURPA and the President's Rural Energy Initiatives[61] have committed public funds to support hydroelectric development or authorized tax expenditures favoring such developing. In terms of developing further public policies which commit public funds to support hydropower development, it is necessary to examine the rationale for these public expenditures.

As discussed earlier, there is a tradition in the United States that hydroelectric sites are *public resources* and should be owned, developed, or administered by public agencies. If one assumes that all hydroelectric sites are a public resource, then it obviously follows that public funds should be used to develop some if not all of these sites.

A second rationale is that hydropower sites are *public goods* or have attributes of public goods. Microeconomic theory teaches that public goods cannot be allocated by a market and, accordingly, must be allocated by the government. Most certainly, especially in the case where an impoundment provides recreational benefits and flood-control benefits, a dam or impoundment is a public good or has significant attributes of a public good.

A third rationale supporting public expenditures in favor of hydroelectric power development acknowledges that hydroelectric power development confers external benefits on society. Again, microeconomic theory holds that market transactions will fail to take into account the external benefits of a particular project or human activity. In the case where the external benefits are not taken into account, microeconomic theory holds also that the resources of society will be misallocated. To correct for the misallocation created by the market failure, it has been suggested that public funds and expenditures be made on behalf of the activity which creates the external benefits.[62]

In the case of hydroelectric power development there can be no question that such development represents exploitation of renewable indigenous energy resources of the United States. To the extent that these renewable resources are developed, the dependency of the United States on foreign oil is reduced. Furthermore, to the extent that dependency on foreign oil is associated with reduced national security, hydropower development will increase national security.[63]

At this point, one is tempted to categorize hydropower development as either development of a public resource, development of a public good, or development of a resource with extensive external benefits. No single categorization, however, is helpful to policy analysis and a single categorization will ignore the accuracies implicit in the other two categorizations.

As noted briefly earlier, there have been various types of public expenditures to foster the development of hydropower in the United States. Federal development of hydropower sites under various pieces of water resource development legislation involves appropriating funds raised substantially through federal taxation. Under water resources development procedures, the Congress of the United States traditionally appropriates the necessary funds for the study, design, and construction of Corps and Burec projects. Notwithstanding that there is a local or state contribution to the project from the locality or the state to be immediately benefited by the project and that the power output of the hydropower project is to be marketed on the basis of the "lowest cost consistent with sound business principles," there is a substantial commitment of public funds to such projects. Insofar as these projects provide pure public goods, i.e., flood control benefits and recreational benefits, it is highly probable that the federal government (or a state government) would be the only possible developer of the project and provider of these public goods.

With respect to the *hydropower* aspects of multipurpose federal water projects, it may be argued that there is a ready market for the development of that part of the project. The justification for federal development of the hydropower aspects of the project is, of course, that the hydropower project confers external benefits on society which would go unrecognized in any market transaction. If, however, public funds are already being spent to support hydropower development by the nonfederal sector, then there may be serious doubts concerning the commitment of federal funds to federal development.[64] Most recently, with burgeoning interest in small-scale hydroelectric development, there has been an increase in public expenditures supporting development of small-scale projects at existing sites. To the extent that public expenditures are supporting nonfederal development of small-scale projects at existing sites, expenditures of federal development funds for small-scale federal development may be unnecessary to capture the external benefits of small-scale hydropower and may even overcompensate such benefits.

The first commitment of public funds to small-scale hydroelectric development occurred with the passage of Title IV of PURPA.[65] Under provisions of that statute, low interest loans for feasility studies and licensing applications are made available by DOE to developers of hydroelectric projects at existing sites of 30 MW or less capacity.[66] Title IV of PURPA also contains a provision for low interest construction loans for hydroelectric projects at existing sites of 30 MW or less. However, funding for these programs has not been approved since 1981 and the programs have, for all intents and purposes, been dismantled. Similarly, the Rural Energy Initiatives have also been dismantled.[67]

The most significant commitment of public funds to small-scale hydroelectric development occurred with the passage of the Crude Oil Windfall Profit Tax Act of 1980 (COWPTA). Under the provisions of that Act, an 11 percent energy tax credit was made available to developers who would undertake to develop hydroelectric projects at existing sites of 125 MW or less. The full 11 percent tax credit was available for projects of 25 MW or less at existing sites and a declining percentage of the 11 percent energy tax credit was made available for projects between 25 and 125 MW. The Act also contained a provision which made fish ladders and other fish passageway investments at hydroelectric facilities eligible for the basic investment tax credit of 10 percent and the 11 percent energy tax credit. The energy tax credit under COWPTA is available for small-scale hydroelectric facilities through 1988. COWPTA also contained provisions which secured tax exempt status for publicly issued debt instruments to support small-scale hydroelectric development at sites owned by public entities and municipalities.

There is one other public expenditure, which is represented by the National Hydropower Study and a number of the activities of the Corps and DOE, which in effect, provides a public good. This public expenditure is designed to provide broadly disseminated information about all aspects of hydropower development to the community or persons interested in hydropower development in the United States. Broadly disseminated information about complex systems and processes is probably a classical public good. To reiterate, markets will fail to provide such a public good.

In the last 2 to 3 years several national and regional conferences have been held on various aspects of hydroelectric development. In addition, DOE has funded several consulting firms, established outreach programs in regional offices, and established model commercialization programs in two of its regional offices. Given the complexity of the licensing process, the system of environmental regulation, the systems of marketing the output of hydroelectric power, and the various financing assistance mechanisms available to hydroelectric developers, public dissemination of information concerning hydroelectric development is most certainly warranted. Whether the expenditures on this form of information transfer are adequate, inadequate, or more than adequate is unknown and probably unknowable.

In terms of development of large, new sites or new multipurpose sites, there is a strong justification for public expenditures in this regard. Presumably, these projects will be multipurpose projects in the tradition of federal water resource project development. Undoubtedly the projects will have flood-control and recreational purposes. These purposes will not be served by market transactions which focus solely on hydroelectric development.

With respect to existing sites, the development of which will involve large environmental impacts (external costs), public development may be appropriate for the reason the markets will not support projects which would have to incur the capital costs necessary to internalize the environmental impact. To the extent that the systems of environmental regulation presently in place over-regulate; i.e., they require the developer to bear the costs of mitigating environmental impacts to which society as a whole is indifferent, public development of these sites may be appropriate. It should be noted that the need for *public* development of large multipurpose sites does not dispose of the question of whether federal or regional, state, or local entities develop a large or environmentally intensive project. In this regard, given that there are in place federal agencies which have the expertise and experience in building such projects, *federal* development of large multistate or regional projects may be appropriate.

With respect to existing sites where the environmental impacts are not great, it would appear that development by market choices to invest would readily occur. There are in place the public expenditure programs of COWPTA to support nonfederal development.

9.3 POLICIES AND TRENDS IN AMERICAN WATER POLICY AND ENERGY DEVELOPMENT

In the preceding part of this chapter, hydroelectric development in the United States has been examined in the context of the systems of resource allocation, environmental regulation, market regulation of the output of hydro-electric plants, and public expenditures tending to favor hydropower development. From that discussion certain policies and trends in policies are discernible. However, this section will make explicit some of the major policy trends which will have an effect on and which have affected hydropower development in the United States. In light of the preceding discussions two trends in policy need only briefly be noted here.

The first trend is the apparent stalemate in large federal water projects and the lack of consensus concerning the role of federal development in large-scale water projects. In 1982, federally sponsored water projects were taking longer and longer periods of time to proceed from study through construction and operation. Documentation of the Corps and Burec indicates that water projects now take approximately 12 to 16 years and even longer from study through actual construction.

The second trend has been the emerging concern of the effect of human activity on the environment. This concern is perhaps prompted by greater understanding of environmental impacts fostered by increased technology and methods of communication. It is also prompted by a better understanding of how the market economy of the United States often fails to take into account the adverse effects of particular activities on persons not parties to the market transactions. To the extent that activities such as federal water development projects are perceived as developing *public resources,* individuals and groups have insisted and will insist that their preferences for development of public resources be considered in any decision-making process. To a large extent environmental regulation has been used by these individuals and groups to express their preferences concerning allocation of "public" resources.

A third development in American society at this time is an increasing disillusionment with government and its ability to substitute for or regulate economic markets. Between 1935 and 1980, the federal government experimented with various forms of government regulation of economic markets. Recently, however, several initiatives have been undertaken to deregulate the domestic passenger airline industry,[68] the motor carrier industry,[69] portions of the electric utility industry,[70] the railroad industry,[71] and the communications industry.[72] Economic literature has increasingly exposed the failures of government economic regulation and underscored the belief that unregulated economic markets best allocate scarce societal resources.

The fourth development was precipitated by the series of shocks to the U.S. economy occasioned by two energy crises of the 1970s. These crises were caused by the increasing dependency of the United States on oil imported from a few countries located in the Middle East and South America and the ability of those countries to form and maintain a cartel to control the price and supply of oil to the world. Between 1973 and 1980 the United States imported roughly 40 percent of its oil needs from these countries or approximately 8 million barrels per day. During this period also, the real price of a barrel of oil increased by almost 400 percent.[73] Moreover, the oil supplying nations embrace political ideologies hostile to American views and are deeply opposed to certain American policies (the continued existence of the state of Israel) in the Middle East. During this same period there were two major wars in the region with the second war between Iraq and Iran presently in progress.

These events prompted a major reexamination of the energy policy of the United States. The principal goal of that energy policy is to reduce substantially American dependency on imported oil and to substitute oil conservation and energy from indigenous sources. Substantial evidence of this policy is found in several pieces of legislation which have been enacted in the last 6 to 7 years. The principal legislation is the National Energy Act of 1978, the COWPTA, and the Energy Security Act of 1980.[74] Very little legislation has addressed hydroelectric development, and what few provisions there are address the problems of nonfederal development of small-scale hydroelectric sites. The Public Utility Regulatory Policies Act of 1978 (PURPA) exhorted the FERC to simplify its licensing process for small-scale facilities and provided, as noted earlier, low interest loans for feasibility studies, licensing activities, and construction at those sites. COWPTA provided tax incentives for nonfederal development, and the Energy Security Act permits FERC to exempt sites of 5 MW or less capacity from federal licensing requirements.

Each of these policy trends will affect hydroelectric development in major ways. The environmental movement for the most part is opposed to the development of nuclear generating plants. Its position is substantially aided by the nuclear accident at Three Mile Island. A part of the environmental movement, and groups which support private initiatives on a small scale, favor development of small renewable dispersed energy sources. The environmental movement continues its opposition to large-scale federally developed water projects including large-scale hydroelectric plants.

The reform of federal economic regulation and greater reliance on the economic market to make investment decisions also affect energy policy. After a relatively brief experiment with price and supply regulation of domestic oil markets and a longer term experiment with price regulation of natural gas production, federal energy policy has recently abandoned these approaches for the market. The consensus is that prices and supplies of oil and natural gas will increase and in fact that is what is happening. Of course, removal of price ceilings on oil and natural gas have made alternative energy resources more attractive.

Regulation of the electric utility industry, long considered one of the most appropriate subjects of pervasive economic regulation because of its perceived *natural monopoly* characteristics, has recently been changed. By the National Energy Act of 1978 (Title II of PURPA), entry into the electric generation market is assured for cogenerators and developers of small generating plants which utilize renewable energy resources.

The two attributes of hydroelectric development favored by American energy policy at this time are, of course, that hydropower is a *renewable* and an *indigenous* energy source. A third characteristic of hydropower development attributable to rehabilitation or expansion of capacity at existing dams is that these projects are comparatively *environmentally benign.* This last characteristic of certain forms of hydropower development reduces substantially opposition to these projects by environmental organizations. Moreover, small-scale hydropower development at existing sites has engendered support by groups and organizations favoring development of small dispersed renewable energy resources. There is evidence that these trends and initiatives have prompted a response by the market. In the last 2 years, FERC has literally experienced an explosion of permit and licensing applications by various organizations, agencies, local governments, and private entrepreneurs. The bulk of these license applications and permit applications are for sites with less than 25 MW of installed capacity presently in existence. A somewhat lesser percentage of these permit and license applications but still greater than 50 percent of the total, are for existing sites of 4 MW capacity or less.[75]

9.4 REFERENCES

1. See *Preliminary Inventory of Hydropower Resources,* vols. 1–6, U.S. Army Corps of Engineers, July 1979.
2. For purposes of this discussion several related microeconomic concepts are used. An *external cost* is a cost borne by a third party who is not a party to the market transaction which imposes the cost. An *external benefit* is a benefit conferred on a third party who is not a party to the market transaction. The economics literature recognizes that market transactions fail to take these *externalities* into account. The result is that third parties who bear the external costs, in fact, subsidize the project. If an activity confers external benefits on third parties, markets will tend to underallocate resources to the activity (i.e., less development than is desired by society will occur). See generally *Public Finance and the Price System,* Edgar K. Browning and Jacquelene M. Browning, Macmillan, New York, 1979, pp. 1–54 (hereinafter *Public Finance*).
3. The most recent, authoritative statement on federal reserved water rights is found in *U.S. v. New Mexico,* 438 U.S. 696 (1978). Federal reserved water rights arise

when the federal government has reserved federal lands for certain purposes. Courts have held that the federal government has reserved water rights necessary to carry out the purposes of the reservation. The appropriation of these water rights dates back to the establishment of the reservation. See *Cappaert v. U.S.*, 426 U.S. 128 (1976). The implications of the federal reserved-water-rights doctrine to western state *appropriation* systems are serious. Under the doctrine, the federal government may be deemed a prior appropriator even though it never perfected its water rights under state law.

4. These attempts are part of the Sagebrush Rebellion. With the enactment of the Federal Land Policies Management Act of 1976 (FLPMA), several western states passed or considered legislation which attempted to assert state title to all public lands and minerals not previously appropriated. These states include Nevada, New Mexico, Utah, Washington, and Wyoming.
5. *Winters v. U.S.*, 207 U.S. 564 (1908) is the seminal decision. The case of *Arizona v. California*, 373 U.S. 546 (1963) expanded Indian reserved water rights.
6. There are numerous treatises on water law. For a general reference to the two systems, see R. E. Clark, *Water and Water Rights: A Treatise on the Law of Waters and Allied Problems*, Allen Smith Co., Indianapolis, Ind., 1967–1972.
7. For an early history, see Jerome G. Kerwin, *Federal Water Power Legislation* (Columbia University Studies in the Social Sciences: No. 274, 1926; repr. AMS Press, Inc., New York), and for a somewhat biased view see Pinchot, *The Long Struggle for Effective Federal Water Legislation*, 14 Geo. Wash. 2 Rev. 9 (1945).
8. 16 U.S.C. §§791 *et seq.* (1976).
9. *See generally* 18 C.F.R. §4.30 *et seq.*, and 45 Fed. Reg. 58371 (1980).
10. 16 U.S.C. §800(a) (1976).
11. 16 U.S.C. §803(a) (1976).
12. 16 U.S.C. §814 (1976).
13. 16 U.S.C. §803(e) (1976).
14. 16 U.S.C. §797(e) (1976).
15. *Id.*
16. 16 U.S.C.A. §824 (1979).
17. P.L. 96-294, 94 Stat. 611.
18. The view of hydropower sites being *public resources* is based on the history of the FPA and federal water projects development policies. It implies that there has been sufficient sentiment among persons inside and outside of government over a long period of time that hydropower sites are owned by the *people* and should be allocated by the servants of the people.
19. A *public good* is defined by economists as a good the consumption of which by one person does not diminish another person's consumption of it. This condition also means that a person cannot practically exclude another person from consuming the good. The flood-control benefits of a dam are, of course, a public good when they inure to the benefit of a large number of people. Economic theory holds that markets fail to allocate *public goods*. See *Public Finance*, pp. 1–54.

20. 16 U.S.C. §§831Y-1 *et seq.* (1978).

21. The Water Resources Planning Act of 1965, 42 U.S.C. §1962 (1974).

22. *Principles* and *Standards:* 38 Fed. Reg. 24,788 (1973). *Procedures:* 44 Fed. Reg. 72,892 (1979).

23. The National Environmental Policy Act of 1969, 42 U.S.C. §§4321 *et seq.* (1976).

24. See *Public Finance,* pp. 1–54.

25. 42 U.S.C. §§4321 *et seq.* (1976) (NEPA); 33 U.S.C. §§1251 *et seq.* (1978) (Clean Water); 16 U.S.C. §661 (1976) (Fish and Wildlife Coordination); 16 U.S.C. §757a (Supp. 1978) (Anadromous Fish Conservation Act); 16 U.S.C. 470–470m (1976) (Historic Preservation); 43 U.S.C. §§1701 *et seq.* (1976) (FLPMA).

26. 16 U.S.C. §1131–1136 (1976) (Wilderness Act); 16 U.S.C. §1271–1281 (1976) (Wild and Scenic); 16 U.S.C. §§1531 *et seq.* (1976) (Endangered Species).

27. 5 U.S.C. §551 *et seq.* (1977).

28. See *United States v. Students Challenging Regulatory Agency Procedures* (SCRAP), 412 U.S. 669 (1973) in which the Supreme Court gave expanded status to groups to challenge administrative action.

29. 43 Fed. Reg. 55,978 (1978) (CEQ regulations).

30. The practice of FERC under NEPA has been discerned as a result of interviews with FERC staff and a number of small-scale hydropower developer-license applicants, and the FERC regulations themselves.

31. 42 U.S.C. §4332(c) (1976).

32. 33 U.S.C. §1251 *et seq.* (1978).

33. 18 C.F.R. §4.51 (1979).

34. 33 U.S.C. 1344.

35. 16 U.S.C. §§661 *et seq.* (1976).

36. 16 U.S.C. §§1531 *et seq.* (1976).

37. *Hill v. TVA,* 419 F. Supp. 753, (E.D. Tenn. 1976), *rev'd* 549 F.2d 1064 (6th Cir. 1977), *aff'd* 437 U.S. 153 (1978).

38. 16 U.S.C. §470–470m (1976).

39. 16 U.S.C. §1241–1249 (1976).

40. 16 U.S.C. §1278(a) (1976).

41. 18 C.F.R. §131.6.

42. 41 Fed. Reg. 2053 (1976).

43. The early economic literature developed the *natural monopoly* theory; for the classical treatment of this theory see F. Zeuthen, *Problems of Monopoly and Economic Welfare, Routledge, London, 1930.*

44. Any number of general treatises discuss electric utility regulation. See, e.g., Priest, *Principles of Public Utility Regulation,* Michie, 1969.

45. 16 U.S.C. §824(a)–824(h) (1976).

46. *FPC v. Southern California Edison Co.,* 376 U.S. 205 (1964).
47. About 90 percent of all revenues from sales of electric power are at retail. See *Statistics for Privately Owned Electric Utilities in the United States-1978,* DOE/EIA–0044(78), 1979, p. 24.
48. 16 U.S.C. §824(b).
49. 7 U.S.C. §§901 *et seq.*
50. In *FPC v. Conway Corp.,* 426 U.S. 271 (1976), the Supreme Court held that the FPC had to examine the anti-competitive effects of higher wholesale rates for public utilities.
51. In *Otter Tail Power Co. v. U.S.,* 410 U.S. 366 (1973), the Supreme Court held, under circumstances of flagrantly anticompetitive practices, Otter Tail could be forced to wheel power.
52. Recent work in the broad area of production function specification has produced mixed results as to the economies of scale in production with large fixed cost.
53. 16 U.S.C. §825s (1976).
54. Bonneville Project Act, 16 U.S.C.§§ *et seq.* See also 41 Op. Att'y Gen. 236 (July 15, 1955).
55. Under these circumstances the manufacturer-developer confronted the classic *monopsony market,* i.e., a market dominated by a single purchaser. See M. Ringo, *Monopsony and the Supply of Power from Small Generating Stations,* Energy Law Institute, Concord, N.H. 1980.
56. P.L. 95-617, 92 Stat. 3117 (1978) (hereinafter PURPA).
57. PURPA, §201.
58. 45 Fed. Reg. 17,965 (1980).
59. 45 Fed. Reg. 17,965 (1980) and 45 Fed. Reg. 12,236 (1980).
60. Title II of the Crude Oil Windfall Profit Tax Act of 1980 (COWPTA) contains the provisions amending the Internal Revenue Code (P.L. 96-817).
61. See *Energy for Rural America, Rural Development Initiatives,* the White House, May, 1979.
62. See *Public Finance,* pp. 1–54.
63. In the past all of the large, successful federal projects conferred the additional benefit of providing a basic physical infrastructure (water supply, flood control, and power) for the economic development of certain regions of the country. Such economic development at the time was perceived to be in the national interest.
64. This argument, of course, does not address the serious question of whether the *federal,* as opposed to some other governmental entity (regional, state, or local) should continue developing water projects even when the water project confers flood-control and recreational benefits.
65. 16 U.S.C. §§2701–2708 (1978).
66. DOE has promulgated regulations [45 Fed. Reg. 3544 (1980)] for the feasibility study and licensing application processes. Loans issued under these regulations are forgivable if the project is infeasible or if the developer cannot obtain a license.

67. For a general discussion of Title IV of PURPA and the REI see, *Federal Obstacles and Incentives to the Development of the Small Scale Hydroelectric Potential of the Nineteen Northeastern United States,* ELI, July 1980, (Federal Report), pp. 221–234.
68. 49 U.S.C. §1301 *et seq.* (Supp. 1980).
69. P.L. 96-296; 93 Stat. 793.
70. Title II, PURPA.
71. 49 U.S.C. §1 *et seq.* (Supp. 1978).
72. The deregulation of the communications industry has largely been by administrative and judicial fiat of the Federal Communications Commission with an assist from the Courts. *Re: The Carterfore Device,* 13 F.C.C.2d 420 (1968); *Re: The Specialized Common Carrier Services,* 29 F.C.C.2d 870 (1971) and the consent decree in *U.S. v. AT&T.*
73. The nominal price of crude went from around $3.00 in 1973 to around $32.00 in mid-1980. Over that period the GNP deflator increased from 110 to 240. Hence, the real price rose about 385 percent, $\$32.00/[(240/110) \times \$3.00] = 4.85$ or a 385 percent increase.
74. 42 U.S.C. 8701 *et seq.* (Supp. 1980).
75. Interview with Ron Corso, Director, Division of Hydroelectric Licensing, Federal Energy Regulatory Commission, Washington, D.C., Oct. 9, 1980. As of June 1980, FERC had received 33 license applications and 152 permit applications for the year. By comparison, in all of 1978, there were 12 license applications and 37 permit applications.

10

Institutional and Policy Environment in Developing Countries

Robert F. Ichord, Jr.

Jack J. Fritz

10.1 ENERGY FOR RURAL DEVELOPMENT: A NEW POLICY ARENA

Current discussion of the role of small and mini hydropower in developing countries is occurring within the context of providing energy for rural development. Until recently, energy in rural areas meant merely the extension of urban electricity distribution and petroleum marketing systems. In many developing countries, these two functions were in different hands—the government operated the centralized electricity system and foreign oil companies, often through private entrepreneurs, owned and operated a network of oil refineries, oil product transportation facilities (i.e., pipelines, trucks, railroad tank cars), distribution depots, and retail outlets.

During the 1970s, the international oil cost and supply crisis and its convergence with the fuel-wood and traditional energy crisis in certain countries focused attention on the need to develop indigenous energy resources. It also highlighted the issue of how to meet both the basic energy needs of expanding rural populations and the energy requirements of rural agricultural and industrial development. The political response was a growing government intervention in virtually all phases of conventional energy production, processing, and distribution. National oil companies and central ministries of energy were established, and national energy planning groups began to assess energy needs, resources, uses, and options from a more comprehensive perspective.

The prime focus of these nascent efforts remains on conventional energy development and utilization in the urban-modern sector; nevertheless, there seems to be a wave of interest in the rural energy situation and the political implications of current trends in this sector. National energy plans are beginning to stress the need for the development of *localized* energy resources, i.e., solar, biomass, wind, and hydro, to meet rural needs and to prevent further environmental degradation from deforestation. International donors are increasing assistance in these areas. The potential of these energy sources received world attention at the August 1981 U.N. Conference on new and renewable sources of energy held in Nairobi, Kenya.

10.2 GROWING INTEREST IN MINI HYDROPOWER

One of the most promising options for meeting rural electrical energy needs is mini and micro hydro, particularly in the tropical highland areas of Latin America, Asia, and Africa. In these regions, it is often the lack of electric power and the dependence on diesel fuel that is limiting the pace of economic development.

As discussed in previous chapters, small waterwheels and turbines have a long history and can be used to provide mechanical as well as electrical power.[1] Francis, Pelton, and propeller turbines are simple in design and require relatively little maintenance. Small dams of earth, masonry, or logs may be constructed in rural areas of developing countries with local resources and labor. Except for speed-control governors, most developing countries have the capacity to build and fabricate micro turbines and civil works.

Although mini hydro units are often more expensive than diesel-powered generators, the life-cycle economic picture is usually more attractive. Estimates of total installed costs (1978) for micro and mini hydro units in Asia range from $400 per kilowatt to approximately $2,000 per kilowatt, depending on scale, origin of equipment, complexity of civil works, and cost of labor. In Papua New Guinea, it was reported that the installed capital costs of a 5-kW micro hydro unit was $438 per kilowatt compared to $580 per kilowatt for a diesel generator of the same size.[2] The economics of mini hydro can be enhanced if additional benefits accrue such as irrigation, water supply, or flood control. The potential for use in existing irrigation projects is largely untapped. An Indian survey estimates that around 10,000 MW of mini hydro capacity could be installed on the canal systems.[3]

Operating costs, of course, depend heavily on capacity factor, which may be only 15 to 20 percent. The low demand for electricity in most remote rural villages, e.g., in Nepal, is considered a serious obstacle to the cost-effective use of mini hydro units. Ways must be sought to utilize the electric power on a 24-hour basis by light industry, for agricultural processing, or municipal services. However, the operating costs for petroleum-fueled engines, often dependent on shipments of diesel or gasoline fuels by trucks over difficult terrain or poor roads, are also very high. To the extent that micro and mini hydro units can replace the fuels used in diesel generators, kerosene lamps, and other petroleum-consuming devices, they can serve to marginally reduce the foreign exchange costs of imported petroleum. This is especially true for more expensive (than crude oil) middle distillate products which are frequently imported because of the inability of local refineries to meet the heavy demand for these products. Additional foreign exchange savings can be realized if the hydropower units are built locally with zero or minimal foreign imports.

Since the technology is proven and if the economics are increasingly attractive for remote sites off the main grid, mini hydro technology is also attractive for social and political reasons. Government leaders in developing countries view with increasing urgency the need to expand rural productivity and employment. Political instability is often associated with the failure to bring modernization to rural areas and with the increasing migration of unemployed and underemployed rural people to overcrowded urban centers. Micro and mini hydro systems offer a way for rural villages to begin electrification with a modest capital outlay, as well as to increase the efficiency of traditional tasks (e.g., grain grinding) and to stimulate the development of new rural industries (e.g., machine shops making agricultural implements). The importance of electricity is summed up by a Chinese expert in mini hydro, "Without electricity, it is impossible for a country or a people's commune to make any long-term planning and development."[4]

Although politicians in many developing countries, e.g., Bangladesh, have elevated the provision of electricity to rural villages to a national commitment, the costs of constructing and maintaining conventional transmission and distribution networks are high, particularly given the characteristically low-load factors in rural areas. A Government of India report concludes:

> The need to draw transmission and distribution lines over long distances from the central electric supply systems makes the cost of supplying electricity to villages very high. This cost is further aggravated because the characteristics of the load in rural areas is such that the losses in transmission and distribution are also high. . . . The present tariff rates involve a very large subsidy to electricity uses in rural areas.[5]

Subsidization of electricity costs in rural areas is usually necessary, particularly to cover the cost of initial connection which could be as high as $120 (Bolivia). Rural subsidization by the urban consumer reverses the traditional food price subsidization of the urban dweller by the rural farmer. The history of rural electrification in the United States is one of massive government programs to bring electricity to farms. This approach worked in the United States because an economic base existed which was able to benefit from the availability of inexpensive electricity. This precondition does not exist in many developing countries.

Mini hydro technology does represent a possible approach, however, in some countries to reduce costs and provide service to remote areas at an earlier time than would have been possible through the expansion of the centralized grid. The problem of high transmission losses and theft may exist for mini hydro based systems as well. Losses in rural systems of the People's Republic of China are running as high as 25 percent, a factor encouraging the building of larger stations that are integrated into local or regional networks.

10.3 ELITE COMPETITION AND GOVERNMENT DECISION MAKING

Developing countries generally exhibit a high degree of competition among elites for access to government authority and influence over government policies and programs in key sectors. The nature of this competition or conflict among elites serves to determine the type of political system (i.e., authoritarian, theocratic, democratic) and the stability and efficacy of the government in achieving the goals of development and order.[6] The values of competing elite groups and the degree to which they represent and support various rural interests is therefore of critical importance to our consideration of the institutional context of rural energy and mini hydro policy and programs.

The following discussions highlight several basic policy issues relating to the demand for electricity and the role of mini hydro technology, the choice of mini hydro systems, and their financing, organization, and management.

10.4 DEMAND FOR ELECTRICITY AND THE ROLE OF MINI HYDROPOWER

Most rural areas of developing countries do not have electricity. Energy consumption is primarily in the form of traditional fuels such as wood or kerosene, predominantly for cooking. Given this pattern of consumption and the increasing difficulty of obtaining these fuels, a view is emerging in some developing countries that the first priority of rural energy policy should not be electricity, but rather augmentation of fuel-wood supplies. The results of a growing number of studies suggest that demand for electricity for productive uses will not develop unless there are a range of other inputs available. Roger Revelle notes some of these factors.[7]

1. The actual or potential existence of adequate transportation facilities to and from markets and sources of raw materials
2. Availability of an adequate supply of other forms of energy needed for a particular industry or industrial complex
3. Sufficient investment capital, either in the form of credit or savings
4. The presence of trained technicians and of managers familiar with problems of marketing, cost accounting, personnel management, and business operation
5. The existence of a market town or small city in which a complex of mutually-supporting industries can be established
6. The availability of skills, disciplines, and comparative costs of labor in the rural areas vs. the labor force in large cities

In addition, experience suggests that successful rural mini hydro systems have three principal characteristics: involvement of local people and organizations, integration into the cash economy, and the existence of a previous or adjacent electrical distribution system. Local decision making in planning, construction, and operation will promote the interests of the community over those of capital-bound policy makers. This is particularly crucial after start-up when technical help is being withdrawn. The notion of ownership, local benefits, and consequences must reside in the community. There must be surplus disposable income which can be used for investment in new production facilities which have not been built because of the lack of available electric power. There must be a familiarity with electricity and its benefits demonstrated through the existence of a diesel-generation system nearby. Clearly, if the benefits have not accrued to the majority in that community, local support for a mini hydro project will not manifest itself.

Where the provision of electricity is deemed appropriate, the issue of centralized or decentralized generation may be important. Competition may develop between the national organization responsible for large-scale power generation and the entity or entities at the national or local level involved in providing energy to rural users. International donors may support one or both groups, further complicating the problem of determining an optimum electric power investment plan for the country.

The emphasis on developing decentralized options such as mini hydro may be viewed as a threat by the centralized national power organization over generation and the development of an integrated national grid. In some countries, however, the unreliability of power from the national grid provides an incentive for businesses and local communities to have their own captive source of power, whether diesel or small hydro. Thus, economics may not always be the dominant concern in government or local community considerations of the mini hydro option.

10.5 ORGANIZATION AND MANAGEMENT

The configuration of institutions involved in rural electrification and mini hydro projects varies from country to country. Four basic approaches can be characterized.

Single Utility Approach

In countries with a socialist tradition, a single national utility may be involved in generation, transmission, and distribution of electricity to both rural and urban areas. This situation is present, for example, in Indonesia, Peru, and Sri Lanka, where the Indonesian National Electric Company (PLN), ELECTRO-PERU, and the Ceylon Electricity Board monopolize government electricity activities. In Indonesia, though, considerable autogeneration capacity exists outside the PLN network. PLN is taking an increased interest in mini hydro and has a full-scale testing laboratory for mini hydro systems in Bogor.

Separate Central and Rural Electrification Authorities Approach

Many developing countries have separated the national power corporation from the organization that distributes power to rural areas. This bifurcation, in part, reflects the large subsidies that are sunk into rural electrification programs. These rural electrification boards (Bangladesh), corporations (India), or administrations (Thailand, Ecuador, Bolivia) may or may not own and manage the distribution system. In Bangladesh, India, Bolivia and the Philippines, cooperatives have been established for this purpose, with technical, managerial, and financial support from the government authority. Some of these rural systems have local mini grids into which power is fed from diesel units. The high operating costs of these units are forcing rural electric authorities and cooperatives to turn to mini hydro if an adequate hydro resource exists near these grids.

Autonomous Mini Hydro Authority Approach

While institutional responsibility for mini hydro generation usually resides in the national power organization or the rural electricity corporation, there are instances, e.g., in Nepal and Peru, where a separate mini hydro board has been established. This board is involved in the construction of mini hydro stations in the remote mountainous regions and is separate from the Rural Electrification Division of the Department of Electricity, which also works with villages in the construction of small hydro plants.[8] In Peru, an ad hoc commission serves to coordinate the planning and execution of mini hydro projects for example.

Local Community Approach

In a growing number of cases, micro hydro units are being installed by local communities, sometimes with the assistance of faculty and students from a nearby technical university. In Papua New Guinea, the Appropriate Technology Development Unit of the University of Technology at Lae has adapted a Pelton runner design to high-head sites near Lae.[9] The Butwal Institute of Technology in Nepal has for many years assisted in local hydro development. Church groups have installed units to pump water and provide electricity to schools and hospitals in Nepal and India. The people's communes in the People's Republic of China have installed thousands of hydro units, generally less than 20 kW in capacity with commune members being trained at college and vocational institutes in all aspects of hydraulic machines.[10] The Peace Corps in Liberia has also been active in the promotion of micro hydro with installation of a 30-kW unit in a remote village. In Pakistan, a similar approach has also been used to install village-level micro hydro systems with a minimum of outside help.

10.6 TYPE AND ORIGIN OF EQUIPMENT

A policy issue that is the subject of intense debate in some countries is whether mini hydro technology should be imported. The 1979 UNIDO/ESCAP Workshop on Mini/Micro Hydro concluded: "It was agreed that the development of technology in the field of mini-hydro plants should take place within the developing countries themselves."[11] While this is certainly occurring in Thailand, India, Nepal, Papua New Guinea, Indonesia, Columbia, Peru, and the Philippines, mini hydro systems from Norway, Sweden, Japan, Canada, the United Kingdom, Germany, the People's Republic of China, and the United States are being purchased by developing countries, generally on concessional terms as part of foreign assistance or export promotion programs. In Asia, the People's Republic of China has provided credits to the Philippines for a large number of mini hydro units.

The issue of imported vs. locally made equipment often revolves around the question of reliability and efficiency. The national utilities want to install systems that will have minimal problems, especially in sites that are remote from centers of technical expertise. The cost differential is considerable, however, between a locally made unit and the more efficient imported foreign technology. The Chinese mini hydro units are among the cheapest foreign units available. Yet at approximately $1,000 per kilowatt for a 5-kW unit, they are over twice as costly as the locally made unit in Papua New Guinea, even before installation. Generally, it is not cost-effective to use imported equipment to gain a few percentage points of efficiency.

The Chinese have succeeded in standardizing units of various sizes. The

People's Republic of China had around 90,000 units in operation at the start of 1980, with an average capacity of 70 kW.

10.7 PRICING AND CREDIT POLICIES

While mini hydro may offer a cheaper alternative than diesel in off-grid locations, the issue of pricing of electricity and its affordability by the local community remains fundamental. The cost of wiring a house may itself be beyond the reach of most families without credit, and the expense of electric appliances may prevent the use of electricity for anything except lighting. Yet, with mini hydro technology there is a built-in incentive to increase capacity factors, since there is no increased fuel cost with additional use. Pricing strategies with mini hydro can be designed to encourage, up to a limit, the consumption of electricity which in turn lowers the cost per kilowatt hour. In Papua New Guinea, the 6-kW mini hydro units near the village of Baindoang supply a 3.6-kW heater element at a community washhouse. This baseload approach eliminates the need for an expensive governor.[12] The villagers pay a fixed charge for the lighting of their houses.

Although some governments have established mechanisms to provide low-interest loans to local communities and cooperatives for energy production projects, programs to encourage the productive uses of electricity have been inadequate. Until recently, rural electrification programs have emphasized lighting of households and have not been closely coordinated with agricultural development projects.

10.8 DETERMINING RURAL DEMAND STRUCTURE

If a community or number of villages have had electricity available from the grid or a diesel generator, forecasting future needs of a nearby nonelectrified community can be determined. However, if no demand history is available from neighboring communities, an analysis needs to be carried out, either through a direct survey and enumeration technique and/or through topographical investigations commonly employed in geophysical resource assessments. However, an in-situ household survey can lead to questionable results since most respondents desire electricity but may not have the means to pay.

The community selling price will be influenced by total demand and capacity factor in the absence of government subsidies. Since the household and small industry loads may be the most significant, estimates can be made based upon similar experience in other countries. However, additional factors must be considered such as provisions for future growth and available energy to meet agricultural and larger industrial needs. The following basic information should be gathered for each community.

1. Current number of inhabitants
2. Average household size
3. Average household energy use and energy expenditures
4. Number of projected connections
5. Number and characteristics of small industries
6. Number of commercial establishments
7. Characterization of public services
8. Projected growth of the community
9. Projected monthly kilowatthour consumption for users

One approach often applied in forecasting is to use per capita installed capacity of electricity generation. This can vary widely depending on the economic level of the community; however, installation capacities of 50 to 60 watts per capita are common.

Fundamental differences in demand structure become evident between urban and small rural communities as electricity becomes available. Table 10.1 illustrates these differences. Urban centers clearly show higher capacity factors, therefore, low peak-power rates and higher annual production rates. The combination of continuous service, a higher degree of innovation, and diverse and industrial loads make electricity increasingly economical as the size of the

TABLE 10.1 Load Comparison: Small Rural vs. Urban Centers During Initial Connection and After 5 Years' Service

Domestic load, excluding cooking and air conditioning	Small rural centers, intermittent supply, 4 hours per day	Urban centers, continuous supply
	Initial	
Peak power:		
per consumer, W	250 to 270	260 to 275
per inhabitant, W	3 to 4.5	3.5 to 5.5
Annual production:		
per consumer, kWh	About 240	325 to 370
per inhabitant, kWh	About 3 to 4	About 4.5 to 7.5
Annual load factor, %	10.5 to 11	14 to 15
	After 5 Years	
Peak power:		
per consumer, W	About 215	About 260
per inhabitant, W	4.7 to 6.2	9 to 14
Annual production:		
per consumer, kWh	About 270	320 to 350
per inhabitant, kWh	6 to 8	11 to 19
Annual load factor, %	About 14	4 to 15

SOURCE: Adapted from Small Scale Power Generation, UN, 1967.

TABLE 10.2 Demand Load for Small-Scale Enterprises in Developing Countries

Activity	Required capacity, kW	Activity	Required capacity, kW
Carpentry	3–15	Brick factory	1–5
Bakery	2–5	Restaurant	1–2
Crafts	1–2	Vegetable canning	5–20
Small sawmills	15–30	Dairy products	2–10
Grinders	10–20	Milk (cooling and preevaporation)	5–20
Flour mill	3–20		
Looms	0.5–6	Electrical and mechanical workshops (repairs)	5–15
Coffee grinder (commercial)	5–30		
Quarry mill	6–30	Gas pumps	2–100
Ice making	6–60	Silos	3–5
Irrigation pump	2–100		

SOURCE: Programa Regional de Pequenas Centrales Hidroelectricas de Olade, *Requerimientos y Metodologías Para La Implementación Masiva de Pequeñas Centrales Hidroeléctricas en Latino America,* Documento de Trabajo, Quito, 26 de Junio de 1980, Anexo VIII.

community becomes larger. Growth in demand as electricity becomes commonplace must be considered in the planning process. Those initially hesitant will later seek to be connected and rapid growth can be expected during the initial 5 years of electrification. As light industrial loads are connected, the capacity factor should improve, especially for larger communities. Peak household loads can be expected to increase 10 percent annually, whereas in urban centers, it may be between 25 and 30 percent. Also, population on the urban periphery is expected to grow much more rapidly than in small rural communities which often lose population to those urban areas.

To determine hydroelectric capacity required for light industrial loads, Table 10.2 is provided presenting typical demand required for a number of small machines commonly found in rural communities. Table 10.3 presents similar information for household appliances. Note the differences between values in Table 10.3 and those given in Chap. 7 for a typical U.S. residence. Com-

TABLE 10.3 Typical Demand for Household Appliances Found in Developing Countries

Appliance	Required capacity, W	Appliance	Required capacity, W
Light bulb	25, 50,	Refrigerator	180
	75, 100	Blender	100
Fluorescent lamp	20	Fan	35
Radio	5	Sewing machine	100
Television	15	Clothes iron	500

bining light industrial loads, household demand, and line loss, a peak demand for an entire community can be determined.

10.9 COMPARATIVE CASE EXPERIENCES

This section examines the institutional development of mini hydro programs in two Asian and two Latin American countries from a comparative perspective.

Philippines and Thailand

The Philippines and Thailand are among the top 10 oil-importing developing countries. Both have approximately the same per capita income levels and have fairly advanced rural electrification programs. These programs involve the extensive use of small diesel generators. In Thailand, there are approximately 400 diesels ranging in size from 50 to 6,000 kW.[13] Both countries are desperately trying to reduce their dependence on imported oil and want to substitute mini hydro units where possible.

Each country has a central rural electrification authority—the National Electrification Administration (NEA) in the Philippines and the Provincial Electricity Authority (PEA) in Thailand. A fundamental difference in approach, however, is found in the development of rural electric cooperatives in the Philippines. As of February 1980, 101 cooperatives had been electrified under the NEA program, representing 27 percent of the barrios and 20 percent of the households. The national objective is to electrify all barrios by 1985. In general, the cooperatives in the Philippines are dominated by the wealthier members of the rural communities—government employees, businesspeople, sugar planters, and professionals.[14] On the other hand, in Thailand, cooperatives have not fared well and the government has resisted this approach in its rural electrification program. Thailand and the Philippines both have substantial potential for mini hydro development. In Thailand, several hundred sites have been identified, especially in the north, and one source estimates a potential capacity of 800 MW of firm, dry-season, unregulated power.[15] Table 10.4

TABLE 10.4 Philippine Mini Hydro Potential

Region	Sites with data, MW	Sites without data, MW	Total sites, MW
Luzon	478	1,248	1,726
Visayas	207	1,327	1,534
Mindanao	88	1,191	1,279
Total	773	3,766	4,539

SOURCE: Government of the Philippines, Ministry of Energy, *Ten-Year Energy Program,* 1980, p. 60.

indicates a confirmed capacity of almost 800 MW in the Philippines and an even larger potential at sites without data.

Responsibility for mini hydro development is dispersed in Thailand, whereas in the Philippines a Presidential Directive in 1979 gave NEA sole jurisdiction in this area. This directive was related to an overall policy decision to promote energy self-sufficiency in all rural cooperatives. Under this policy, NEA has begun working with rural cooperatives in planning for the introduction of either mini hydro systems or *dendro-thermal* units (wood-burning or gas-fired power plants). NEA has signed agreements with the People's Republic of China and France providing credits for hydro units and gas-turbine power plants. NEA will provide loans and technical assistance to cooperatives in the adoption of these technologies.

In Thailand, both the National Energy Administration and PEA are currently installing mini hydro units in rural areas. The Royal Forestry Department and the Royal Irrigation Department also operate small, locally made units. While PEA, which reports to the Ministry of Interior, is the agency of the government responsible for distributing electricity to rural consumers, the National Energy Administration, which is under the Ministry of Science, Technology, and Energy, has a broad mandate in the energy planning and supply field and views itself as playing a catalytic role in the development and demonstration of alternative energy technologies for rural areas.

The National Energy Administration and PEA officials hold somewhat different philosophies toward the development of mini hydro in Thailand. The National Energy Administration is concerned with developing an indigenous low-cost system that is appropriate in rural villages, particularly those that are remote from the main grid or from district centers. Officials also believe the local cooperatives can effectively build and manage micro hydro systems at significantly lower costs than with imported units. The National Energy Administration has obtained assistance from the UNDP, USAID, and Japan for micro hydro projects. PEA, on the other hand, is seeking to provide reliable power to rural district grids and to replace high-cost diesel generators as quickly as possible. PEA looks to proven, imported equipment and is receiving assistance from Norway, as part of the World Bank's second rural electrification loan, as well as from Finland for projects involving units with a capacity of about 500 kW.

The main entity responsible for economic planning in Thailand is the National Economic and Social Development Board (NESBD). The NESBD is also considering the program and structure of the mini hydro sector in Thailand. An overall Energy Master Plan is in preparation with assistance from the Asian Development Bank. This plan will assess the potential from mini hydro and other alternative technologies in meeting future energy needs.

The pace of development in mini hydro in both the Philippines and Thailand is clearly accelerating. In the Philippines, NEA plans to install about 200 MW of mini-hydro by 1987. This growth of decentralized generating capacity is necessitating increased investment coordination in the power sector. Foreign donors have expressed to the Philippines government the importance of coordination between NEA and the National Power Corporation (NPC), which is embarking on a major program to build large generating plants based on geothermal energy and domestic and imported coal. Similarly in Thailand, the Electricity Generating Authority of Thailand is planning new generating capacity based on lignite and imported coal. At this point, neither country has a firm understanding of the optimal future mix of centralized and decentralized capacity. However, it will be interesting to compare the relative cost/benefit of the two approaches, in particular to see whether the Philippine policy of rural energy self-sufficiency stands the economic test of time.

Peru and Ecuador

In contrast, Peru and Ecuador are currently oil-exporting countries whose petroleum resources will be severely diminished by 1990 from high domestic consumption resulting from government price subsidies during the 1970s. However, policy-makers in both countries realize that electricity in rural areas must be provided through mini hydro plants instead of traditional diesel sets.

The theoretical hydroelectric potential for Peru as calculated by the Ministry of Energy and Mines (MEM) in 1979 was approximately 200,000 MW. Approximately 30 percent is exploitable, or some 60,000 MW, of which only 3 percent has been developed so far. The electricity sector is organized through ELECTROPERU, a public-sector enterprise charged with system operation and project execution. ELECTROPERU is composed of five operating regions but with 87 percent of the demand concentrated in the Lima area. As of 1979, approximately 73 percent of Peru's electricity generation or 5,470 GWh of production from 1,397 MW of capacity, came from hydropower, primarily large plants.

Within ELECTROPERU, small or mini hyro plants are within the purview of the Office of Applied Technology (OPTA) under its Programa de Pequeñas Centrales Hidroeléctricas. OPTA began its program of mini hydro development in 1978, initially carrying out studies of power systems with 50 to 1,000 kW capacity to meet the electricity needs of rural people. By 1980 the budget was up to $1.85 million with several plants under construction. Additional plans call for 50 plants to be built between 1980 and 1985, many financed through foreign assistance programs.

In carrying out its programs to dot the countryside with mini hydro plants, ELECTROPERU generally seeks local participation by the municipality, particularly during the construction phase. Subsequent system operation is also carried out by the local authorities after training is provided by ELECTROPERU. This approach has proved successful in bringing about cooperation between rural people and central utility decision makers.

Ecuador, on the other hand, has only begun to exploit its hydroelectric potential, estimated to be 22,000 MW. Total installed capacity in 1977 was 612 MW. Thermal and hydropower plants were developed on a large-scale basis only during the seventies. By 1980 roughly 60 percent of Ecuador's population remained without electric power.

Aside from a small number of regional power companies, the Instituto Ecuatoriano de Electrification (INECEL), under the jurisdiction of the Ministry of National Resources and Energy (MNRE), is responsible for generating and distributing electricity in Ecuador. The development of energy resources in Ecuador had not been coordinated until the creation of the Instituto Nacional de Energía (INE) in 1978.

Because Ecuador's oil reserves were rapidly being depleted and oil importation was expected to begin during the late eighties, a national energy policy needed to be established. Planning efforts followed by conservation and exploitation of the vast hydropower base are going forward during the early eighties.

Development of hydroelectric power is one of the cornerstones of the national energy plan. The contribution to total electric generation from large and small hydro plants is expected to be increased to 80 percent by 1985 in contrast to 32 percent in 1978. INECEL will continue to execute and manage the grid and the large electric generating projects, whereas regional and municipal power companies will manage local systems. It is expected that these efforts will provide power to 60 percent of the rural inhabitants. The grid will be extended to larger communities, principally those with more than 500 inhabitants, near major roads, and with high agricultural productivity. Since this program will bypass many isolated villages a Mini Micro Hydroelectric Plan was formulated.

Prior to construction of a national grid, INECEL had constructed some 50 mini micro hydropower systems, some of which were subsequently abandoned. Under the new plan retired plants will be brought back into operation or the equipment moved to more attractive load centers. The management of new mini micro hydro systems is expected to be through cooperatives, local power companies, or municipalities. Each management structure will depend upon the characteristics of existing local organizations. Management expertise and training will be provided by INECEL. This type of organizational structure is similar to that found in Peru with a central technical focus determining how energy technology will be applied in rural areas but with administration left to the local level. This type of situation could lead to misperceived needs or concentration on the larger urban centers. Therefore, a balanced program meeting the needs of both urban and rural peoples is necessary.

10.10 CONCLUSIONS

The issue of the role of mini hydro in developing countries is caught up in current policy debate over the merits of rural electrification. In most developing countries, the balance of power continues to rest with the proponents of total national electrification. Policymakers view the provision of electricity to rural areas as much in political as in social or economic terms. Electrification is a tool of political integration and provides tangible proof of the commitment of the government to improving the quality of rural life.

Yet, these proponents are increasingly having to face some uncomfortable truths about rural electrification. Douglas Smith summarizes several of these realities:[16]

1. The benefits of rural electrification flow primarily to the already better off;
2. The most most urgent energy needs, for example fuel-wood replacement to slow deforestation, are neglected through reliance on rural electrification to power villages;
3. Rural electrification only occasionally stimulates industrial enterprise sufficiently to result in new economic activities in rural areas; and
4. The inefficiencies of rural electrification combined with rapidly escalating costs of central station power plants are demanding millions of dollars of subsidy to rural electrification.

The development of decentralized approaches to electricity generation, such as mini hydro, wind, photovoltaics, or biomass, may not fundamentally alter these facts. Certainly, the potential for the development of a locally based and managed mini hydro industry exists in many developing countries. However, Chinese experience has demonstrated that mini hydro technology can serve to mobilize the community and to enhance productivity of rural agriculture and small industry. Nevertheless, planners should be extremely sensitive to the overall problem of village *energization* (as contrasted with village electrification) and to the social impact of various energy supply and delivery systems. Only in this way can appropriate policies, programs, and institutions be designed to meet both the basic energy needs of the poor for cooking and heating and the requirements of rural agriculture, industry, and transportation for mechanical, electrical, and thermal energy.

Note: Material in this chapter represents the personal views of the authors and does not reflect AID policy.

10.11 REFERENCES

1. National Academy of Sciences, *Energy for Rural Development: Renewable Resources and Alternative Technologies for Developing Countries,* NAS, Washington, D.C., 1976. See especially pp. 37–40 and 137–164.
2. Allen R. Inversin, "A Pelton Micro-Hydro Prototype Design," Lae, Papua New Guinea, Appropriate Technology Development Unit, 1980, pp. 23–260.
3. Government of India, Planning Commission, *Report of the Working Group on Energy Policy,* New Delhi, 1979, p. 93.
4. Mao Wen Jing and Deng Bing Li, "An Introduction to the Development of Small

Hydro-Power Generation in China," Paper presented at UNIDO/ESCAP/RCTT Seminar on the Exchange of Experiences and Technology Transfer on Micro-Hydro Electric Generation Units, September 1979, p. 4.

5. Government of India, *op. cit.*, p. 91
6. David Apter, *Choice and the Politics of Allocation,* New Haven, Yale Univ. Press, 1971.
7. Roger Revelle, "Future Energy Resources and Use in Asia: The Needs for Research and Development," in *Proceedings of the USAID Asia Bureau Conference on Energy, Forestry, and Environment, Manila, Nov. 12–16, 1979,* p. 46.
8. UNIDO, *Draft Report on UNIDO/ESCAP/RCTT Joint Meeting on the Exchange of Experiences and Technology Transfer on Micro-Hydro Electric Generation Units, September 1979,* p. 8.
9. Inversin, op. cit.
10. Mao Wen Jing and Deng Bing Li, op. cit., p. 20, and UNIDO report, op. cit., p. 29.
11. UNIDO report, op. cit., p. 20.
12. Allen R. Inversin, "Technical Notes on the Baindoang Micro-Hydro and Water Supply Scheme," Lae, Papua New Guinea, Appropriate Technology Development Unit, February 1980.
13. Morton Gorden, "Institutional Analysis of the Small Hydro Program in Thailand," *Report to AID, March 10, 1980,* p. 13.
14. Agency for International Development, *The Philippines: Rural Electrification,* AID Project Impact Evaluation Report No. 15, December 1980, p. C-1.
15. Meta Systems, *Report to AID on Renewable Non-Conventional Energy,* July 1979, annex IX, p. 1.
16. Douglas Smith, "Rural Electrification or Village Energization?" *Interciencia,* vol. 5, no. 2, March–April 1980, p. 86.

11

Economic and Financial Feasibility

Jack J. Fritz
Jerry W. Knapp
Herschel Jones

11.1 BACKGROUND

The economic and financial feasibility of small hydropower must be determined relative to reasonable alternatives. The principal economic characteristics of small hydropower are high initial capital cost and low operating costs. After allowances are made for escalation, future annual fixed costs such as operation, maintenance, and replacements represent a predictable cost of power compared to fossil-fuel sources. Before 1973, the low cost of electricity generated from oil-fired plants made the construction of new hydropower facilities unattractive on the basis of traditional economic analysis. In the United States, during the past 40 years, thermal plants tended to replace small hydro facilities as they were retired through obsolescence. The current cost in the United States of a large oil- or gas-fired plant is approximately $700 (1980) per installed kilowatt. Thermal plants now in the planning stages may cost $1,200 per kilowatt by the time construction is complete in the mid-1980s. These costs are as much as 10 times the cost of similar plants constructed 20 years ago. A substantial part of the increased costs is attributable to the air-quality controls now required for fuel-burning plants. Even more costly, the economics of nuclear power remains uncertain. Costs of constructing hydroelectric plants have also increased but at a lesser rate.

Since hydropower is capital intensive, costs are very sensitive to interest rates. Armstrong demonstrated (Fig. 11.1) the value of a hydroelectric kilowatt over the value of a kilowatt from a thermal plant.[1] If the cost of oil was $30 per barrel when the thermal plant started operations, and if a 5 percent per year increase in the cost of oil is assumed, then the value of a hydro kilowatt would be $4,000 more than that of an oil-fired kilowatt. A 35-year life span was assumed for the thermal plant and a 50-year life span for the hydro plant with a 10 percent interest rate for both. Clearly, the key element making thermal electricity uneconomical in the long run is the probable increasing cost of fossil fuels.

Further cost data were noted by Moore, who compared system capital costs in 97 less developed countries (LDCs).[2] Table 11.1 gives installed cost per kilowatt of generating capacity.

Although initial hydro capital costs are higher than for thermal plants, it

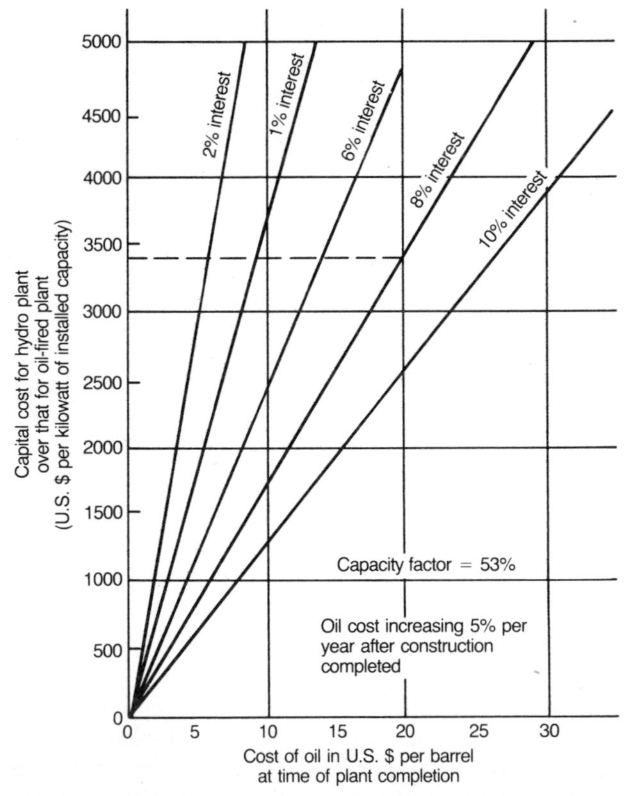

FIG. 11.1 Additional capital cost of a hydro plant over oil-fired plants as a function of oil cost.

TABLE 11.1 Average System Costs per Kilowatt of Generating Capacity in 97 Developing Countries

System type	$US, 1979
Hydro	1,296
Geothermal	1,564
Nuclear	1,436
Thermal	836

must be stressed that hydro systems have a built-in check against inflation because energy costs are nil. Only maintenance and labor costs will increase annually at a rate depending on local economic conditions. Also, where thermal systems exist, the marginal cost of expanding that electrical power capacity becomes an issue. In the United States, with the balance-of-payments already strained by imported-oil costs, increased use of fossil fuels should be discouraged. Future social and political as well as economic circumstances may make hydropower more attractive.

In rural areas of developing countries, a useful contrast to minihydropower costs is the application of small diesel-electric sets. Installed costs in 1979 ranged from $1,200 to $2,100 per kilowatt. Generally, capital costs for simple small hydro plants are in the upper end of the diesel-electric capital cost range. However, in serving other loads, the displacement of fossil fuel is realized in such applications as diesel pumping, kerosene refrigerators, and private diesel-electric generators for lighting and small industry. These applications of electricity in a rural community may have a high economic multiplier effect. Productive uses of available power have the most potential for increasing the wealth of the community if used by small industry or for agricultural processing and irrigation.

A comparative analysis of the cost of generating electrical energy by small hydro and diesel installations indicates that, despite the high initial cost of investment for small hydro plants, the cost of generation per kilowatthour is relatively low, as shown in Table 11.2. When both capital cost and annual fixed costs over the project life are considered, the hydro-generated electricity may be more economical.

In addition to fuel cost being an important consideration, load factor is also critical. The typical average cost of generation per kilowatthour for hydro and diesel power stations with average annual load factors varying from 25 to 65 percent in developing countries is indicated in Table 11.3. Note that these figures are based on conservative estimates. In practice, the useful life of hydro civil works, which account for 50 to 70 percent of total investment, extends well beyond the estimated plant life, and figures of 50 to 60 years are often used. These cost estimates have been computed on average costs at mid-1981 constant price levels.

TABLE 11.2 Comparison of Hydropower and Diesel Generated Electricity Costs (1,000 kW Facility)

Cost type	Hydropower	Diesel
1. Investment cost per kW,	2,000.00*	700.00†
2. Useful life of equipment, years	35	15
3. Capital recovery factor for an interest rate of 10 percent per annum	0.1037	0.13145
4. Power generated per year at a load factor of 45 percent, kWh	3,942	3,942
5. *Annual fixed costs:*		
Debt service, $	207.40	92.00
Annual cost for maintenance at 1 percent and 2.5 percent of capital investment, $	20.00	17.50
Annual cost for operation at 0.5 percent of investment, $	10.00	3.50
Annual charges for administration and overhead at 0.5 percent of investment, $	10.00	3.50
Total annual fixed cost, $	247.40	116.50
6. Annual fixed cost per kWh, $	0.0627	0.030
7. *Variable cost:*		
Cost of fuel per kWh in remote localities, $	–	0.18‡
8. Total cost of generation per kWh at 45 percent load factor, $	0.0627	0.21

SOURCE: Adapted from Ref. 3.
*Assumption for mid-1981 price level.
†Assumption on the basis of average investment cost of diesel power stations up to 1,000 kW capacity.
‡Based on the price of diesel fuel as $0.55 per liter delivered at site.

TABLE 11.3 Load Factor and Average Generation Cost for Hydro and Diesel Power Stations

Average annual plant load factor, %	Average cost of generation per kWh, U.S. cents	
	Hydro	Diesel*
25	11.3	23.3
35	8.0	21.8
45	6.27	21.0
55	5.13	20.4
65	4.34	20.0

SOURCE: Adapted from Ref. 3.
*Diesel calculated at U.S. $.55 per liter.

11.2 PROJECT ANALYSIS

The economic and financial feasibility must be demonstrated for any hydroelectric project under consideration. The treatment of economic and financial feasibility as herein presented is not meant to be comprehensive, but simply to serve as an introduction. In-depth treatment of the subject can be found in Van der Tak[4] and Grant et al.[5]

Economic feasibility, ideally the determination of project benefits and costs from the viewpoint of society as a whole, should include both tangible and intangible benefits and costs in the analysis. However, only tangible benefits and costs can be dealt with accurately from an accounting point of view. Economic feasibility as dealt with in this chapter is considered from the standpoint of the sponsoring enterprise in contrast to the nation as a whole. When measures of total benefits accruing from the project exceed the total costs incurred, the project is regarded as economically feasible.

Financial feasibility, however, states whether the tangible value of the output of the project will be sufficient to amortize the project loan, pay operation and maintenance costs, and meet interest and other financial obligations. Or it can be simply construed as the cash flow situation. Financial feasibility is also viewed from the perspective of the project sponsor.

An economically feasible project may not be financially feasible. For example, a hydroelectric project may have considerable intangible benefits due to flood control or recreation for which the sponsor cannot collect revenues. These intangibles make it economically feasible, but the revenue from sale of the project's electrical output may be insufficient to support the required debt service. Conversely, a project may be sound financially but not economically because social costs outweigh social benefits.

This chapter considers both financial and economic feasibility. Economic and financial data as well as technical data form the basis for decision making. Alternative courses of action, such as choice of site, installed capacity, size of impoundment, and method of financing, must be evaluated in terms of costs and benefits to the project sponsor. Such financial and economic analyses are generally included in reconnaissance studies but in more detail in feasibility studies.

For projects in the United States, a comprehensive approach to both economic and financial analysis includes a market analysis, detailed system cost estimates of several possible physical site alternatives, cash flow analysis during and after construction, attendant regulatory and licensing requirements which may influence project feasibility, and alternative methods of financing.

If the sponsor or power purchaser is a large electric utility, the market analysis may need to include the investigation of methods to integrate hydro-generated power into the regional grid so that it complements existing thermal power production. Proper integration depends on such factors as the annual and seasonal energy production of the project, determination of this demand profile, and whether the project structures can store water for daily, seasonal, or cyclical periods. In contrast, a small, isolated plant supplying power to rural communities may require only a local power survey with provisions for future increases in load growth.

A variety of methods can be used for the economic analysis of capital projects. In the analysis of small hydropower projects, specific techniques are considered universal practice and have traditionally yielded an objective economic picture. However, increases in fossil-fuel prices, uncertain inflation, and escalating construction costs have made some traditional methods of economic analysis less than satisfactory. Today, in spite of high interest rates, some capital intensive projects may be more attractive compared to projects having low capital costs and high annual operating costs. For this reason traditional analysis techniques need to be selectively applied.

Various site-specific hydropower schemes requiring separate economic and financial analysis may fall into one of several categories:

1. Site-specific analysis to determine an optimal plant configuration or capacity for given resource and demand profiles. Examples include specification of multiple turbine-generator units of different sizes in situations where severe fluctuations in demand or water availability exist, reservoir size, or dam height, etc.
2. Analysis to compare a hydropower project with a thermal or alternative-energy plant.
3. Analysis to determine an optimal site for a hydropower project given diverse resource, demand, and geographic constraints.
4. The application of alternative financial schemes.
5. Multipurpose analysis to determine costs and benefits in conjunction with other projects such as irrigation, flood control, and water supply.

The basic principle for carrying out these studies is to compare alternatives on an equivalent basis, such as a fixed energy output, time frame, and constant dollar values. In the analysis, the quantifiable assumptions must be first established as follows:

1. Determine all base line project assumptions, such as the period of analysis, discount rate, cost of capital, and other economic and financial variables.
2. Estimate project costs which include capital costs; one-time costs, such as permits; annual operating and maintenance costs; and provisions for renewals and replacements. Estimates of engineering fees, excavation, construction, labor and materials, legal fees, and interest during construction must all be determined and placed within the desired construction schedule.
3. Ascertain project benefits, principally the revenue stream generated by the sale of electricity (or potable or irrigation water if the project is multipurpose).
4. Determine the source of financing and the specific terms of the loan. If the project is to be constructed by an agency other than the user of its output, the contract for the sale of the power may be the key to its financial feasibility.
5. Choose an appropriate economic analysis methodology, and determine economic and financial feasibility.

6. Perform a sensitivity analysis to determine how costs and benefits react to variations in such factors as plant capacity, discount rate, financing, storage capacity, load factor, and demand forecasts.

The common approach to economic analyses has been to compare costs and revenues over a consistent time period on a ratio basis or net positive benefit basis. Several measures using discounted cash flow techniques can be employed: internal rate of return (IRR), benefit/cost ratio (B/C), net present value (NPV), and life-cycle costs (LCC). Each technique has its advantages, disadvantages, and appropriate applications. Those aspects are described in the following section.

11.3 BASIC TOOLS OF ECONOMIC ANALYSIS

In analysis of projects on the basis of current monetary value, a fundamental approach is to apply discounting to a series of future payments to determine the value of those payments in current dollars. To develop the basic mathematics of analysis, we first introduce some conceptual definitions.

Time Frame

The useful life of the project is a key parameter in establishing economic and financial feasibility. Project life span may not be equivalent to the financing period. For example, for hydropower plants, the project life span may be in excess of 50 years, and in some cases up to 75 years, whereas the financing period is 50 years. At this point significant repair work or replacement of the turbine may be necessary. Useful lifetime for small-scale thermal plants is between 20 and 30 years at which time they usually require a major overhaul involving replacement of boiler tubes, rotating equipment, ash handling equipment, steam turbine replacement, and possibly generator parts. It therefore can be assumed that the useful life of a thermal plant is always less than that for a hydro plant. In the case of micro sets, up to 100 kW, diesel-generator useful life is often less than 10 years, especially under adverse conditions found in many rural applications.

Interest Rate

The interest is the fee that must be paid by the user for the lender's capital. The rate is set in the capital markets and fluctuates with changes in the health of the economy and government fiscal and monetary policies. Interest rate will depend on the source of the capital, lower rates being available from public lenders and higher rates required by commercial lenders. The interest rate is used in determining the debt service.

Discount Rate

The discount rate is used for determining economic feasibility whereas the interest rate is used to ascertain financial feasibility. The proper rate to use for testing economic feasibility is the opportunity cost of capital to society. This is the rate of return that could be earned by investing the capital cost of the project in a venture of similar risk or an alternative marginal project.

The appropriate interest rate for determining financial feasibility is equivalent to the interest rate paid on the bonds or other securities sold to finance the project. This will vary for different organizations. On the other hand, the discount rate for a regulated utility should be the rate of return it is earning or expects to earn on its rate base. For a public agency in the United States it will be equivalent to the interest rate on the tax-exempt bonds the agency will sell to obtain construction funds. The discount rate will depend upon its source of funds including the Rural Electrification Administration (REA) and the Cooperative Finance Corporation for an REA-financed electric cooperative.

Discounted Cash Flow

One of the basic tools for determining the economic feasibility is discounted cash flow. All cash expenditures are tabulated during the chosen period for comparison each year. The total cash expenditure for each year is then discounted to the present and cumulatively added to a single sum. This sum is then compared with similar sums of discounted expenditures for alternatives. The alternative with the smallest sum is clearly the least costly. A similar comparison is made with cash revenues or receipts for the same period. The ratio between the sum of the discounted receipts and the sum of the discounted expenditures yields the benefit/cost ratio, (B/C), for example.

Certain rules must be followed in making discounted cash flow analyses:

1. The same period of years must be used for each alternative set of cash expenditures and each alternative set of cash receipts.
2. The alternatives must have the same energy production and capacity. In some cases this may require adjustments to the costs of the lowest cost alternative. For example, a small hydroelectric plant may produce energy on an intermittent or seasonal basis and thus have no saleable peak-capacity output. To make a valid comparison with a diesel-generating plant which produces both firm energy and has saleable peaking capacity, the cost of the diesel plant must be adjusted by subtracting the market value of its peaking capacity.
3. Cash expenditures will include renewals and replacements, but if the years in which these will be made cannot be accurately predicted, then an estimated average annual cash expenditure for renewals and replacements as well as an accelerated depreciation schedule can be used since those costs will occur far in the future.

4. If the period chosen is shorter than the estimated service life of the hydroelectric plant or one of the alternatives, the discounted salvage value at the end of the period should be deducted from the sum of the discounted cash expenditures. Ideally, two or three diesel life cycles should correspond to one hydro life for simplicity.
5. If the period chosen exceeds the estimated service life of an alternative plant, the costs of replacing the plant should be shown in the year or years preceding the end of its life. Of course, if the second plant's service life extends beyond the life of the hydroelectric plant, the discounted salvage value should be deducted from the sum of the discounted cash expenditures of the alternative plant. It is important to remember that capital and operating costs are exclusive of any payments for interest, interest during construction, amortization of debt, or charges for depreciation.

Discounting transforms all future costs and revenues into the present time frame so they can be compared on a current monetary basis. These sums are simply called the *present worth* or *present value.* All future expenditures and revenues are modified or discounted by a factor which provides escalation due to opportunity costs and resource depletion. The following expression illustrates the concept:

$$P_0 = S_n/(1 + k)^n \qquad (11.1)$$

where P_0 is a sum of money today, S_n is a sum of money at year n, and k is the discount rate. The reciprocal of the term $(1 + k)^n$ or quotient P_0/S_n is the *present worth factor.*

As an example, if we have a sum of \$100 in the future and discount it at the rate of 10 percent for 5 years, our future \$100 is worth \$62 today. The corollary is clearly that if we invest today's \$62 at 10 percent interest, we have \$100 at the end of 5 years.

11.4 ECONOMIC ANALYSIS TECHNIQUES

By definition, the economic analysis compares all benefits and costs associated with a project during its useful life. Costs include initial and recurring annual expenditures, whereas the benefits include revenues from the sale of power. Intangible or nonmonetary benefits accruing to the community may also be included if appropriate, as in a developing country. The project may be single-purpose, that is, to supply electricity, or multipurpose, for example, flood control, irrigation, or water supply. If these objectives are included in the project, then associated costs and tangible benefits must be included in the analysis. Cost escalation may also be considered in such an analysis, depending on the expected rises in labor, materials, fuels, and other consumables.

To carry out in-depth economic analysis, all project parameters must be defined beginning with the following:

1. Installed capacity
2. Annual energy production
3. Load factor
4. Capital cost
5. Operating and maintenance cost
6. Expected interest rate
7. Discount rate
8. Construction period
9. Financing period
10. Value of energy
11. Cost of distribution system
12. Quantifiable community benefits

Items 1 through 12 are sufficient to carry out the type of analysis required in both prefeasibility or feasibility studies. However, when comparing a mini hydro plant to a diesel generator, clearly the price of fuel and its escalation over future years will be required. In addition, periodic rebuilding of the diesel engine (every 5 to 7 years) is necessary and will require a single cash outlay.

Benefit/Cost Ratio

The benefit/cost (B/C) ratio technique is most frequently applied in analyzing capital projects. The method compares the present worth of plant costs and benefits on a ratio basis. Projects with a ratio of less than 1 are generally discarded. Mathematically, the present worth of project benefits are divided by the present worth of costs as follows:

$$R_{b/c} = \frac{P_b}{P_c} = \frac{\sum_{j=0}^{n} \frac{B_j}{(1+k)^j}}{\sum_{j=0}^{n} \frac{C_j}{(1+k)^j}} \tag{11.2}$$

where $R_{b/c}$ is the benefit/cost ratio (B/C), P_b is the sum of present worth of benefits, P_c is the sum of present worth of costs, B_j is the benefit for year j, C_j is the costs for year j, and n is the number of years of analysis.

This approach should be applied with caution because the project having the largest ratio may not yield the largest benefit, since the ratio is not indicative of project magnitude. Projects should be analyzed incrementally, that is, the benefit/cost ratio should be computed for projects which may have larger costs but which yield a somewhat higher ratio. An analysis of this type must also include a view of the absolute quantities as well as the ratios.

An example of benefit/cost analysis of a mini hydro plant is illustrated with the following parameters:[6]

1. Installed capacity, 500 kW
2. Annual energy production, 2.45 $\times$ 10^6 kWh per year
3. Load factor, 56 percent
4. Capital cost, $375,000
5. O&M costs, $15,000 per year
6. Financing cost, 12.5 percent
7. Construction period, 1 year
8. Financing period, 15 years
9. Escalation rate, 10 percent
10. Value of energy, 2.5¢/kWh

Table 11.4 shows how typical benefit/cost computations are carried out. The *cost stream* consists of a single capital outlay in year 0 and recurring expenditures for operation and maintenance which increase at an annual rate of 10 percent. The *benefit stream* is simply annual revenues from the sale of

TABLE 11.4 Example of Benefit/Cost Ratio Computation*

	Cost streams				Present worth	
Year	Capital	O&M, 10%	Benefit stream, 10%	Present worth factor, 12.5%	Cost	Benefit
0	375,000			1.00000		
1		16,500	67,375	.88889	14,667	59,889
2		18,150	74,112	.79012	14,341	58,558
3		19,965	81,524	.70233	14,022	57,257
4		21,961	89,676	.62430	13,710	55,984
5		24,158	98,644	.55493	13,406	54,740
6		26,573	108,508	.49327	13,108	53,524
7		29,231	119,359	.43846	12,817	52,334
8		32,154	131,295	.38974	12,523	51,171
9		35,369	144,424	.34644	12,253	50,034
10		38,906	158,867	.30795	11,981	48,922
11		42,797	174,753	.27373	11,715	47,835
12		47,076	192,229	.24332	11,454	46,772
13		51,784	211,452	.21628	11,200	45,733
14		56,962	232,597	.19225	10,951	44,717
15		62,659	255,856	.17089	10,708	43,723
TOTALS					188,864	771,194

Capital cost = $375,000
Present value O&M = $188,864
Present value of benefits = $771,194

$$\text{Benefit/cost ratio (B/C)} = \frac{\$771{,}194}{\$375{,}000 + \$188{,}864} = 1.3677$$

Net present value (NPV) = $771,194 − ($375,000 + $188,864) = $207,330

SOURCE: Adapted from Ref. 6.
*All figures in U.S. dollars.

energy production at its sale price, 2.5¢/kWh, increasing also at 10 percent annually. The present worth of costs and benefits is shown in the last column. These quantities are the product of the present worth factor and benefits or costs for each year beginning with year 1. To compute the benefit/cost ratio Eq. (11.2) is applied. A ratio larger than unity makes the project economically feasible.

Net Present Value

Another approach, particularly useful for ranking multiple projects is to compute the net present value (NPV). This procedure is simply to compute the difference between discounted benefits and costs for each year, beginning with outlays for construction. Mathematically, this can be shown as:

$$N_{pv} = \sum_{j=0}^{n} \frac{B_j - C_j}{(1 + k)^j} \tag{11.3}$$

where N_{pv} is net present value (NPV).

The net present value of our example project is $207,330 as computed with the above expression. For the project to be economically feasible, N_{pv} must be positive. If $B_j - C_j$ is constant throughout the life of the project exclusive of capital cost, then:

$$N_{pv} = -S_0 + (B_j - C_j) \left[\frac{(1 + k)^n - 1}{k(1 + k)^n} \right] \tag{11.4}$$

where S_0 is the capital cost which is paid during year 0.*

This technique allows the decision maker to reduce economic and financial information of a project to a single figure. Such a procedure can be used when evaluating a number of projects to establish a priority. Or it can be used as a screening technique as well as decision tool to decide among a number of alternative cost schemes at a single plant. Projects with a negative N_{pv} are rejected. The major disadvantage of this technique is that the magnitude of the decision variable N_{pv} yields no information about the ratio of costs to benefits.

Internal Rate of Return

Simply stated, the internal rate of return (IRR) is that discount rate at which the net present worth is equal to zero. Computations usually require an iterative procedure the results of which can also be used as a ranking or screening tool. The essential element is to reject all projects which have an internal rate of return less than the opportunity cost of capital. This type of analysis lends itself well to computer application when choosing among either a large number

*In traditional economic analysis, capital costs are not considered a separate item(s), but form part of the cost stream (C_j). Equation 11.4 and subsequent formulations assume capital costs are paid within the first year of the project, otherwise they must be discounted to the present as with other future costs.

of projects or specific project alternatives. This method should be applied carefully since it represents the aggregate, revealing little about future project economics, particularly if at some future time costs begin to exceed benefits.

Life-Cycle Costs

Life-cycle costing (a variation of simple discounted cash flow) is particularly useful for comparing the economics of mini hydropower to a thermal energy system alternative. The rationale for adopting the life-cycle costs (LCC) approach over other methods is related to rising prices of fossil fuels as well as to the need to consider all costs, operating and capital together, over the entire project lifetime.

The life-cycle cost of an energy system can be defined as the present value sum of all expenditures related to capital, operation, debt service, and maintenance over its useful life. Life-cycle costing is not unique, yet there is currently more interest in the method because of the need to compare energy systems with different cost profiles, e.g., comparing systems with relatively high front-end capital costs with those characterized by high operating costs. It is a well-known fact that high initial costs associated with certain energy systems, hydro for example, may be justifiable in the face of higher operating costs of fossil-fuel systems due to continuing real price increases.

The LCC approach considers the present value of all costs of two or more alternatives at a specific site. Here it is considered useful in appraising low capital cost diesel-generation with a high capital cost mini hydro plant. A trade-off situation exists in that after a period of time a break-even point is reached where low capital cost and accumulated high cost of fuel of the diesel are equivalent to the high capital cost and low accumulated O&M costs of the mini hydro plant. Clearly, it is beyond this point that the mini hydro system becomes more cost effective.

For example, if we propose two systems, d designated for diesel and h for hydro, and assume the life of the hydro plant is equivalent to three diesel lives, total expenditures for each could be characterized as follows:

$$E_d = S_{d0} + \frac{S_{d1} - s_{d1}}{(1 + k)^{n1}} + \frac{S_{d2} - s_{d2}}{(1 + k)^{n2}} - \frac{s_{d3}}{(1 + k)^{n3}} + \sum_{j=1}^{n3} \frac{C_{dj}}{(1 + k)^j} \tag{11.5a}$$

$$E_h = S_{h0} + \sum_{j=1}^{n3} \frac{C_{hj}}{(1 + k)^j} \tag{11.5b}$$

where

E_d, E_h = life-cycle expenditures for diesel and hydro alternatives
S_{d0}, S_{d1}, S_{d2} = capital costs for diesel sets in years 0, n_1, n_2
S_{h0} = capital cost for hydro plant in year 0
s_{d1}, s_{d2}, s_{d3} = salvage values for diesel sets in years n_1, n_2, n_3

n_1, n_2, n_3 = replacement years for the diesel set
C_{dj}, C_{hj} = O&M costs for diesel and hydro for each year

The above expressions have been simplified with zero assumed salvage value for the mini hydro plant. They can easily be modified to represent an actual situation.

11.5 FINANCIAL OR CASH FLOW ANALYSIS

A key element in the feasibility study is the financial analysis or determination of the project's cash flow position over the period of the loan. This analysis will show whether the project is self-supporting or whether deficits are likely to develop. Depending on the policies of the funding organization, provisions can be made to meet payback commitments during deficit periods which might occur during initial start-up.

The determination of financial feasibility is simply to compare revenues and expenditures, including loan repayment, on an annual cash basis. An example of a cash flow statement is given in Table 11.5 which uses the data from Table 11.4. Elements which were not considered in the example were taxes and depreciation. The specific magnitude of these items would depend on the loan agreement, local financial practices, and government policies. From Table 11.5, it is evident that during the first 2 years of the project, negative cash flows occur which must be covered from other financial resources. Generally short term loans can be negotiated to cover such a period.

TABLE 11.5 Example of Financial Analysis*

	Costs				
Year	Debt service†	O&M	Total	Benefits	Net cash flow (benefits − costs)
1	56,536	16,500	73,036	67,375	−5,661
2	56,536	18,150	74,686	74,112	−574
3	56,536	19,965	76,501	81,524	5,023
4	56,536	21,961	78,497	89,676	11,179
5	56,536	24,158	80,694	98,644	17,950
6	56,536	26,573	83,109	108,508	25,399
7	56,536	29,231	85,767	119,359	33,592
8	56,536	32,154	88,690	131,295	42,605
9	56,536	35,369	91,905	144,424	52,516
10	56,536	38,906	95,442	158,867	63,425
11	56,536	42,797	99,333	174,753	75,420
12	56,536	47,076	103,612	192,229	88,617
13	56,536	51,784	108,320	211,452	103,132
14	56,536	56,962	113,498	232,597	119,099
15	56,536	62,659	119,195	255,856	136,661

SOURCE: Adapted from Ref. 6.
*All figures in U.S. dollars.
†$375,000 at 12.5 percent for 15 years. Capital recovery factor equals .15076.

11.6 DETERMINING BENEFITS

The benefit of electricity production is assumed to be the price the user is willing to pay for it. That price is determined by a market survey or established tariff structure but must be no more than the price of electricity produced by the lowest cost alternative means. Reasonably accurate consumption trends and forecasts of monthly demand patterns are required. For small and mini hydropower systems, a popular approach is to consider the economic value equal to the cost of diesel-generated power. The selling price is that price at which the user is willing to buy and the price at which the seller is willing to sell. However, the selling price must be higher than the cost of production or the project will not be financially feasible.

In February 1980 the U.S. Federal Regulatory Commission issued regulations implementing Section 210 of the Public Utilities Regulatory Policy Act (PURPA) of 1978. Under these regulations, developers of small hydroelectric projects (under 30 MW capacity) can sell their output to publicly-owned or investor-owned utilities, subject to PURPA, at their avoided cost for equal quantities of energy and capacity. Avoided cost is defined as the incremental (not average) cost the utility would otherwise have to pay for purchased power or the incremental cost of operating its highest cost generating facility to produce an equal amount of energy (e.g., gas turbines). Avoided capacity cost is the cost of new peaking capacity from a small hydroelectric project. In some cases, these avoided costs are based on generation from older oil-fired plants and range from 4.9 to 6.1 cents per kilowatthour.[7]

In the analysis of small rural systems in developing countries, both financial and economic feasibility will depend on maximizing the benefits. For example, available electricity from a mini hydro plant may reduce costs other than the cost of electricity for those who are connected to the system. These cost-reduction benefits might include avoidance of kerosene for lighting, diesel fuel used for irrigation pumping, or other resource savings. Additional indirect benefits might include (1) the increased agricultural production due to an electric irrigation pump, (2) improved health and human productivity because of available potable water made possible through groundwater pumping, (3) increased job opportunities and a more positive balance-of-payments for the community due to increased rural industry made possible by electricity, (4) the availability of night school classes, and (5) public safety improvements because of public lighting. Although these economic benefits are difficult to quantify, they significantly improve the quality of life in rural communities of developing countries.

An approach to quantification of these benefits can be shown mathematically. Total benefits for some year j can be represented by B_j as follows:

$$B_j = \underbrace{t_{1j}d_{1j} + t_{2j}d_{2j}}_{\text{[Revenues]}} + \underbrace{s_j\,p_j + P_cAy}_{\left[\substack{\text{Cost savings through} \\ \text{increased productivity}}\right]} \tag{11.6}$$

where

B_j = benefits accrued during year j, $

t_{1j} = normalized residential tariff for community, $/kWh

d_{1j} = aggregate annual residential demand, kWh

t_{2j} = normalized industrial/commercial tariff for community, $/kWh

d_{2j} = aggregate annual industrial/commercial demand, kWh

s_j = normalized annual cost savings for lighting and potable water pumping, diesel fuel, $

p_j = number of irrigation pumps

P_c = crop selling price, $/ton

A = irrigated area planted in crop, hectares

y = annual increase in productivity, tons/hectare

The above expression assumes that an existing diesel set is in place to pump potable water and provide power for lighting. The hydropower set would displace the diesel providing electricity for those purposes in addition to power for irrigation pumping, possibly increasing agricultural productivity. Expression (11.6) is specific to the assumed circumstances; similar approaches can be developed for other situations.

11.7 DETERMINING COSTS

Capital costs for hydro plants are difficult to accurately estimate in the absence of site-specific information. Capital costs may be high in comparison with various types of thermal plants. However, as pointed out, electricity production costs of hydro plants tend to be dominated by financial (debt service) instead of operating (energy) costs. Little is known regarding future costs without an accurate estimate of the capital costs by an engineering firm.

Some general guidelines on costs are available. Capital costs of new or renovated plant in the United States can be determined using accepted estimation techniques and recently published sources.[8] However, outside the United States, the equipment and labor market varies widely over time and place. Similar plants in different countries may have altogether different project cost profiles. Costs which are quoted must be carefully examined in light of local currency strength. A problem which continues to plague such efforts is that in economies with severe inflation economic feasibility could be endangered by the time the plant is constructed, which may be several years after the analysis.

Project capital costs are defined as the stream of disbursements required during preparation, construction, and start-up. This stream of expenditures must be carefully planned so that initial funding is sufficient to cover the period. Estimates must be in constant dollars and annual disbursements should be expressed as a percentage of total capital costs or in absolute amounts considering annual escalation. As an example, the capital cost ranges for a typical high-head site is given in Table 11.6.

TABLE 11.6 Capital Costs for Small High-Head Hydro Plants in the United States (1982)

Capital expenditure	Approximate 1980 costs, $/kW
Turbine-generator	400–800
Influent civil works, canal, sand trap, headworks, penstock	300–500
Powerhouse	200–300
Electrical equipment	100
Other mechanical equipment	*
Service road, outlet works	100–200
Cost of land	*
Distribution system	*
Start-Up	*
Engineering	100–200
Interest during construction	*
Taxes, duties, levies	*
Contingencies	100–200

*Site specific.

Clearly, all the costs in Table 11.6 are site specific, however as a rule of thumb, equipment costs are half or less than half of total capital costs. Civil costs vary widely depending on plant configuration, cost of materials, and local labor. Contingencies, engineering, and start-up costs are calculated as fixed percentages of total cost. Other factors to consider include whether the plant is in existence and or needs only to be upgraded. Often, the equipment is moved to another site or a turbine-generator is installed in an existing irrigation control structure. These site-specific situations require very specific cost and economic analyses. Under these situations there are no "rough" cost guidelines.

Some generalized cost relationships have been developed to accelerate the estimation process for "typical" sites. One of the most practical approaches is outlined in *Hydropower Cost Estimating Manual* published by the U.S. Army Corps of Engineers.[8] Some discussion of component costs can be found in previous chapters of this handbook. Another approach as suggested by Gordon and Penman is given below for low-head sites:[9]

Item	Cost expression (1980)	Comments
Turbine-generator equipment, $	$C_T = 40{,}000(P/H)^{0.53}$	Units below 1.5 MW, Swedish experience, 1978
Turbine-generator equipment, $	$C_T = 4{,}000P^{0.7}H^{-0.35}$	Units below 5 MW, U.S. experience

Note: C_T is cost in dollars, P is power capacity in kW, and H is head in meters.

Electricity production costs are a function of annual costs. Key determinants of the magnitude of these costs are initial capital charges, the construction period influencing interest paid during construction, the useful life of the plant which may influence the interest rate and size of loan payments, the load factor or percentage of full operating capacity utilized, and other operating and maintenance costs.

Most energy production costs are primarily fixed and can be broken into the following categories:

1. Debt service
2. Salaries or wages of operating personnel
3. Taxes
4. Water-right payments
5. Depreciation of facilities
6. Appreciation on land
7. Regular maintenance expenses
8. Unplanned repairs (renewals and replacements)
9. General and administrative

As the list suggests, virtually all significant expenses can be forecast. General and administrative costs are fixed costs. However, unplanned repairs could be a significant item during emergency periods; for example, in the event of a flood there would be significant clean-up expense and repair work. Obviously, these costs cannot be predicted, but funds should be available in the event of emergencies.

The annual cost of power generation will increase each year because of escalation of operation and maintenance costs. Often, a levelized or average annual generation cost can be determined in dollars per kilowatthour over n years by discounting all costs forward on a present worth basis using the following expression:

$$L = R\left(\frac{a^n - 1}{i\,a^n}\right) + A\left(\frac{b}{i - r}\right)\left(\frac{\frac{a}{b} - 1}{\frac{a^n}{b}}\right) \tag{11.7}$$

where $a = 1 + i$, $b = 1 + r$, and L is the total annual expenditure; R is the annual debt service; A is the annual O&M costs including interim replacements, insurance, taxes, wages, G&A; r is the inflation rate of O&M costs; n is the number of years; and i is the interest rate. The cost of power is simply:

$$\$/\text{kWh} = \frac{L}{\text{Annual energy production}}$$

The cost of energy at U.S. plants are often forecast as a fraction of capital cost as suggested by Gordon and Penman.[9] They assume that annual costs are 12.5 percent of capital costs.

11.8 U.S. FINANCING

As shown earlier, the cost of producing electricity is dependent upon plant operations, regulation policies, and financing. In the United States, two financing alternatives were available: government loans administered through the Department of Energy (DOE) until 1983 and the sale of bonds. The Public Utility Regulation Act of 1978 made funds available through DOE for up to 75 percent of project costs. These loans were for 30 years with an interest rate determined at the time of application. The conventional approach to financing by public agencies is to sell tax-exempt revenue bonds to be retired in 40 years or less. An entire project could be financed this way, or, until 1983, with DOE financing and the remaining 25 percent financed privately.

11.9 SOME CASE STUDY EXAMPLES

Mora Canal Drop[10]

In 1978, the Tudor Engineering Company of San Francisco, California, carried out a feasibility study on the hydroelectric potential of the Mora Canal Drop located near Boise, Idaho. The study was requested by the Boise Project Board of Control, an irrigation control authority. Recommendations included installation of a 1,900-kW tubular-type turbine-generator operating under a head of approximately 12 m (39.36 ft).

Table 11.7 presents costs and computed energy production costs under a bond financing arrangement. Two costs are given because two flows were used

TABLE 11.7 Energy Production Cost Computation for Mora Canal Drop

Initial capital cost	$1,787,750
Annual expense:	
Repayment: $1,000,000 7%, 40 years	75,091
$ 787,750 10%, 40 years	80,555
Operation and maintenance	24,700
Total annual expenses	$ 180,346

Energy production cost:

Average annual energy = 8,113,000 kWh

$$\text{EPC} = \frac{180{,}346}{8{,}113{,}000} = \$0.022/\text{kWh}$$

Average annual energy = 6,612,000 kWh

$$\text{EPC} = \frac{180{,}346}{6{,}612{,}000} = \$0.027/\text{kWh}$$

TABLE 11.8 Benefit/Cost Ratios with Conventional Funding Option Mora Canal Drop

	Discount Rate, %		
	6	8	10
Energy			
8,113,000 kWh/Yr			
Present worth benefits	$3,324,784	2,580,510	2,091,408
Present worth costs	2,745,190	2,158,178	1,766,975
Benefit/cost ratio	1.21	1.20	1.18
Energy			
6,612,000 kWh/yr			
Present worth benefits	$2,709,660	2,103,085	1,704,473
Present worth costs	2,745,190	2,158,178	1,766,975
Benefit/cost ratio	0.99	0.97	0.96

in the original computations. These costs must be met by revenues if the project is to be economically viable. Table 11.8 presents benefit/cost ratios for three discount rates. Only the higher flow condition will yield a positive benefit/cost ratio at all three discount rates. Further study could be warranted since escalation was not considered in this model. For example, if electricity cost escalation is included, the increase in benefits will outstrip increases in O&M costs thereby raising the benefit/cost ratio. This example demonstrates the sensitivity of energy production costs to the amount of energy production and the benefit/cost ratio to discount rate.

IWRRI Parametric Study[11]

When the decision maker is faced with a number of possible technical alternatives as well as with uncertainty regarding interest rates and cost escalation in fossil-fuel prices, a detailed sensitivity analysis of energy production costs to these factors must be carried out. In a sensitivity analysis or a parametric study, several variables are given a range of values, and their effects on the objective function noted. A typical hydropower parametric study would focus on the cost of electricity subject to variations in plant size, discount rate, load factor, capital cost, O&M costs, and plant life.

A recent (1978) study by the Idaho Water Resources Research Institute (IWRRI) illustrates the technique. Their hypothetical example is reproduced in part here. The basic assumptions are

1. Installed generating capacity is 6 MW.
2. Capital costs vary from $1,000 to $3,500 per installed kilowatt, in $500 increments.
3. The plant is funded through a 40-year loan at 10 percent interest.
4. Operating and maintenance (O&M) costs are $50,000 the first year.

5. Energy production is 20×10^6 kWh/year with a 100 percent plant factor.
6. The value of electric power varies between 20 and 50 mills*/kWh in 5-mill increments.
7. Escalation is 5 percent per year.

The benefit/cost ratio was the preferred method of analysis. The present value of costs and benefits are compared using discounted cash flow as previously presented. Table 11.9 shows a typical calculation sheet assuming $1,500/kW installed capital cost. Computations of this type can be easily carried out on a microcomputer. The benefit/cost ratio for this scenario is 1.37. Figure 11.2 shows the value of energy produced under various values of capital cost and first year energy values for power purchased from nonhydro suppliers. Note that a large capital investment decision made early has an advantage over future uncertainty regarding fossil-fuel prices. Figure 11.3 shows the benefit/cost ratio as a function of capital cost and the energy value of non-hydro-generated power. As the cost of purchased power increases, the benefit/cost ratio increases for an investment decision currently made. Figure 11.4 shows the energy value as function of capital cost when the benefit/cost ratio is 1.

A parametric study of this type allows the decision maker to examine a large number of possible alternatives under conditions of uncertainty. A determination can be made as to which specific scenario makes the hydropower system economically feasible. This type of sensitivity analysis is being developed in conjunction with efforts to select economically viable sites from a large number of possibilities.[12,13]

Swaziland Micro Hydro Project[14]

Life-cycle costs for a diesel-pump project are compared with a micro hydro electric pump combination. In conjunction with a World Health Organization (WHO) Integrated Water Supply and Sanitation Demonstration Project, located in the Shiselweni district of Swaziland, seven diesel pumps are to be installed to supply potable water to approximately 1,000 homesteads. Located in the vicinity of one of the boreholes is a small stream, the Matimatima, where a micro hydro plant is to be installed to provide electricity for an electric pump to replace one of the proposed diesels.

Life-cycle cost analysis as described in Section 11.4 is used to compare alternatives. The basic economic assumptions are given in Table 11.10. Project life for small diesels is usually 8 to 10 years and for micro hydro plants 15 to 20 years. However, for simplicity, 10 years is used for both. Discount rates of 10, 20, and 30 percent are applied to cover a range of funding possibilities, including private. For very high discount rates, choosing a long time horizon may be irrelevant since expenditures after a certain period would be inconsequential.

*A mill is the popular definition of $0.001.

TABLE 11.9 IWRRI Hydro Example—Calculation Sheet

Year	Repay, $ (40-year/10%)	O&M, $	Total, $	Hydro mill rate, mills/kWh	Energy purchase, $
1	920,335.00	50,000.00	970,335.00	48.52	800,000.00
2	920,335.00	52,500.00	972,834.94	48.64	839,999.94
3	920,335.00	55,124.99	975,459.94	48.77	881,999.88
4	920,335.00	57,881.24	978,216.19	48.91	926,099.81
5	920,335.00	60,775.30	981,110.25	49.06	972,404.75
⋮	⋮	⋮	⋮	⋮	⋮
36	920,335.00	275,789.50	1,196,133.00	59.81	4,412,776.00
37	920,335.00	289,588.38	1,209,923.00	60.50	4,633,414.00
38	920,335.00	304,067.75	1,224,402.00	61.22	4,865,084.00
39	920,335.00	319,271.13	1,239,606.00	61.98	5,108,338,00
40	920,335.00	335,234.63	1,255,569.00	62.78	5,363,754.00

Benefit/cost ratio = 1.37
6 MW/20,000,000 kWh annual energy
40 mill/kWh thermal energy purchase alternatives
Capital cost: $1500/kW installed
Assumed interest rate: 10%
Assumed 5% annual inflationary rate on O&M and purchase energy

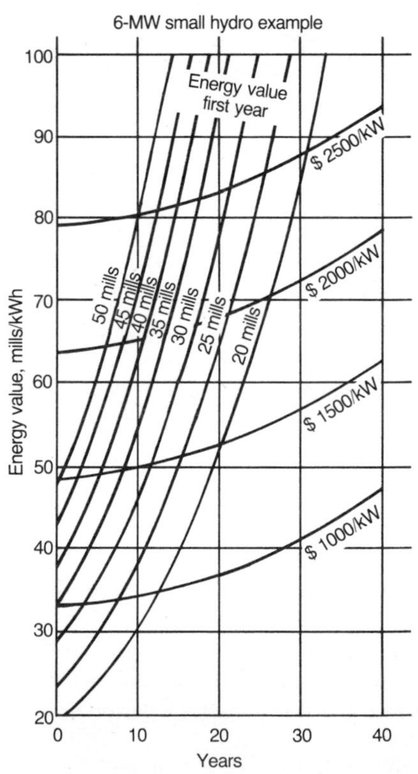

FIG. 11.2 Value of energy under various capital costs.

Purchase mill rate, mills/kWh	Present worth factor (PWF) 10%	Present worth (PW) cost, $	Present worth (PW) benefit, $	Present worth (PW) surplus, $
40.00	0.9090914	882,123.13	727,273.06	−154,850.06
42.00	0.8264475	803,997.00	694,215.88	−109,781.13
44.10	0.7513167	732,879.31	662,661.25	−70,218.06
46.30	0.6830157	668,137.00	632,540.69	−35,596.31
48.62	0.6209239	609,194.81	603,789.38	−5,405.44
⋮	⋮	⋮	⋮	⋮
220.64	0.0323502	838,695.17	142,754.25	104,059.06
231.67	0.0294093	35,583.02	136,265.56	100,682.50
243.26	0.0267358	32,735.32	130,071.75	97,336.38
255.42	0.0243053	30,128.96	124,159.50	94,030.50
268.19	0.0220957	27,742.69	118,515.94	90,773.19

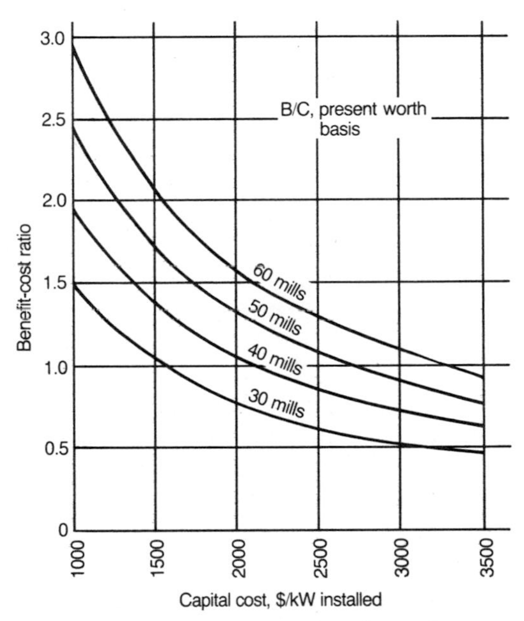

FIG. 11.3 B/C ratio as a function of capital cost.

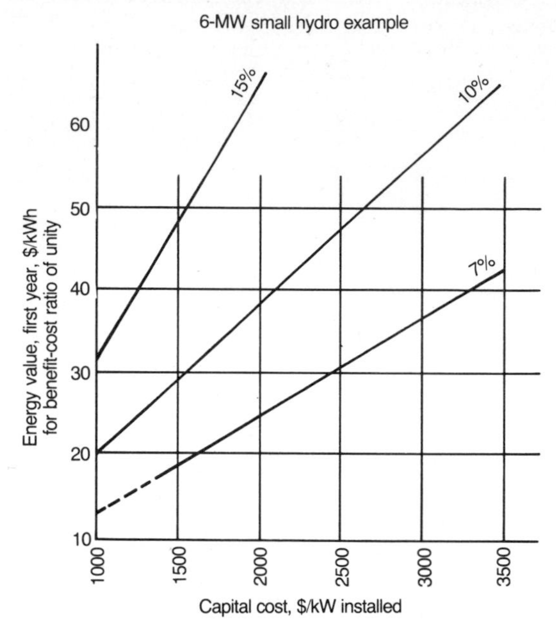

FIG. 11.4 Energy value as a function of capital cost and a B/C ratio of unity at three interest rates.

Residual or scrap value is difficult to determine; however, 10 percent for mechanical equipment is not uncommon.

Diesel-fuel prices will continue to rise. Swaziland diesel-fuel prices were approximately $1.50 per gallon in 1979, the year of this investigation. Considerable uncertainty continues to exist in future long-term diesel-fuel prices. It is therefore convenient to present a wide band of price increases and then conduct a sensitivity analysis. Three fuel price scenarios were used as shown in Table 11.10: high, intermediate, and low. Since the intermediate projection is no more likely to occur than the high or low, it was important to use all three projections to determine the sensitivity of life cycle costs to different levels of fuel prices.

TABLE 11.10 Life-Cycle Costing Basic Assumptions

Economic:	
Project life	10 years
Discount rate	10, 20, 30%
Residual value	10% of capital cost
Diesel-fuel price increase*	5, 10, 20%

*Base price, $1.50/gal, 1979. Diesel fuel price for any year can be computed by $P = P_b (1 + f)^t$, where P is price year t, P_b is base price, here $1.50, f is annual increase, here 10%, and t is years, here 5. The diesel fuel price increase does not include inflation since a constant-dollar analysis is used.

TABLE 11.11 Diesel-Pump Costs, December 1979

Capital Costs	
Small diesel, 10 kW*	$1,000
Centrifugal pump, 500-ft (152.4-m) head	600
Frame, controls, miscellaneous equipment	400
	$2,000
Annual Operation and Maintenance Cost	
Diesel fuel, 0.50 L/kWh $1.50/gal, pumping for 4 h per day yields $16.48 per day, or	$2,370
Maintenance, 10% of capital cost	200
	$2,570

*Total head is 163 m (535 ft) at Matimatima which includes a well 42.6 m (139.7 ft) deep. No pressure drop is assumed due to friction since pipe size has not been determined and the pump will not be drawing from the bottom of the well but from the drawdown located above the bottom. To pump 3 L/s (approximately 50 gal/min) over a 163-m (525-ft) head requires 4.7 kW or with a 50 percent efficient engine/pump will require a 10-kW machine.

TABLE 11.12 Hydro-electric Pump Costs, December 1979

Capital Costs	
Electro-mechanical equipment:	
10-kW turbine-generation set	$ 6,000
Extension line, 900 V, 1 km (0.621 mi)	4,000
Civil works impoundment, penstock, powerhouse	3,000
Electric motor pump, 500-ft (152-m) head, 50 gal/min (189 L/min)	1,000
	$14,000
Annual Operation and Maintenance Costs	
6% of capital costs	$ 840

Table 11.11 presents diesel-pump capital and operation and maintenance (O&M) costs and Table 11.12 shows the same costs for the micro hydro electric pump combination. As expected, capital costs are higher for the micro hydro but operating costs are less because of nil fuel expenses. It should be noted that the civil cost of $3,000 is approximate, based on a preliminary design, since detailed engineering is necessary to make an exact determination.

Computations were carried out to determine the sensitivity of life-cycle costs to:

1. Three reference diesel-fuel annual real price increases, 5, 10, and 20 percent.
2. Three discount rates, 10, 20, and 30 percent.

Figure 11.5 illustrates the chronological constant dollar cash flows for the two projects. Note the effect of diesel-fuel price increases. On the basis of constant dollar cash flows, the diesel pump is cheaper for the first few years. However, the high capital cost of the micro hydro installation is overshadowed after approximately 3 years by the escalating fuel price increases which led to a large cumulative expenditure regardless of discount rate. Since hydropower is capital intensive, life-cycle costs are particularly sensitive to discount rates as we see below.

Figure 11.6 shows life-cycle costs or present worth of all future expenditures

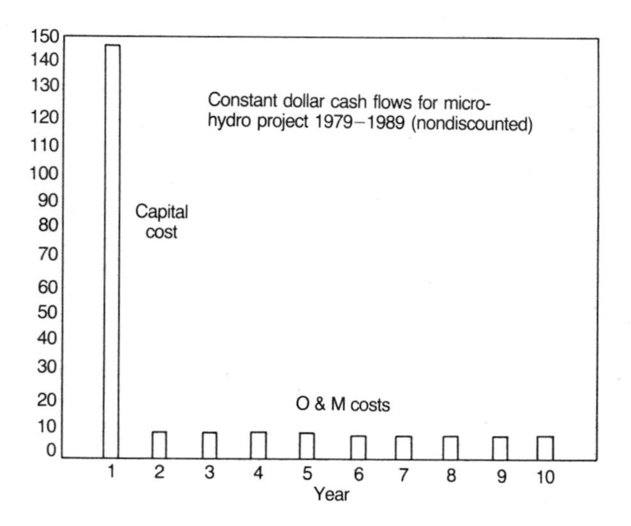

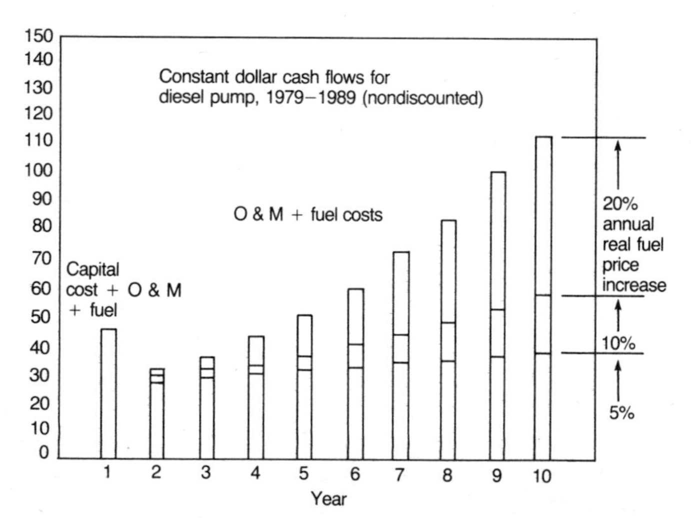

FIG. 11.5 Constant dollar cash flows for diesel pump and micro hydro set.

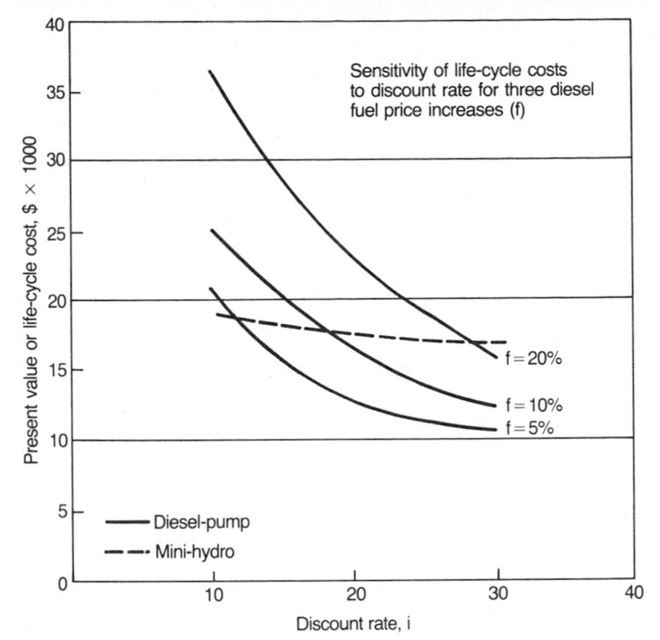

FIG. 11.6 Sensitivity of life-cycle costs to discount rate for three diesel fuel price increases *f*.

as a function of discount rate. The rather interesting result is that, with high discount rates, the diesel pump becomes competitive with the hydro scheme. However, it is somewhat doubtful that the dual effect of a real fuel price increase of only 5 or 10 percent will occur simultaneously with discount rates going above 20 percent. It is assumed that the discount rate is informally related to energy price increases. A discount rate of 15 percent and an annual real fuel price increase of 10 percent appear more likely. Under those circumstances, the micro hydro scheme is favorable.

An additional factor not considered in this analysis is that generally the project life for the micro hydro exceeds 10 years, whereas for the diesel pump, it is usually less than 10 years. Replacement costs were also not considered. If they were, the micro hydro project would appear clearly more attractive.

The expressions which were used to compute the present worth of all future expenditures for the two schemes are given below. The present value of the micro hydro life-cycle costs are limited to current capital cost and O&M costs. The present value of the diesel-pump system costs include capital costs, and O&M costs as well as fuel costs subject to annual escalation.

Micro Hydro

$$E_h = 14{,}000 + 840 \sum_{j=1}^{10} \frac{1}{(1 + k)^j}$$

Diesel Pump

$$E_d = 2{,}000 + 200 \sum_{j=1}^{10} \frac{1}{(1+k)^j} + 2{,}370 \left[\frac{1}{(1+k)}\right] + \ldots$$
$$+ 2{,}370\,(1+f)^{10} \left[\frac{1}{(1+k)^{10}}\right]$$

where, f is the diesel-fuel price increase exclusive of inflation.

It would appear from the analysis that the micro hydro is cost competitive based on the assumptions. If we use only constant dollar cash flow, the micro hydro is competitive after about 3-years' service. By choosing life-cycle cost analysis, the magnitude of the discount rate has affected the economic viability of the project. However, since we assume that diesel-fuel prices will continue to climb and the discount rate is highly variable, it certainly appears prudent to spend more of today's dollars to avoid uncertain expenses in the future.

Mbulu Hydro Project[3]

In conjunction with their energy analysis efforts, the United Nations Department of Technical Cooperation for Development sent a reconnaissance mission to the Arusha region of Tanzania in 1981 to evaluate the feasibility of small hydropower. A site on the confluence of the Hainu and Nambisi rivers near the town of Mbulu was selected for analysis.

The project consists of a 1500-kW capacity run-of-river plant utilizing three 500-kW turbine-generators operating under a head of 420 m (1377.6 ft). A

TABLE 11.13 Capital Cost of Mbulu Project*

Item	Depreciation period, years	Capital investment, $ US 000	Capital recovery factor, 10% annual interest	Annual cost, $ US 000
Generation:				
Civil works	35	2,250.00		
Plant and equipment	35	750.00		
Subtotal		3,000.00		
Interest during construction		210.00		
Subtotal generation:		3,210.00	0.1037	330
Transmission, transformation,		1,107.00		
and distribution	25			
Interest during				
construction		77.50		
Subtotal		1,184.50	0.1102	130
GRAND TOTAL		4,394.50		

SOURCE: Adapted from Ref. 3.
*U.S. dollars.

TABLE 11.14 Annual Fixed Costs of the Mbulu Project*

Cost item	Generation Works	Transmission and distribution
Maintenance, 1% for generation and 2% for transmission and distribution	30,000	22,000
Operation, 0.5%	15,000	5,500
Overhead, 0.5%	15,000	5,500
Capital Investment	330,000	130,000
Total	390,000	163,000

SOURCE: Adapted from Ref. 3.
*U.S. dollars

distribution system supplying power to two communities, Mbulu and Babati was also planned. It is anticipated that 8.0 GWh of energy at a 60 percent load factor will be available. Construction is estimated to take 3 years with capital costs expended at 10, 60, and 100 percent cumulatively.

Tables 11.13 and 11.14 show capital cost and annual fixed costs for the project. The cost of generation was computed as follows:

$$\frac{\text{Annual fixed costs}}{\text{Annual energy production}} = \frac{\$556{,}000}{8.0\text{ GWh}} = \$0.070/\text{kWh}$$

If we assume a loss of 10 percent during transmission, leaving 7.2 GWh, then the cost to the consumer would be $0.077/kWh.

11.10 SUMMARY

The material and case study examples given in this chapter were presented to illustrate the variety of economic analysis techniques which can be applied to determine both economic and financial feasibility. However, informed judgment and reasoning must be used in developing underlying assumptions and choice of analysis techniques. Clearly, an analyst could influence the outcome through the choice of techniques and assumptions. There are no universal approaches, but only rough guidelines. Care must, therefore, be exercised in developing the economic and financial rationale for a particular project.

In addition, economic and financial analysis of a hydro scheme in a developing country may require the use of shadow economics, determining the effects of subsidization and balance of payments implication. The information presented in this chapter is primarily for use in an unplanned market economy where revenues gained through the sale of electricity must fully support a project.

11.11 REFERENCES

1. E. L. Armstrong, "The Impact of the World's Energy Problems on Low Head Hydroelectric Power," in J. S. Gladwell and C. C. Warnick (eds.), *Low Head Hydro,* Idaho Water Resources Research Institute, Moscow, Idaho, 1978.
2. E. A. Moore, "Electricity Supply and Demand Forecasting for Developing Countries," Energy Department, World Bank, Washington, D.C., 1979.
3. United Nations, "Evaluation of Small Hydro-Power Sites in Tanzania," TCD/INT-80-R47/3, U.N. Department of Technical Cooperation for Development, New York, 1982.
4. H. G. Van der Tak, *The Economic Choice Between Hydroelectric and Thermal Power Developments,* Johns Hopkins Univ. Press, Baltimore, Md., 1966.
5. E. L. Grant, W. G. Ireson, and R. S. Leavenworth, *Principles of Engineering Economy,* 7th ed., John Wiley & Sons, New York, 1982.
6. M. Henwood, "Economic and Financial Feasibility Study Methodologies," *Proc. Small Hydroelectric Powerplants,* NRECA, Washington, D.C., 1980.
7. Announcement of Pacific Gas and Electric Company, February 1980.
8. U.S. Army Corps of Engineers, *Hydropower Cost Estimating Manual,* Institute of Water Resources, National Hydropower Study, May 1979.
9. J. L. Gordon and A. C. Penman, "Quick Estimating Techniques for Small Hydro Potential," *Water Power and Dam Construction,* September 1980.
10. Tudor Engineering Co., "Final Report on Potential Hydroelectric Power, Mora Canal Drop," DOE No. EW-78-F-07-1760, San Francisco, December 1978.
11. J. S. Gladwell, "Small Hydro: Some Practical Planning and Design Considerations," IWWRI, University of Idaho, Moscow, Idaho, April 1980.
12. S. R. Harper, *Assessment of Small Hydropower Resources in the Republic of Korea,* Olympic Assoc. Co., Seattle, Wash., 1982.
13. *Methodology for Regional Assessment of Small Scale Hydropower,* Tudor Engineering Co., San Francisco, 1983.
14. J. Fritz, "Brief Comments on the Cost Aspects of Mini-Hydropower Systems," *Small Scale Hydropower in Africa,* NRECA, Washington, D.C., 1982.

12

System Design and Case Studies

Jack J. Fritz

12.1 PLANNING A SMALL HYDROPOWER PROJECT

Implementing a hydropower project requires a series of sequential steps which can broadly be categorized as planning activities, procurement, construction, and start-up. A typical project schedule as described by the U.S. Army Corps of Engineers is shown in Fig. 12.1.[1]

Planning studies include all types of investigations performed to determine the desirability of carrying out a project. These studies are initiated when a proposal for a site is deemed worthy of interest and should be completed when construction starts (on Fig. 12.1, planning studies therefore would extend from time 0 to approximately the 24th month).

Planning studies vary in scope, detail, depth, and intended audience and lead to various decisions and commitments made during the preconstruction period. Following business and international practice, planning studies are generally grouped in three main categories: (1) reconnaisance studies, also referred to as appraisal and prefeasibility studies, (2) feasibility studies, and (3) detailed design studies or definitive site studies.

Briefly, the objective of *reconnaisance studies* is to determine if the project under investigation merits a full feasibility study (months 0 to 1 on Fig. 12.1). *Feasibility studies* will define the proposed project more concretely and assess its potential as well as determine whether an investment commitment should be made (months 1 to 6 on Fig. 12.1). *Definitive site studies* are final studies performed between the time of the implementation commitment and the beginning of construction (months 6 to 24 on Fig. 12.1). These studies result in permit applications, licensing negotiations, financing arrangements, marketing agreements, and engineering designs and specifications.

As indicated in Fig. 12.1, the budget required to perform all site studies could reach 25 percent of total project cost; reconnaissance and feasibility studies alone may require up to about 10 percent of total cost. That sufficient funding must be available to reach a substantiated "go/no-go" decision must be kept in mind by planners and investors when initiating a proposed project.

The main purpose of this handbook is to provide planners, engineers, and decision makers responsible for the development of hydropower resources with sufficient background and understanding of the tasks and elements included in the design and analysis process. Emphasis is placed on reconnaissance and feasibility analysis during which the role of the planner or project initiator is crucial. Once a decision to implement has been reached, the detailed site studies can be performed by specialized teams or individuals. At that point, the services of a project engineer are required in coordinating the activities of several specialists.

12.2 RECONNAISSANCE STUDIES

The reconnaissance study is a small-scale or abridged feasibility study intended to determine which of several sites should be considered for more in-depth analysis. Therefore, it is broad and designed to highlight the critical issues, taking a first cut at presenting the principal technical, economic, and financial factors. It does not include a sensitive analysis but might propose several technical alternatives. Reconnaissance studies are designed to reduce the risk of unfavorable projects by increasing the potential for selecting attractive sites thereby optimizing the investment. It is relatively simple to rank several reconnaissance studies to identify sites which merit priority development. These sites would have favorable indicators such as a positive benefit/cost ratio warranting further in-depth investigation.

The U.S. Army Corps of Engineers defined a reconnaissance study as "a preliminary feasibility study designed to ascertain whether a feasibility study is warranted."[1] A reconnaissance study is, therefore, an abridged feasibility study having as its major objectives a preliminary economic evaluation and identification of the critical issues which will have to be considered during the feasibility study.

Figure 12.2 shows a generalized study flowchart proposed by Tudor Engineering Company for performing a feasibility study.[2] The same basic flowchart can be used for reconnaissance studies with somewhat less depth of investiga-

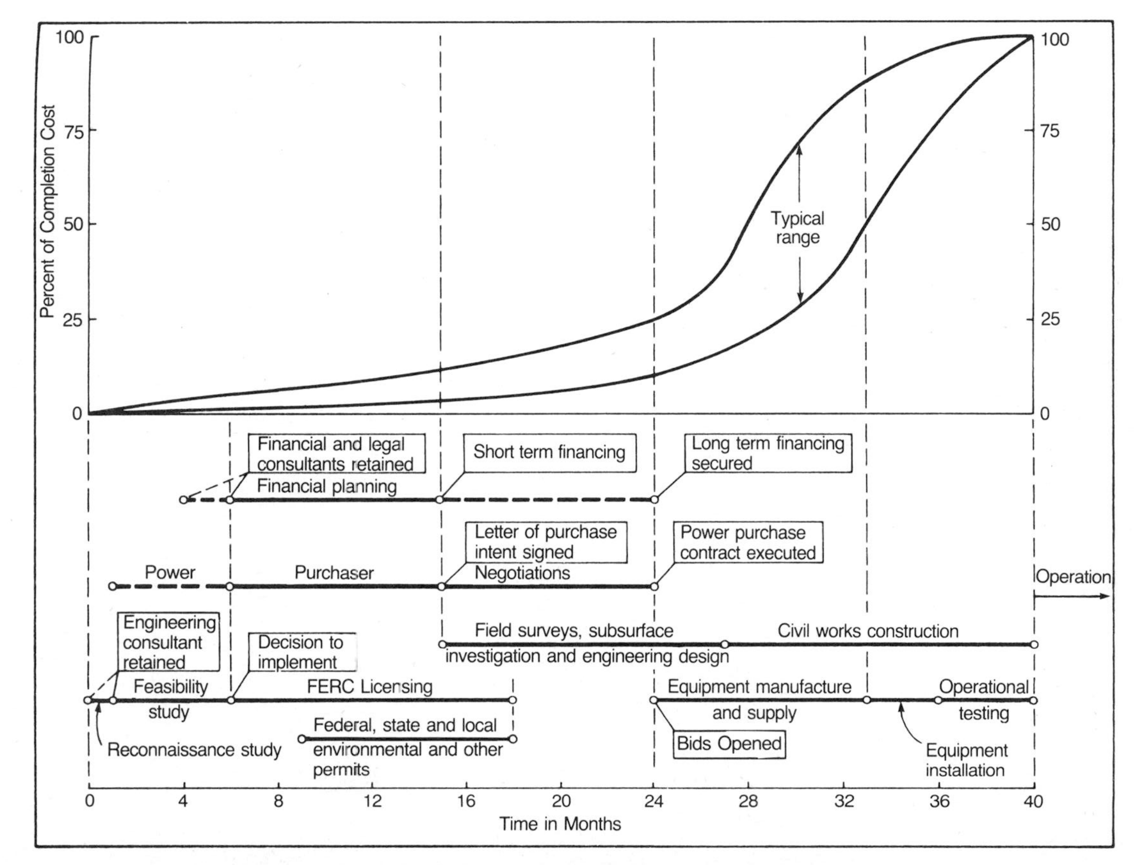

FIG. 12.1 Typical project implementation schedule and expenditure patterns. *(From Ref. 1.)*

tion. The procedure proposed by the U.S. Army Corps of Engineers for conducting a reconnaissance study is also similar to that shown in Fig. 12.2. The steps involved in the reconnaissance analysis are briefly described:

- **Steps 1 to 3:** Project identification, site conditions, and project layout: *Project identification* includes determining location of the project, approximate market for the electricity produced (local demand or interconnection to existing grid), existing facilities, etc. *Site conditions* include securing maps, determining geological conditions, status of water rights, etc. A preliminary arrangement based on site conditions can then be proposed. The *project layout* should enable the designer to determine the available head and take into account other water demands such as irrigation.
- **Steps 4 and 9:** Identify potential environmental factors critical to the development.
- **Step 5:** Institutional factors: This task should evaluate regulatory, institutional, licensing, marketing, and transmission factors as well as identify the agencies or other governing bodies involved in resolving these issues. Potential roadblocks should also be identified.
- **Steps 6 and 11:** Flow data, capacity, and energy output: When available, flow data should be used to evaluate the energy output for a range of power capacities. At some sites approximate flows may have to be generated due to the lack of sufficient historical data.
- **Step 7:** Proposed project layout is made on the basis of preliminary flow data.
- **Step 12:** Project costs: Project costs for various ranges of capacities and plant designs are determined using standard tables of cost versus capacity and plant characteristics (cost versus capacity and head, for example).
- **Steps 8 and 10:** Alternative sources of electrical power: The cost of energy produced from sources such as diesel or grid as well as the potential for sales should be estimated. The most attractive financing mechanism can also be identified. These could include government loans, grants, or other investment strategies.
- **Steps 13 and 14:** System optimization: The economic data generated for various configurations are used to select the capacity of the plant and determine the project's physical arrangement. This task is basically an analysis of the sensitivity of costs and revenues to alternative design options.
- **Steps 15 and 16:** Determine environmental and institutional requirements: Environmental constraints should be identified and a plan developed to meet all institutional requirements.
- **Steps 17 to 20:** Finalize plant characteristics, costs, economics, and financing: These steps will fix project characteristics, economics, and

financing mechanisms. They constitute the *bottom line* of the reconnaissance study on the basis of which recommendations are made.

- **Step 21:** Project schedule: A tentative project schedule should be prepared.
- **Step 22:** Recommendations and final report: This task summarizes the findings with emphasis on the economic feasibility of the project and identification of areas requiring further investigation. A recommendation is made on this basis to proceed with a full scale feasibility study.

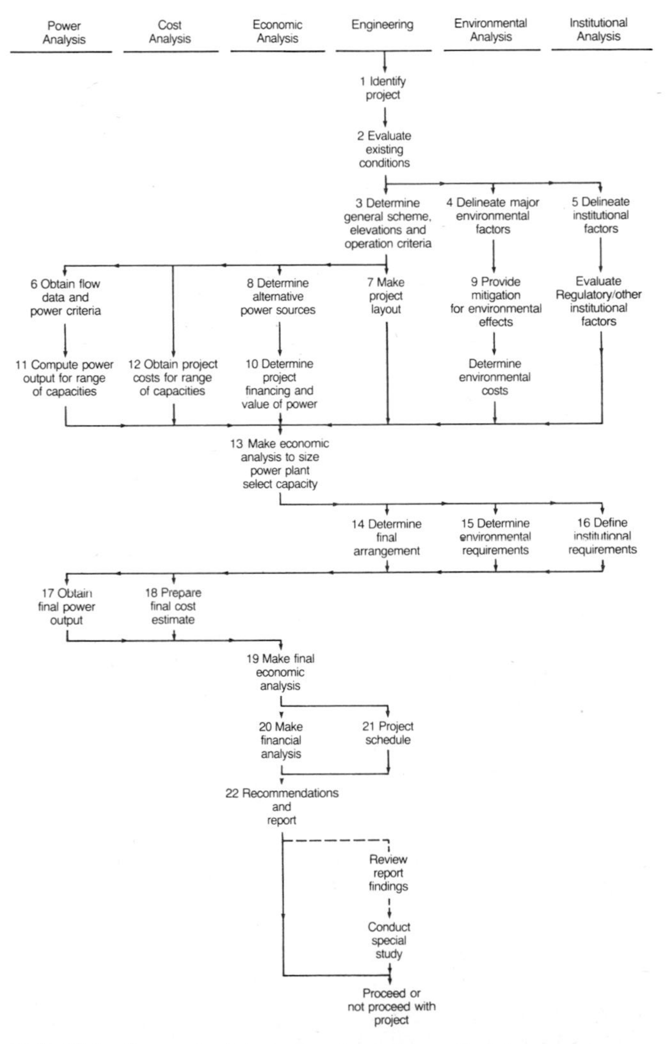

FIG. 12.2 Generalized study flowchart. *(From Ref. 2.)*

The flowchart described is not rigid and should be modified as dictated by site-specific conditions and institutional requirements. Based on U.S. labor and overall project costs, the U.S. Army Corps of Engineers estimated the cost of reconnaissance studies to be from 0.15 to 0.30 percent of the estimated construction cost. These estimates can be improved when project scale is known. A more realistic estimate of the effort involved in reconnaissance studies may be about 10 to 15 worker-days over a 15-to-30-day time frame for projects in the 1-MW range and 45 to 60 worker-days over a 45-to-90-day time period for projects in the 15-MW capacity range.[1]

In the case of mini or micro hydro projects, an experienced engineer with a background in economic analysis should be able to perform the evaluation. Reconnaissance studies for small hydro projects in developing countries may require the participation of an engineer, an economist, and a social scientist. Table 12.1 illustrates the key elements suggested for a reconnaissance study in a developing country.

12.3 FEASIBILITY STUDIES

The aim of reconnaissance studies is to carry out a preliminary economic analysis and to identify the issues critical to the successful implementation of the

TABLE 12.1 Elements of a Reconnaissance Study in a Developing Country

I. Site-development rationale • Economic profile of region • Energy demand	VI. Project benefits • Identify revenue stream • Identify indirect benefits
II. Site description • Community description • Physical site characteristics	VII. Economic feasibility • Cost/benefit analysis or rate-of-return
III. Power and energy potential • Hydraulic resource • Power determination • Annual energy production and profile	VIII. Financial feasibility • Identify funding mechanism • Covenants of loan agreement • Cash flow analysis
IV. Physical configuration • Turbo-generating equipment • Electrical equipment • Distribution system • Civil works	IX. Environmental impact • During construction • During operation
V. Project costs • Capital costs • Annualized O&M costs • Annual investment cost	X. Social analysis • Management and operation • Local participation opportunities • General community impact • Institutional arrangements

project. In contrast, the purpose of a feasibility study is to firm up the economic analysis and to analyze those unresolved issues. The definition of a feasibility study proposed by the U.S. Army Corps of Engineers, "an investigation performed to formulate a hydropower project and definitively assess its desirability for implementation," summarizes the scope of a feasibility study.[1]

A full-scale feasibility study contains all the elements of the reconnaissance study, usually including a sensitivity analysis of various design and finance options, design drawings, and in-depth discussion of all issues raised in previous studies. A team incorporating several disciplines is generally necessary: geotechnical engineering, hydropower engineering, electrical engineering, economics, finance, law, and social sciences. The project is formulated, from beginning to end. Costs and benefits must be clearly defined. Construction and operation management must be determined as well. The feasibility study should guide a contractor in carrying out all details of the project. The in-depth economic and financial analyses normally would include a sensitivity analysis to determine the effect of variable discount rates, inflation, and several tariff structures. The analysis would also address financing strategies—public, private, or quasi-public—loan repayment, and cash flows.

Cursory system design is carried out during the feasibility phase. This generally includes a detailed site plan giving plant configuration and required excavation. It should include overall drawings of the powerhouse and all civil and electrical works; a detailed list of materials, labor, and supervisory services; budgets; cash flows; and a detailed implementation schedule.

In addition, an in-depth determination of the productive uses of electricity should be made for sites in developing countries. Increased agricultural productivity or the establishment of small rural industries become the principal economic justification for development of a site. When computing the benefits of a particular development, the additional income resulting from increased energy availability must be factored into the analysis.

The specific tasks required for a feasibility study were outlined on Fig. 12.2. This approach was developed by the Tudor Engineering Company under contract to the U.S. Department of Interior's Water and Power Resource Service.[2] The tasks are organized in an increasing order of specificity as exemplified by the "engineering" column. For example, tasks 1 to 3 deal with general, broad characteristics of the project, task 7 suggests one or several possible layouts for the project, task 14 focuses on a final design for the project, and task 21 proposes a schedule for implementation. A similar approach to specificity is pursued in other areas of investigation: economic analysis, cost analysis, power analysis, etc. The background provided by the reconnaissance study should reduce the alternatives to be examined in each area of investigation. As an example, the reconnaissance study has identified the critical environmental issues relevant to the project, thereby focusing the scope of the feasibility study in that area.

The tasks included in a feasibility study are described below (refer to Fig. 12.2 for the sequence of tasks):

1. **Project Identification:** The scope of this task is to identify the location and main features of the project. The project may be built around existing structures, such as an irrigation dam and reservoir, or interconnection with an electrical grid may be planned. The near-term market and projected demand pattern for electricity should be determined. This may be a critical factor in developing areas where the capacity factor may be low initially, thereby reducing the economic attractiveness of a project. (Data needed to estimate projected demand were presented in Chaps. 7 and 10.)
2. **Site Conditions:** After a site has been identified it is necessary to evaluate local conditions. This procedure should include the following:
   - Procurement of site plans for existing developments, if any.
   - Obtain local topography map.
   - Determine the legal access to the site for both project construction and operation.
   - Obtain any existing data on site geology and determine the need for additional geologic exploration.
   - Review the history of prior developments at or near the site to determine any possible constraints on this or future developments.
   - Determine the current status of water rights for hydroelectric development.

   (The legal background on water rights and use required to perform this task were presented in Chap. 9.)
3. **Project Layout and Operational Criteria:** A preliminary project layout is made after evaluating site conditions. The layout should specify further evaluations required to determine the expected head through the turbine. It is also necessary to establish if the water releases from the project are to be allocated for other uses. These uses govern plant operation; power production can be significantly different if there can be operational flexibility permitting optimization of the power output. However, at this stage both the project layout and operation criteria are preliminary and subject to later modification as the study proceeds. (Methods for making a preliminary power production estimate were discussed in Chaps. 2, 3, and 4.)
4. **Environmental Factors:** The major environmental issues need to be investigated concurrently while preparing the preliminary layout. These include:
   - Effects of the project on fish and wildlife resources.
   - Changes in the water quality.
   - Determination of recreational benefits.

- Determination if an archeological investigation is required.
- Impact on visual appearances.

(Environmental considerations were reviewed in Chap. 8.)

5. **Institutional Factors:** The investigation of the institutional factors should be done at the same time as the preliminary layout preparation as follows:

- Application process for various required federal and state permits and licenses.
- Negotiations with those governmental agencies that may maintain rights at the proposed site.
- Preliminary negotiations for the sale or transmission of the electric energy.
- Review of special programs and procedures applicable to small hydro projects

(As discussed in Chap. 9, simplified licensing procedures have been proposed by the U.S. government to encourage the development of small hydro—less than 15-MW capacity. These procedures should be carefully reviewed to assess how they impact on the project under investigation.)

6. **Power Capacity and Energy Production:** A determination should be made of the potential capacity and annual energy production. The hydrologic analysis is key for determining the project output and subsequent economic benefits. A first step is the determination of the average monthly streamflows near the proposed site. Sources of data may include international, national, state, regional, and local agencies. Where records do not exist, or for short periods only, it may be necessary to synthesize streamflow by correlation with nearby stream gaging or precipitation stations. (Methods for estimating streamflows are discussed in Chaps. 2 and 3.) Before using the historic streamflow records, it is necessary to determine if there have been changes in the river system as a result of construction of regulating reservoirs or additional consumptive uses. If so, the historic streamflows should be modified to represent the present water uses. Such data are sometimes available as a result of prior work by agencies interested in the watershed.

Another alternative is to select a recent period of streamflows having the same means as the long-term record. This should point out the variability of monthly and yearly data highlighting critical dry periods, if any. The streamflows adopted can then be used to evaluate the power and energy production of the project and the size and effectiveness of any reservoir storage to be provided by a dam. Where a new reservoir and dam are shown to be feasible and effective, a separate cost/benefit analysis may be required to determine the optimal reservoir volume and dam height.

The second aspect of the hydrologic study is determination of flood volumes, peak discharges, and their frequency of occurrence. From this data follows the design of the spillway and other water diversion works. A further requirement is to determine the tailwater curve for the power plant site. It is used generally in its lower range for determining the operating head for the plant and in its upper range for the design of hydraulic structures around the powerhouse and outlet works to mitigate for flood events. (These aspects of the hydrologic study were discussed in Chap. 2.)

7. **Review Preliminary Project Layout:** A cursory review should be made of the flow data and power production so that refinements could be made to the preliminary project layout. Specific project features which affect power generation include:

   - Penstock length and diameter.
   - Size of tailrace excavation.
   - Preliminary estimate of the optimal turbine size and type.
   - Head loss through the waterways.
   - Tailwater calculated as a function of turbine flow.

   (Data needed to perform this task are found in Chaps. 4 and 6.)

8. **Alternative Electricity Sources:** It is necessary to identify the end-user to determine the expected cost of power that will be replaced by the proposed project. If the expected user is a nongenerating utility (or commercial company) the alternative electricity originates through power purchases. If the expected user is a generating utility, then the alternative electricity will likely come from existing or proposed plant facilities. In either case, the expected cost of the alternative electricity should be evaluated for use in estimating the value of energy from the proposed development.

   If the expected market will be new users, such as might be the case in rural areas, the alternative energy could probably come from diesel generation. The cost of alternative energy provided by electricity generation other than hydro should be used to estimate the value of energy from the project. In addition, it will be necessary to estimate the price that the potential market can bear; the lowest price for electricity generated could be too high for rural people. (Comparative cost/benefit analysis as described in Chap. 11 may have to be performed for potential alternative electric power sources to ensure that the hydro approach is the most economically attractive.)

9. **Major Environmental Factors:** On the basis of the refined project layout, a final review should be made of key environmental questions. An assessment should be made to determine which items could cause significant problems. (During construction and operation, a plan should be developed to resolve these problem areas as discussed in Chap. 8.)

10. **Initial Energy Sales and Project Financing:** When the project capacity and annual energy production have been determined, an initial evaluation for sales and project financing can be started. This will include the following items:
    - Review user demand curves if the energy is to be used by the developer.
    - Identify the various potential energy purchasers if any or all of the output is to be sold.
    - Make preliminary contacts, if feasible, with potential purchaser to determine the approximate value of the energy for use in preparation of an economic assessment.
    - Determine if probable project financing will be general obligation or revenue bonds if a domestic project.
    - Determine if there are currently any government programs to assist in project financing, loans, or grants for hydroelectric development.
    - Determine the suitability for each method of financing and, on a preliminary basis, choose the best option.

    (Approaches to perform these tasks were briefly discussed in Chaps. 9 and 11.)
11. **Expected Power and Energy:** Estimates should be made of the expected power capacity and energy production for several plant configurations in a range around the estimated optimal capacity. A minimum of four estimates should be made to establish the relationship between the plant capacity and annual energy production. If it appears suitable to use more than one type of turbine, then power and energy estimates should be made for each type, including multiple turbines. The possibility for developing more power and energy using a longer penstock and lowering the powerhouse (and afterbay) elevation, yielding a higher operating head, should be investigated. (Data required to perform these estimates were discussed in Chaps. 2, 3, and 4.)
12. **Estimated Project Costs:** Estimated project costs should be made for each plant configuration. The estimates need only be approximate since the differences between the estimates will determine the optimal capacity when the economic analysis is carried out. The estimated costs must include costs for any remedial measures taken for mitigation of harmful environmental effects. If detrimental environmental effects occur which cannot be remedied, an annual cost should be added to the annual operating cost. Some of the costs associated with environmental remedies have already been estimated in task 4. Others, specific to some system configurations, will have to be estimated on the basis of data found in Chap. 8. Rough cost estimates for the various components of the project (dam, turbine, switchyard, etc.) can be derived from data provided in the chapters dealing with these components or in Ref. 3.

13. **Optimal Plant Capacity:** Having the total project cost and estimated energy values for each plant configuration, one approach to ascertaining the optimal plant capacity is outlined:
    - Evaluate the annual costs and benefits over the entire project life.
    - Estimate the debt service, on the basis of estimated capital costs and method of financing, for each year for a 30- to 40-year period.
    - Estimate annual operation, maintenance, and replacement, first year's cost, and escalate future years by a forecast rate of escalation.
    - Estimate power and energy benefits throughout the analysis in a similar manner.
    - Convert the future values of costs to present worth values using the assumed discount rate cost of money to the developer.
    - Determine for each configuration and capacity the ratio of present worth benefits to present worth costs—the overall project benefit/cost ratio.

      The optimal project is clearly the one having the greatest difference between the total project benefits and the total project costs. If the benefit/cost ratio is less than 1 for all proposed configurations then the project is not economically feasible.

    (Additional approaches to performing such economic benefit/cost ratio analyses were discussed in Chap. 11.)
14. **Final Arrangement:** On the basis of the optimal project configuration, the final project layout drawings can be prepared. These drawings will be much less detailed than the final design drawings, but they must be sufficient to allow resolution of all space allocations or construction difficulties. Design specifications should be prepared which establish equipment, building, and facility criteria and which include a project design and construction schedule.
15. **Final Environmental Requirements:** A final determination should be made of remaining environmental issues. All required remedial measures should be identified and associated costs estimated. Any remaining adverse environmental effects should be specified for later inclusion in the environmental impact statement.
16. **Final Institutional Requirements:** A final plan for satisfying the institutional requirements should be prepared. A schedule with milestones should be drawn up of all institutional requirements.
17. **Final Energy Output:** A final estimate of the project power capacity and energy production is made. This estimate is prepared in a similar manner as earlier estimates; however, new data on project layout or operating criteria affecting the power output are included in the analysis.
18. **Final Cost Estimate:** A final cost estimate is prepared based on the final project layout. This estimate includes all project costs, construction, development, and other indirect costs. The bond issue or loan requirements can

be determined from this total project cost estimate. Annual cost of operation is also estimated.

19. **Definitive Economic Analysis:** A definitive economic analysis is made based on the final power capacity and energy output and the complete cost estimate. This determination is made in the same manner as the previously described economic analysis.
20. **Financial Analysis:** A review is made of the project on completion of the final economic analysis. The review includes the following:
    - Financing of the project development.
    - Costs associated with sale or transmission of the energy.
    - Suitability of both the project and the developer for the assumed type of financing.
    - Cash flow analysis during both the development period and the initial years of plant operation.

    It is not unusual for a project that is economically feasible to have cash flow requirements that preclude project development.
21. **Project Schedule:** A comprehensive project schedule should be prepared. The schedule should include significant milestones for design, purchasing of major equipment items having a long lead time, construction activities, start-up, and the process of meeting institutional requirements.
22. **Recommendation and Report:** A final report and recommendations are made on the basis of the foregoing analysis. The recommendations should include a decision to continue or not to continue the project and any further studies that should be performed. Often, special follow-up studies have to be made on items of environmental concern.

12.4 STREAMLINING THE STUDIES PROCESS IN A DEVELOPING COUNTRY

It is clear from the previous section that in a national mini hydropower program, a large number of plants would have to be sited and studied. This process, if executed in some detail, is time-consuming and represents a significant fraction of the capital cost. (A typical table of contents for a feasibility study in a developing country is given in Table 12.2.) Ways of streamlining this process are being developed in conjunction with major mini hydropower programs.

Procedures for site selection must be based on the physical, economic, and social realities of the region or country of interest. Initial attempts focus on the use of computer-aided procedures.[4,5] Such an applied methodology has been developed by ELECTROPERU, the national electrical utility of Peru, in conjunction with its mini hydro program.

As pointed out in Chap. 1, Peru is a country of varied topography and cli-

TABLE 12.2 Elements of a Feasibility Study in a Developing Country

I. Introduction
- Mini hydro in national energy strategy
- Rationale for development of site
- Sociocultural profile of community
- Economic/energy profile of community

II. Site description
- Physical features
- Community energy-use profile
- Existing diesel generation

III. Power and energy production
- Description of the hydrologic resource
- Power capacity and annual energy production
- Alternative plant schemes

IV. Physical configuration
- General arrangement of power systems
- Distribution systems
- Powerhouse details
- Details of civil works
- Electromechanical equipment specifications
- Safety equipment
- Excavation plan

V. Project costs
- Engineering and design costs
- Capital cost breakdown
- O&M costs

VI. Project benefits
- Tariff structure
- Direct revenue streams
- Increased community productivity
- Total community benefit

VII. Determine economic feasibility
- Cost/benefit analysis or
- Life-cycle cost analysis of alternative

VIII. Financial feasibility
- Loan or grant covenants
- Schedule of payments
- Financial arrangements

IX. Institutional analysis
- System operation and maintenance
- Field or service arrangements
- Tariff collection and administration

X. Environmental impact
- Description of flora and fauna
- Impact during construction
- Impact during operation

XI. Schedules
- Construction
- Cash flow analysis

XII. Summary recommendations

mate. It is arid along its Pacific coast where most population, agriculture, and industry clings. A short distance from the coast begin the Andes mountains which bisect the country separating the arid coast from the lush lowlands to the east which drain into the Amazon basin. Numerous mountain valleys along the Andes, some of which drain to the Pacific, are often used for irrigation. However, most of the well-watered valleys are on the eastern slopes where precipitation is abundant but population centers are few. Nevertheless, at least 1000 mini hydro sites have been identified which have the necessary hydrologic resource and energy demand. A process to prioritize a large number of sites was therefore developed.

In the ELECTROPERU procedure, several requirements had to be fulfilled before a community would be considered for inclusion in the list of possible sites. The factors considered were (1) reasonable proximity to appropriate water resource, (2) accessibility by road, (3) an area of agricultural, livestock, or other commercial activity, and (4) cooperation and interest in the community. Determination as to whether a community met these conditions could in part be satisfied by examination of topographic maps which show elevation, water resources, community location and population density, principal agricultural activities, and location of existing power grids. Visits to the communities would verify the accuracy of the maps, the degree of local interest, and level of prosperity. Through this procedure a large number of sites were identified.

Prioritization of sites for final study required a quantitative approach that ranked the communities. The specific criteria and their order used by ELECTROPERU follow:

1. Hydrologic resources and geologic characteristics
2. Topography of the hydrologic resource
3. Population benefited
4. Population density
5. Energy demand
6. Existence and type of access roads
7. Existence of drinking-water supplies
8. Rate of population increase
9. Thermal energy system alternatives
10. Accessibility to other power plants
11. Topography for transmission lines
12. General topography of region
13. Number of communities located in the vicinity

Each criterion was assigned a value from 1 to 10 with the sum representing the position that the site would occupy in the ranking scheme. Mathematically, it is simply

$$R = a_1x_1 + a_2x_2 + \cdots + a_nx_n + \cdots + a_{13}x_{13} \qquad (12.1)$$

where

R = ranking number

a_n = assigned value of each criterion, $0 \leq a_n \leq 1.0$ and $a_1 + a_2 + \cdots a_{13} = 1$

x_n = the score attached to each criterion, $1 \leq x_n \leq 10$

Clearly, the highest ranking number is 130 and the lowest 0.0. Once the ranking has been established, additional work to determine the approximate cost of developing specific sites followed. This was done by computing the capital cost and cost of electricity production from existing data at sites which have been developed nearby. Figure 12.3 shows the relationships that emerge from this sort of analysis. As the distance from the hydro plant to the community increases, from L_1 to L_3, unit energy costs increase. Clearly, as system capacity increases, unit energy costs are reduced due to a higher load factor and other economies of scale. Energy cost can then be used as a further ranking mechanism.

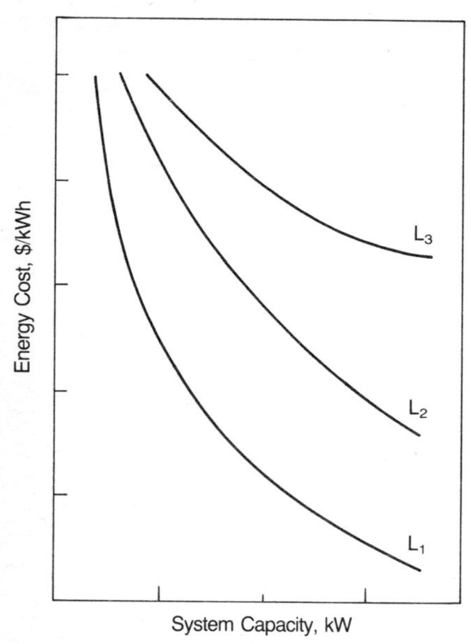

FIG. 12.3 Energy production costs as a function of system capacity and transmission distance.

12.5 CASE STUDIES

In this section, five small or mini hydropower case studies are presented to illustrate how information can be organized into a study format. Each is a summary or abridged version of the original studies. The format for the case studies is shown in Table 12.3. The information was provided with permission by engineering organizations and represents the principal conclusions of investigations carried out at actual sites. The oldest case study (1964), includes somewhat dated cost information, but the analysis is, nevertheless, meaningful.

Each case study presents a unique approach in its analysis. This is primarily due to specific site conditions, local existing power authorities, and different economic scenarios. However, it is evident that in each case sufficient technical and economic information must be given in order for a decision maker to come to a clear conclusion.

Port Arthur Dam, Ladysmith, Wisconsin (1964)[6]

Information courtesy of Barr Engineering Co., Minneapolis, Minnesota. Project Manager: Douglas W. Barr.

Background

The Port Arthur Hydroelectric Power Plant is located on the Flambeau River near Ladysmith, Wisconsin, and owned by the Lake Superior District Power Company. This study was carried out because the existing dam and powerhouse (built in 1906) were in poor condition requiring significant repair. The head is approximately 15 ft (4.57 m) and the power capacity 500 kW. The existing timber crib dam had serious leakage with the spillway apron eroded

TABLE 12.3 Case Study Format

Background	Benefits and costs
• Reason for development	• Capital and annual cost estimates
• Institutional history	• Expected revenues and other benefits
• Broad economic profile	Economic and financial analysis
• Previous development	• Established discount and interest rates
Site description	• Benefit/cost or other analysis
• Community energy demand profile	• Cash flow analysis
• Existing service	Development plan
• Physical description of site	• Schedule
• Hydrology	• Required permits and licenses
Plant characteristics	• Institutional arrangements
• Capacity and energy production	Discussion
• Civil works description	
• Electromechanical equipment description	

away, leading to continued scouring downstream. The powerhouse was in somewhat better condition showing only some wall cracks.

Site Description

Flow measurements of the Flambeau River had been taken since 1915 some 11 mi (18 km) upstream making data adjustment necessary. Since this was a low-head site, tailwater curves also had to be developed. Regulation of both headwater and tailwater had to be considered in determining power output and energy production.

The most feasible headwater pool level was found to be elevation 1,098 ft (334.9 m). The new proposed spillway and powerhouse would be capable of passing 22,000 ft^3/s (623 m^3/s) with a pool level at elevation 1,100 ft (335 m). This was equivalent to a flood which can be expected to reoccur every 50 years.

With discharges greater than 22,000 ft^3/s (623 m^3/s), a new emergency spillway would begin to function, discharging an additional 4,000 ft^3/s (113 m^3/s) when the pool level increased above elevation 1,100 ft (335 m). Thus, the combined capacity of the spillway, powerhouse, and emergency spillway was 26,000 ft^3/s (736 m^3/s). This discharge is slightly greater than that which was expected to reoccur every 100 years. It was also equivalent to the flow which would have been discharged by the existing structure with the pool at the same water level elevation. Therefore, there was no increase in flood hazard due to the proposed new structures with a higher normal pool level.

Plant Characteristics

Various configurations were studied to determine optimal headwater/tailwater elevations and number of turbines. The final analysis was made for both two or three tubular-type turbines, each of 1,200 kW capacity. Power production was run-of-river predictable from the flow-duration curve in Fig. 12.4. With two turbines, annual energy production was between 13.0 GWh per year and 14.5 GWh per year, and with three turbines, it was between 14.68 GWh per year and 16.530 GWh per year, depending on headwater elevation. Renovation of this plant, because of attendant water regulation would create a power loss at the Ladysmith plant, 7 mi (11 km) upstream of between 0.6 GWh per year and 1.5 GWh per year. This resulted in net energy production somewhat less than given in the stated figures. In addition to a new powerhouse, a new earth dam and spillway were recommended. (See Figs. 12.5 and 12.6.)

Benefits and Costs

Estimated costs for this project are given in Table 12.4. A dam and power plant containing two 1,200-kW power units would cost $1,266,700 (1964), including estimating flowage rights and credit for the cost of abandonment. This was

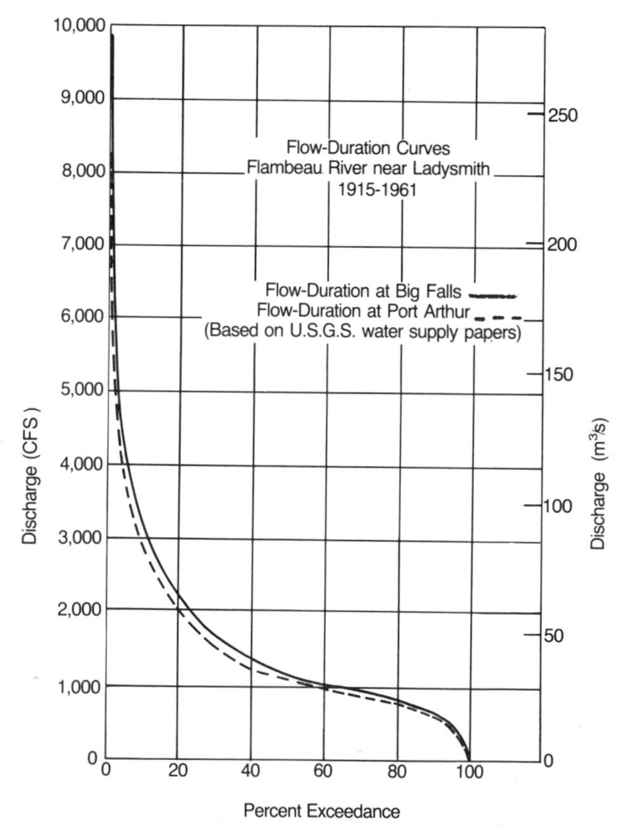

FIG. 12.4 Flow-duration curve for Port Arthur project.

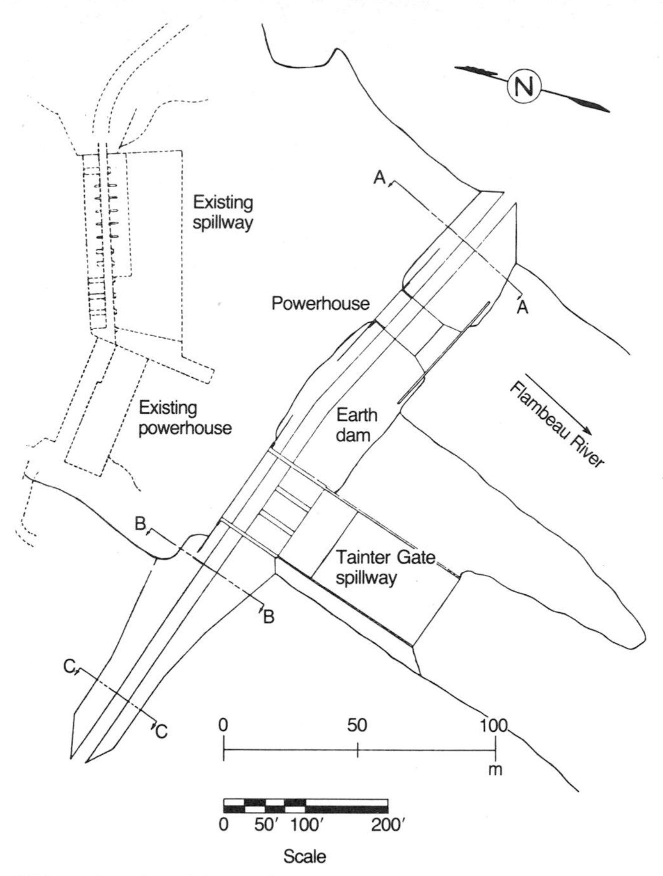

FIG. 12.5 Plan of dam and powerhouse for Port Arthur project.

equivalent to $528 per kilowatt of electric power capacity for generating an average of 12,644,100 kWh per year. If the plant contained three 1,200-kW power units, the cost would have been $1,487,800, including the estimated cost of flowage rights and credit for the cost of abandonment. This was equivalent to $413 per kilowatt of installed capacity with generation at 14,644,100 kWh per year.

Based on an approximate determination of annual expenses, the apparent annual cost of the dam and power plant containing two 1,200-kW units was $148,050 per year. The resulting energy cost was $0.0116 per kilowatthour. With three power units, annual cost was $163,430 per year, resulting in an energy cost of $0.0111 per kilowatthour.

Using $0.01 per kilowatthour as an approximation for the probable maximum value of the energy produced, the average annual increase in power production was converted to average annual increase in income due to the rise in head. The results are summarized in the solid "Benefit" curve on Fig. 12.7.

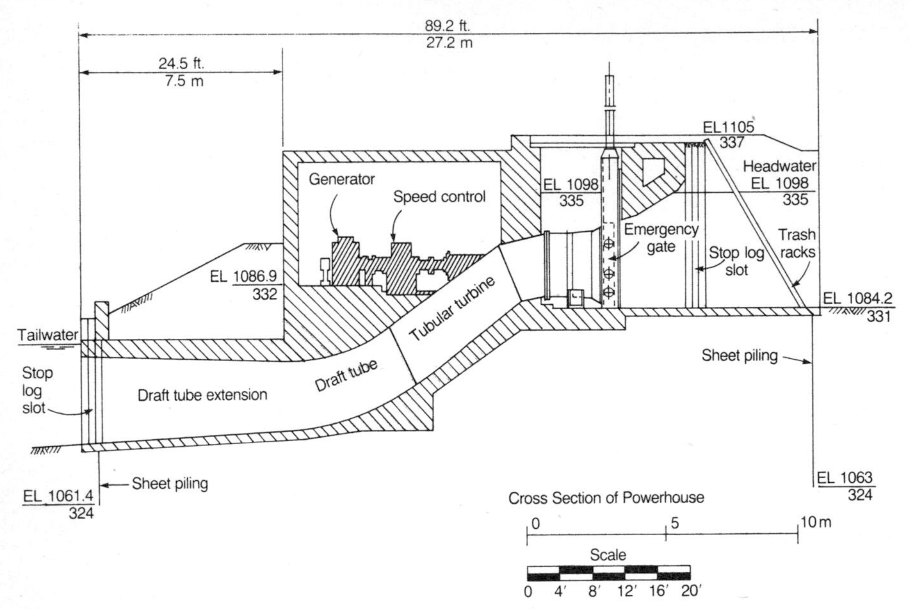

FIG. 12.6 Cross section of Port Arthur powerhouse.

A higher headwater elevation would have caused a correspondingly higher construction cost for the dam and power plant. An estimate was made of the probable increase in cost due to raising the pool. That cost was then converted to an annual cost on the basis of a useful life of 65 years, taxes at 2.4 percent, and financing through bonds at 5.6 percent interest. The resulting annual cost curve is also shown as the solid "Cost" curve.

The distance between the income curve and the cost curve indicates the net increase in income. Both the 1,098-ft (334.9-m) pool and the 1,099-ft (335.2-m) pool gave similar increased net incomes. The former level was selected for the remainder of the feasibility study.

After completing a preliminary design of the dam and power plant, the probable cost of raising the pool elevation was reviewed and the annual cost computed on the basis of 10.4 percent of the construction cost. The revised annual cost is shown by the dotted line in Fig. 12.7. The increased income from the higher pool was also reviewed and computed on the basis of $0.0114 per kilowatthour and plotted as the "revised benefit" curve. Again, an optimum pool elevation of 1,098 ft (334.9 m) is indicated. Specific B/C ratios were not computed; however, Fig. 12.7 indicates they would be positive.

Development Plan

No information was given.

TABLE 12.4 Port Arthur Dam Reconstruction Cost Estimate

Item	Costs 2 Units	3 Units
Spillway	$ 499,670	$ 499,670
Earth dam	138,900	138,900
Cofferdam	56,330	56,330
Emergency spillway	34,000	34,000
Powerhouse (3-turbine capacity)	120,140	162,740
Total	$ 849,040	$ 891,640
Plus 10% contingencies	84,900	89,160
Total	$ 933,940	$ 980,800
Tube, turbine, intake, gate, speed increaser, induction generator—1,200 kW, and necessary switchgear	$ 344,000	$ 513,000
Transformers	15,760	21,000
Total	$1,293,700	$1,514,800
Cost of flowage rights	26,000	26,000
	$1,319,700	$1,540,800
Less cost of abandonment (avoided by reconstruction)	53,000	53,000
Total estimated construction cost	$1,266,700	$1,487,800
Estimated generation, kWh/year		
Energy produced at turbine (Less power loss at Ladysmith)	13,758,250	15,908,700
Energy produced at transformer output (Less power loss at Ladysmith)	12,644,100	14,664,100

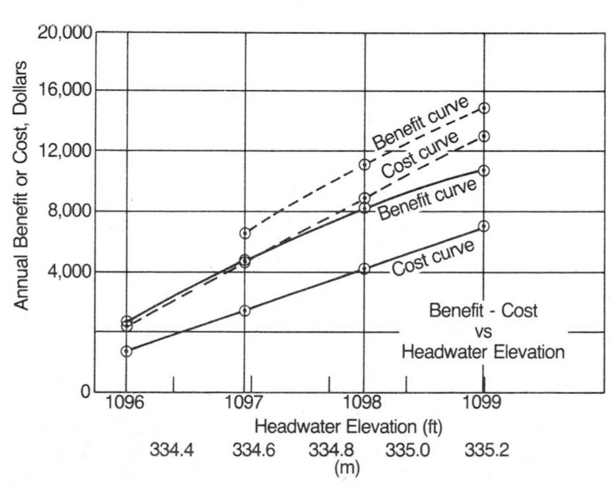

FIG. 12.7 Benefit/cost ratio vs. headwater elevation for Port Arthur project.

Discussion

This project analysis was relatively simple, requiring only determination of the number of turbines, two or three, and a headwater/tailwater elevation analysis. An in-depth benefit/cost analysis was not necessary since an established energy sales price was used. The final choice of headwater elevation, associated costs, and benefits was left to the client.

Mora Canal Drop, Boise, Idaho (1978)[7]

Information courtesy of Tudor Engineering Co., San Francisco, California. Project Manager: Gordon Little.

Background

This study was carried out at the request of the Boise Project Board of Control by the Tudor Engineering Company during the summer of 1978. The majority of the funding for the study was provided through the Department of Energy (DOE).

The Mora Drop structure is located in southwestern Idaho approximately 20 miles (32 km) southwest of Boise on the Mora Canal. The purpose of the existing facility is to drop the flows of the Mora Canal a vertical distance of 36 ft (11 m) from the northern slope of Kuna Butte to the valley below. The canal flows range between 200 and 800 ft^3/s (5.7 and 22.7 m^3/s) and provides irrigation water for approximately 70 mi^2 (181 km^2) of farmland.

The Mora Canal is part of a large canal system known as the Arrowrock Division of the Boise Project, a development of the U.S. Bureau of Reclamation. The Arrowrock Division provides for irrigation of approximately 200,000 acres (81,000 hectares) of farmland located to the west of Boise between the Boise and Snake rivers.

Site Description

The site consists of the Upper Mora Canal with an approximate capacity of 1,100 ft^3/s (31.2 m^3/s), a headworks structure with four gates and a weir that controls the flow of up to 300 ft^3/s (8.5 m^3/s) into the Waldvogel Canal, the Mora Canal Drop structure, a downstream stilling basin, and the Lower Mora Canal with a capacity of 850 ft^3/s (24 m^3/s).

The drop structure is composed of an inlet section with three operating gates and three abandoned gate slots and a covered 13 by 6 ft (4 by 1.8 m) rectangular drop structure 140 ft (42.7 m) in length. Below the drop, the canal continues to the west for 15 mi (24 km) to the south side of Lake Lowell. The canal capacity decreases gradually from about 850 ft^3/s (24 m^3/s) just below the Mora Drop to around 250 ft^3/s (7.1 m^3/s) near Lake Lowell.

Hydrologically, detailed flow records for the Mora Canal were available only for the period of 1972 to 1977. However, long-term average data existed for the area water supplies indicating that the 5-year period of 1972 through 1976 is very close to the long-term average. Therefore it was felt that using this period of record would be indicative of the long-term average values which could reasonably be expected to occur in the future.

Project water deliveries for the entire Boise Project service area for the period from 1926 to 1977 were recorded. Examination of this data shows that annual diversions have varied greatly between 1926 and 1961, but since 1961, they have steadied considerably. The period 1972 through 1976 was very close to the long-term average project water deliveries and can therefore be expected to indicate future average deliveries as well. Also, water from Lake Lowell, which can take several routes, can be conveyed through the Mora Canal when capacity is available.

The average monthly power and energy production for the operating season are shown by 2-week periods in Table 12.5 and are plotted on Figs. 12.8 and 12.9 respectively. Total annual energy production was expected to average 8,113,000 kWh, based on an average operating season from March 16 through November 15 using the historical average flows and the additional flows which can be diverted to Lake Lowell. Average weekly power production ranged from 684 kW in March and November to 1,900 kW during the peak irrigation sea-

TABLE 12.5 Average Monthly Power and Energy Production for the Mora Canal Drop

Period	Historical flows only, average		Historical flows with Lake Lowell flows, average	
	Power, kW	Energy, kWh	Power, kW	Energy, kWh
3/16 to 3/31	0	0	684	263,000
4/1 to 4/15	0	0	905	309,000
4/16 to 4/30	902	324,000	1,083	389,000
5/1 to 5/15	1,610	578,000	1,691	607,000
5/16 to 5/31	1,687	648,000	1,700	653,000
6/1 to 6/15	1,559	562,000	1,653	596,000
6/16 to 6/30	1,777	636,000	1,777	636,000
7/1 to 7/15	1,900	684,000	1,900	684,000
7/16 to 7/31	1,900	730,000	1,900	730,000
8/1 to 8/15	1,900	684,000	1,900	684,000
8/16 to 8/31	1,824	701,000	1,824	701,000
9/1 to 9/15	1,501	544,000	1,501	544,000
9/16 to 9/30	1,111	400,000	1,178	424,000
10/1 to 10/15	336	121,000	1,064	383,000
10/16 to 10/31	0	0	684	263,000
11/1 to 11/15	0	0	684	247,000
Total		6,612,000 kWh		8,113,000 kWh

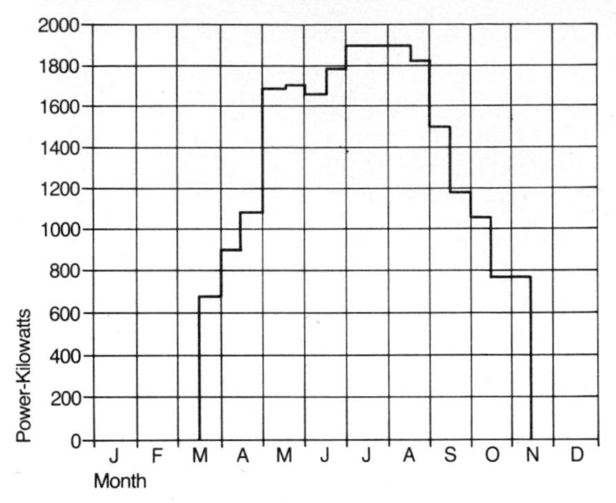

FIG. 12.8 Mora Canal Drop average monthly power capacity potential.

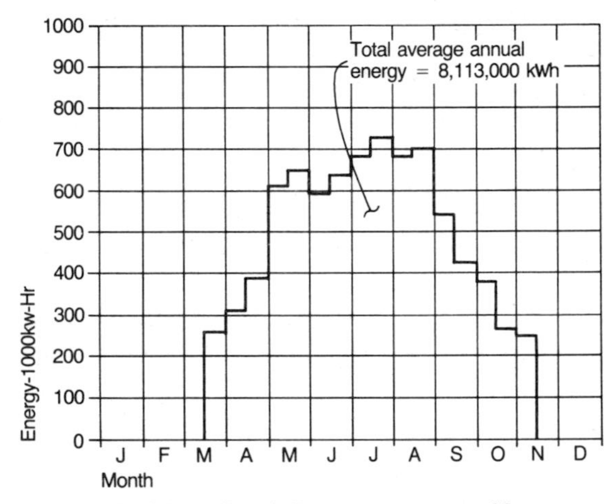

FIG. 12.9 Mora Canal Drop average monthly energy production.

son in July and August. Using the historical flows only, the average annual energy production was 6,612,000 kWh based on an operating season of April 16 through October 15.

The results of a power study, conducted for the 1977 irrigation season indicate that during any rare drought of this magnitude, annual energy production will total about 4,600,000 kWh or approximately 57 percent of the average production. However, it is not likely that another drought as severe as the 1977 season will occur within the economic lifetime of the Mora Drop power plant.

Plant Characteristics

Specifically at the drop, the new facilities will consist of a new headgate structure for the Waldvogel Canal, to be located downstream of its present location; new gates for the existing chute, which will remain in place as an emergency bypass of the turbines; the power plant intake works, with a conduit connecting to the turbine; the power plant itself; and outlet works that tie into the Mora Canal just downstream of the existing stilling basin.

Table 12.6 shows a cost comparison of five alternatives for the power plant configuration. Note Fig. 12.10 for a plan of the site with a tubular turbine and a propeller turbine. A 1,900-kW turbine was recommended with annual energy production at 8,113,000 kWh based on the hydrologic record.

TABLE 12.6 Cost Comparison of Alternative Power Plant Configurations for the Mora Canal Drop*

Configuration	Costs, $		
	Civil	Mech./elec.	Total
Vertical-shaft propeller, open-flume design	460,000	830,000	1,290,000
Vertical-shaft propeller, conc. penstock and scroll case	474,000	799,000	1,273,000
Tubular-type turbine	434,000	855,000	1,289,000
Bulb-type turbine	519,000	1,374,000	1,893,000
Cross-flow type turbine	524,000	853,000	1,576,800

*Fall 1978.

Benefits and Costs

In this case, the *economic value* of hydroelectric power can be taken to be the least cost of alternative power. Since the power produced at Mora Canal is not firm power in the traditional definition (since it cannot be guaranteed to be available 100 percent of the time) it has a value only as nonfirm, or secondary, hydroelectric energy. However, the production pattern closely matches the peak area-demand pattern and a case can therefore be made that the energy from the Mora Canal Drop may be a form of firm energy since equivalent firm energy would have to be supplied if the Mora Canal Drop energy was not utilized. A conservative approach would be to consider the economic value of this energy equal to the cost of fuel oil utilized to generate power at a conventional oil-fired generating station.

The selling price, or *financial value,* of the power is also important. The financial value is the price at which the developer of the power source would sell the power to a power purchaser. The financial value must be greater than the development cost, or the project is economically unfeasible. Similarly, the selling price must be less than the economic value, or the purchaser would have

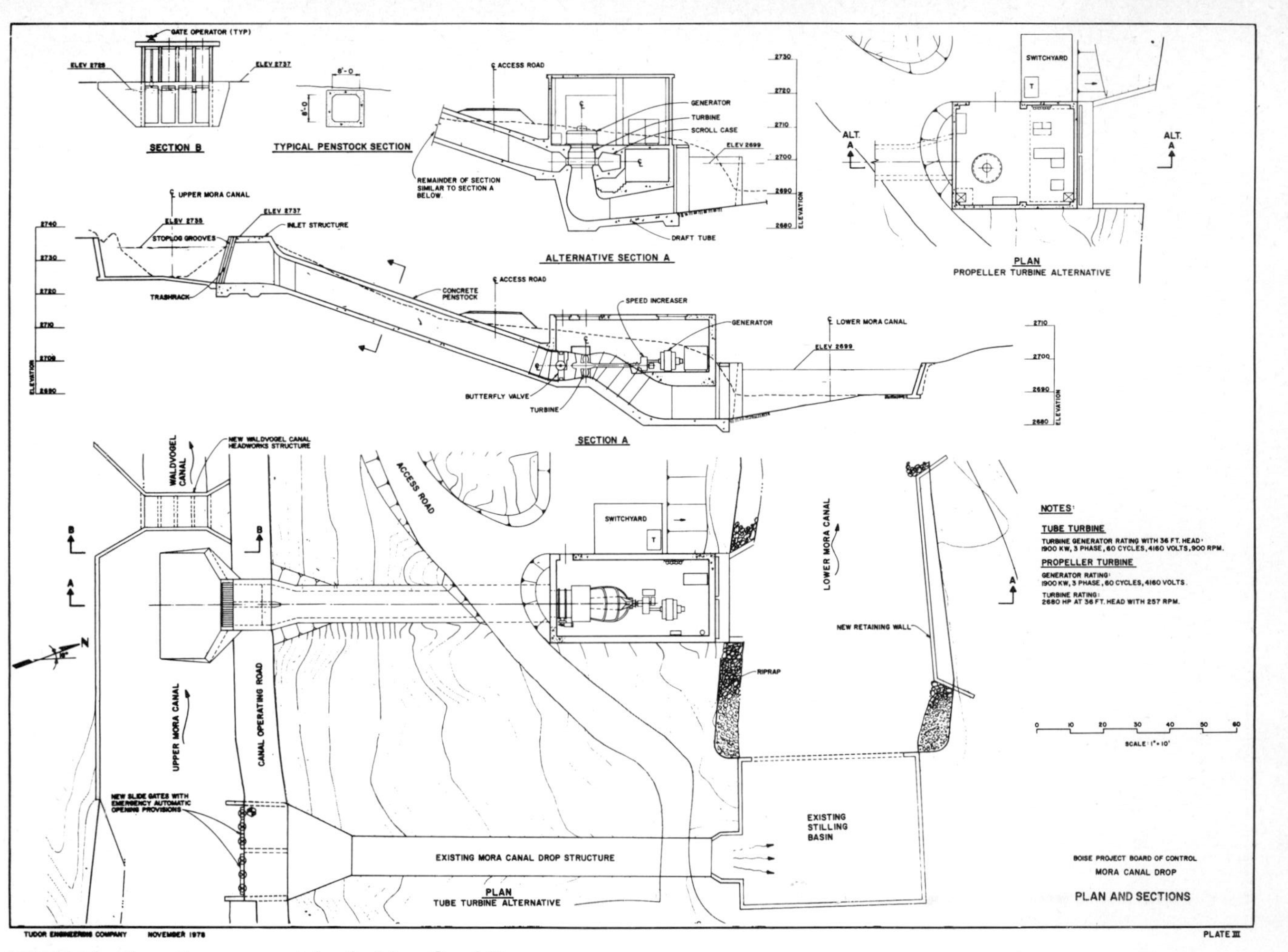

FIG. 12.10 General arrangement for the Mora Canal Drop.

no incentive to buy the power. Thus, if the power is to be sold to a power purchaser, the financial value must be less than the economic value.

During initial discussions held with the Idaho Power Company a price of $0.026 per kilowatthour was indicated as being considered reasonable (about 1978). Since the average annual energy production was estimated to be 8,113,000 kWh, this value yields an estimated average annual revenue of $210,938.

The project costs consist of the initial capital costs and continuing annual costs. The initial capital costs consist of construction costs and the interest during construction. Construction costs are summarized in Table 12.7 for both vertical-shaft propeller turbine and tubular turbine. These cost estimates were based on recent bid prices for similar construction work and direct quotes from various equipment manufacturers.

TABLE 12.7 Construction Cost Estimates for the Mora Canal Drop*

Cost Item	Tubular-type turbine	Propeller-type turbine alternative
Land and land rights	$ 2,000	$ 2,000
Structures and improvements	145,700	166,750
Reservoirs, dams, and waterways	214,000	226,000
Waterwheels, turbines, and generators	629,000	568,000
Accessory electrical equipment	95,000	105,000
Miscellaneous power plant equipment	53,000	53,000
Transmission line	100,000	100,000
Subtotal	$1,238,700	$1,220,750
Contingencies	141,300	153,000
Total construction cost	$1,380,000	$1,373,750

*Fall 1978.

The engineering costs are presented in Table 12.8. The necessary engineering activities consist of an environmental impact assessment, the final FERC application, surveys and soil investigations, miscellaneous permits, contract plans and specifications, and construction management. The administrative costs are also shown in Table 12.8. They consist of administrative efforts by the Boise Board staff and fees for the financial consultant and the bond counsel. The total initial costs are also summarized on the table. As shown, the total initial cost is estimated to be $1,783,750.

The annual costs for the project will consist of a bond or loan repayment cost and the operation and maintenance (O&M) costs. The bond or loan repayment costs depend on the source of the funds, the length of repayment time, and the interest rate.

The estimated annual O&M costs are also shown in Table 12.8. They include insurance, labor costs, interim replacement costs (for miscellaneous

TABLE 12.8 Capital and O&M Cost Summary for the Mora Canal Drop*

Engineering costs	
Environmental impact assessment	$ 20,000
Final FERC application/approval (minor)	15,000
Surveys and soil investigation	10,000
Miscellaneous permits	5,000
Contract plans and specifications	125,000
Construction management	101,000
Total engineering costs	$276,000
Administrative costs	
Administration	$ 14,000
Financial consultants	25,000
Bond counsel	30,000
Total administrative costs	$ 69,000
Total capital costs	
Construction cost (propeller type)	$1,373,750
Engineering cost	276,000
Administrative cost	69,000
Subtotal	$1,718,750
Interest during construction	65,000
Total initial cost	$1,783,750
Total O&M costs	
Insurance	$ 3,500
O&M labor	12,000
Interim replacement	4,000
General expenses	5,200
Total annual O&M cost	$24,700

*Fall 1978.

repairs during the life of the facility), and general expenses. As shown, the annual O&M costs will total $24,700 per year.

The costs of producing energy are directly related to the method of financing employed. This will determine the interest rate and the repayment period and therefore directly influence the annual expenses and energy costs.

There were two potential sources of funding recommended to the Boise Project Board of Control to finance the Mora Canal Drop Project (about 1978): government loans through the U.S. Department of Energy (DOE) and the direct sale of general obligation or revenue bonds.

1. Obtain a loan from DOE for the first 75 percent of the project cost at an interest rate of 6⅞ percent and a term of 30 years. Sell tax exempt revenue

bonds for the remaining 25 percent of the project cost at an interest rate of 7 percent and a term of 40 years. This funding method will be called the "DOE Loan Option" in the ensuing discussions.

2. Sell tax exempt revenue bonds in the amount of $1,000,000 at a nontaxable interest rate of 7 percent. Sell bonds sufficient to cover the remaining project cost at a taxable rate of 10 percent. The term of both bond issues would be 40 years. This funding method will be called the "Conventional Funding Option" in the ensuing discussion.

With the project costs, financing options, and power production established, the project energy production cost and the cost per installed kilowatt can now be determined.

Table 12.9 presents the production costs for the Mora Canal Drop using the DOE Loan Option. As shown, the annual expense of $164,607 is a combination of bond and loan payments and the annual operation and maintenance cost. Under this financial option, the costs would be $0.0203 per kilowatthour with an average annual energy production of 8,113,000 kWh (with Lake Lowell flow) and $0.0249 per kilowatthour with an average annual energy production of 6,612,000 kWh (using historical flows only).

Table 12.9 also presents the production costs using the Conventional Funding Option. As shown, the annual expenses are a combination of two bond payments and the annual operation and maintenance cost. Under this option the costs would be $0.0222 per kilowatthour and $0.0273 per kilowatthour with an average annual energy production of 6,612,000 kWh.

As can be seen, the lowest cost is $0.0203 per kilowatthour using the DOE Loan Option and assuming that the historical Mora Canal Drop flows are augmented as previously discussed with water for Lake Lowell.

With an installed capacity of 1,900 kW, and an initial cost of $1,783,750, the project has a cost per installed kilowatt of $937. This figure is independent of the financing option used.

The economic evaluation of this project was a comparison of the benefits and costs over the life of the project using the benefit/cost (B/C) ratio. Private utilities generally use the rate of return on investment, but public agencies generally utilize the B/C ratio.

The economic life was taken as 50 years. This was the estimated length of time the project will be in service producing power and accruing benefits.

As previously discussed, the projected costs depend on the method of financing employed. Table 12.10 presents the annual costs over the 50-year project life for the DOE Loan Option and Conventional Funding Option methods of financing. These cost streams show the repayment costs ending after 30 or 40 years, when complete repayment has been made, but with the annual O&M costs continuing over the entire 50-year project life.

The project benefits depend on the average annual energy generated. Table 12.10 also shows the annual project benefits with 6,612,000 kWh annually (his-

TABLE 12.9 Energy Production Costs for the Mora Canal Drop*

DOE Loan Option		Conventional Funding Option	
Initial cost	$1,783,750	Initial cost	$1,787,750
Annual expenses:		Annual expenses:	
Repayment: 0.75 × $1,783,750 at 6⅞%, 30 years	106,458	Repayment: $1,000,000 @ 7%, 40 years	75,091
0.25 × $1,783,750 at 7%, 40 years	33,449	$787,750 @ 10%, 40 years	80,555
Operation and maintenance	24,700	Operation and maintenance	24,700
Total annual expenses	$164,607	Total annual expenses	$180,346
Energy production cost (EPC):		Energy production cost (EPC):	
(a) Average annual energy = 8,113,000 kWh:		(a) Average annual energy = 8,113,000 kWh	
$EPC = \frac{164{,}607 \times 1{,}000}{8{,}113{,}000} = \$0.0203/kWh$		$EPC = \frac{180{,}346 \times 1{,}000}{8{,}113{,}000} = \$0.0222\ kWh$	
(b) Average annual energy = 6,612,000 kWh:		(b) Average annual energy = 6,612,000 kWh	
$EPC = \frac{164{,}607 \times 1{,}000}{6{,}612{,}000} = \$0.0249/kWh$		$EPC = \frac{180{,}346 \times 1{,}000}{6{,}612{,}000} = \$0.0273/kWh$	

*Fall 1978.

TABLE 12.10 Cost and Benefit Streams for the Mora Canal Drop*

	Cost stream		Benefit stream	
Year	DOE loan	Conventional funding	At 6.6 × 10^6 kWh	At 8.1 × 10^6 kWh
1	$164,607	$180,346	$171,912	$210,938
↓	↓	↓	↓	↓
30	164,607			
31	58,149			
↓	↓	↓	↓	↓
40	58,149	180,346		
41	24,700	24,700		
↓	↓	↓	↓	↓
50	24,700	24,700	171,912	210,938

*Fall 1978.

torical flows only) and 8,113,000 kWh annually (historical flows augmented with Lake Lowell flows). It is assumed that the added Lake Lowell flows would also be available under normal conditions, but for comparative purposes, both conditions are presented.

The resulting B/C ratios for the DOE Loan Option are shown in Table 12.11. Since interest rates are currently in a considerable state of flux, the present worth for the benefits and costs have been computed using discount rates of 6 percent, 8 percent, and 10 percent. With average annual energy of 8,113,000 kWh the B/C ratio ranges from 1.41 to 1.33 depending on the discount rate. With average annual energy of 6,612,000 kWh the B/C ranges from 1.15 to 1.08.

The B/C ratios for the Conventional Funding Option are also shown in Table 12.11. With average annual energy of 8,113,000 kWh the B/C ratio ranges from 1.21 to 1.18 and with an average annual energy of 6,612,000 kWh, the B/C ratio ranges from 0.99 to 0.96. These tables show that the B/C ratios are not very dependent on the discount rate used, but they are quite sensitive to the average annual energy value used and the method of financing chosen.

It should be noted that inflation has not been taken into account in the analysis, and the results are therefore conservative. The only component of the annual costs that can escalate is the $24,700 O&M component since the bond or loan repayment costs of $139,907 or $155,646 will be fixed. However, the annual benefits of $171,912 and $210,938 will almost certainly increase since part of the power sales agreement will provide for the selling price of the project power to be tied to a fuel-cost index which is expected to increase.

TABLE 12.11 Benefit/Cost Ratios for the Mora Canal Drop*

	Discount rate		
	6%	8%	10%
DOE Loan Option			
Energy: 8,113,000 kWh/yr			
Present worth benefits	$3,324,784	$2,580,510	$2,091,408
Present worth costs	$2,357,964	$1,899,520	$1,575,560
Benefit/cost ratio	1.41	1.36	1.33
Energy: 6,612,000 kWh/yr			
Present worth benefits	$2,709,660	$2,103,085	$1,704,473
Present worth costs	$2,357,964	$1,899,520	$1,575,560
Benefit/cost ratio	1.15	1.11	1.08
Conventional Funding Option			
Energy: 8,113,000 kWh/yr			
Present worth benefits	$3,324,784	$2,580,510	$2,091,408
Present worth costs	$2,745,190	$2,158,178	$1,766,975
Benefit/cost ratio	1.21	1.20	1.18
Energy: 6,612,000 kWh/yr			
Present worth benefits	$2,709,660	$2,103,085	$1,704,473
Present worth costs	$2,745,190	$2,158,178	$1,766,975
Benefit/cost ratio	0.99	0.97	0.96

*Fall 1978.

From the given data and analysis it can be concluded that the Mora Canal Drop is an economically feasible project with the DOE loan option.

Development Plan

The first activity in developing this site involves negotiations with a power purchaser. The end result of the negotiations will be a *memorandum of understanding* in which the Boise Project Board of Control will pledge to sell and the power purchaser will pledge to buy all power produced at the Mora Canal site. This document will stipulate all details, including a clause in which the selling price of the electric energy will be tied to a suitable escalation index. Following the final design of the project when all costs are known, the document will be revised as required to form the final power purchase contract.

The other activities follow the start of power negotiations. Assuming some bond sales will be necessary, legal services and professional advice regarding financial assistance is obtained during the power sales agreement and continued until all bonds are sold or until the necessary loans are obtained from DOE.

The FERC application and design and construction activities will start immediately after the conclusion of the power negotiations. A recently passed energy bill directed the FERC to prepare revised and simplified guidelines for small hydroelectric projects. However, 1 year should be allowed for the FERC license application approval process. A part of the FERC application is the preparation of an environmental impact statement (EIS). It is a major activity and is often also required by other agencies as well.

The process to obtain all other necessary permits and agreements will also be initiated following the completion of the power negotiations. In the case of the Mora Canal Power Project, an agreement between the U.S. Bureau of Reclamation for the use of Boise Project lands and facilities must be completed. Also, a water-use permit for power generation must be obtained from the Idaho Department of Water Resources.

The entire permit, licensing, and construction process will take approximately 2 years.

Discussion

The analysis performed for the project analysis is typical, having basic elements of a full feasibility study. The principal considerations are the choice of propeller- or tubular-type turbine and the financing option. The financial analysis was carried out using both a DOE loan and conventional bond financing. The federal-funding option was clearly more attractive; however, the discount rates were not firmly established. The B/C ratios were somewhat low, indicating the possibility of an economically infeasible project should discount rates and costs increase without a compensating increase in the electricity sales price. It is evident that the conclusions of the study are fragile, requiring revision if the Boise Project Board of Control seeks to implement the project.

Scotts Flat and Lower Scotts Flat, Nevada County, California (1981)[8]

Information courtcsy of Tudor Engineering Company, San Francisco, California. Project Manager: Candice King.

Background

This case study presents the results of an investigation of the feasibility of installing two hydroelectric power plants at existing dams which form Scotts Flat Reservoir and Lower Scotts Flat Reservoir (also known as Deer Creek Division) in Nevada County, California. The dams are owned and operated by the Nevada Irrigation District to provide storage for irrigation and domestic use. The power plants would utilize flows which presently pass through the dams' outlet works. Neither dam has existing hydroelectric-generating facilities. Energy produced by the project would be sold to one of several potential power purchasers in the area.

The feasibility study was authorized by an agreement, in 1981, between the Nevada Irrigation District (NID), the proposed owner and operator of the power development, and Tudor Engineering Company. The study was made possible by a loan obtained by NID from the U.S. Department of Energy (DOE).

The Scotts Flat Dam is located on Deer Creek, approximately 5 mi (8 km) east of the town of Nevada City, California. Scotts Flat Reservoir provides storage for downstream irrigation and domestic use. An earthfill dam and the existing outlet structure would be utilized for site development. One side of the existing bifurcation would be connected to a horizontal Francis turbine and the other side would be used for bypass capabilities.

About ½ mi (0.8 km) downstream of Scotts Flat Dam is Lower Scotts Flat Dam, a concrete-arch structure. It is approximately 4 mi (6.4 km) east of the town of Nevada City. Lower Scotts Flat Dam functions as the diversion point for D-S canal water stored in Scotts Flat Reservoir. There is no suitable access into this site and roads would have to be constructed. The project consists of a penstock running from the existing outlet to the base of the dam where a horizontal Francis turbine would be located.

Site Description

1. Scotts Flat Reservoir: Scotts Flat Reservoir is formed by a 175-ft (53.4-m) high earthfill dam initially constructed to a height of 140 ft (42.7 m) in 1949 and raised to its present height in 1964. The crest length is 960 ft (292.8 m) at elevation 3,085 ft (940.9 m). The spillway crest is at elevation 1,075 ft (327.9 m) and is 239-ft (72.9-m) long. There are 48,500 acre-feet (59.8×10^6 m^3) of usable storage between the bottom of the intake (elevation 2,925 ft, or 892 m) and the spillway crest with no dead storage. The maximum water surface area is 715 acres (2.9×10^6 m^2). The Scotts Flat outlet works consist of two 36-in (0.9-m) inlet pipes joined in a wye to a 751-ft (229-m) long, 2-m (6.56-ft) diameter lined tunnel. At the axis of the original dam, a concrete plug has been installed and the water flows into a 600-ft (183-m) long, 1-m (3.28-ft) diameter steel pipe, then through a bifurcation to two 30-in (0.76-m) outlet pipes. The two outlet pipes have butterfly valves and the inlet has hydraulically operated gates.

2. Lower Scotts Flat Reservoir: Lower Scotts Flat Reservoir is formed by a 92-ft (28-m) high, 325-ft (99-m) long concrete arch dam on Deer Creek. Releases from Scotts Flat Reservoir flow down Deer Creek for approximately ½ mi (0.8 km) to Lower Scotts Flat Reservoir. The maximum capacity of the reservoir is 1,400 acre-feet (1.73 m^3) at the spillway elevation of 2,896.6 ft (883.5 m). The spillway is a 240-ft (73-m) overflow section in the center of the dam. The headworks for the D-S canal are located on the south side of the dam. Because of the high floor elevation of the canal 2,888 ft (881 m), the reservoir is normally kept full (elevation 2,896 or 883 m) to ease diversion into the canal. The capacity-elevation curve for the Lower Scotts Flat Reservoir is shown in Fig. 12.11.

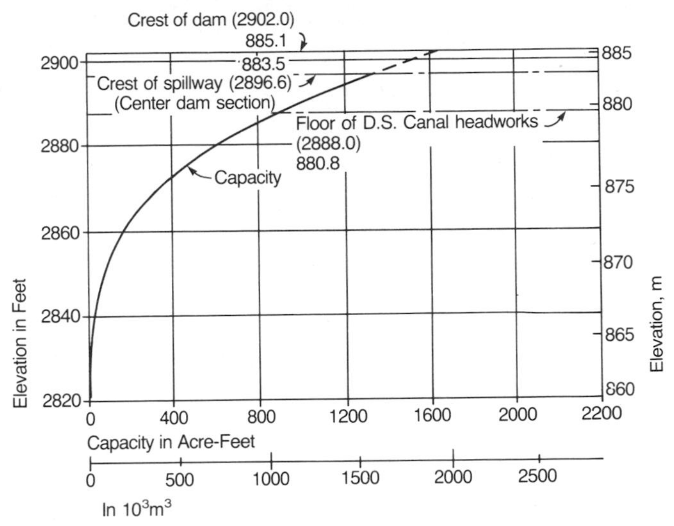

FIG. 12.11 Capacity-elevation curve for Lower Scotts Flat.

3. Hydrology: There are no streamflow gages on Deer Creek near Scotts Flat Dam. Daily reservoir elevations along with records of diversions to and from Deer Creek and an estimation of natural streamflow based on a correlation with a streamflow gage from a similar adjacent basin were used in a computer model to reconstruct the apparent inflow and outflow from Scotts Flat Reservoir and Lower Scotts Flat Reservoir.

Nearby Oregon Creek was used as an index of probable natural flow in Deer Creek. It is located 12 mi (19.3 km) north of Deer Creek at similar elevation, precipitation, and exposure. Because records at this gage were only available from 1968 through 1980, this period was chosen as the study period. The assumption was made that the District would continue to operate the reservoirs in the future as they did during the period used for the operation studies.

The flow-duration curve for 1968 through 1980 is shown in Fig. 12.12, and the average, maximum, and minimum monthly discharges for Lower Scotts Flat Reservoirs are shown in Fig. 12.13. The same figures for Scotts Flat are in the feasibility study but are not reproduced.

4. Spillway and Tailwater Analysis: At Scotts Flat, the spillway was designed for a maximum discharge of 10,000 ft^3/s (283.2 m^3/s) with 5 ft (1.5 m) of freeboard. The California Division of Safety of Dams stated that the probable maximum flood at Scotts Flat Dam would be 24,000 ft^3/s (680 m^3/s) and the spillway would be able to pass the flood with 1 ft (0.3 m) of freeboard at the dam.

There was no calculation as to the capability of Lower Scotts Flat Dam to pass this probable maximum flood. The 1,000-year flood of 8,600 ft^3/s (243 m^3/s) was calculated to pass at elevation 2,902 ft (885 m), a depth of 5.4 ft

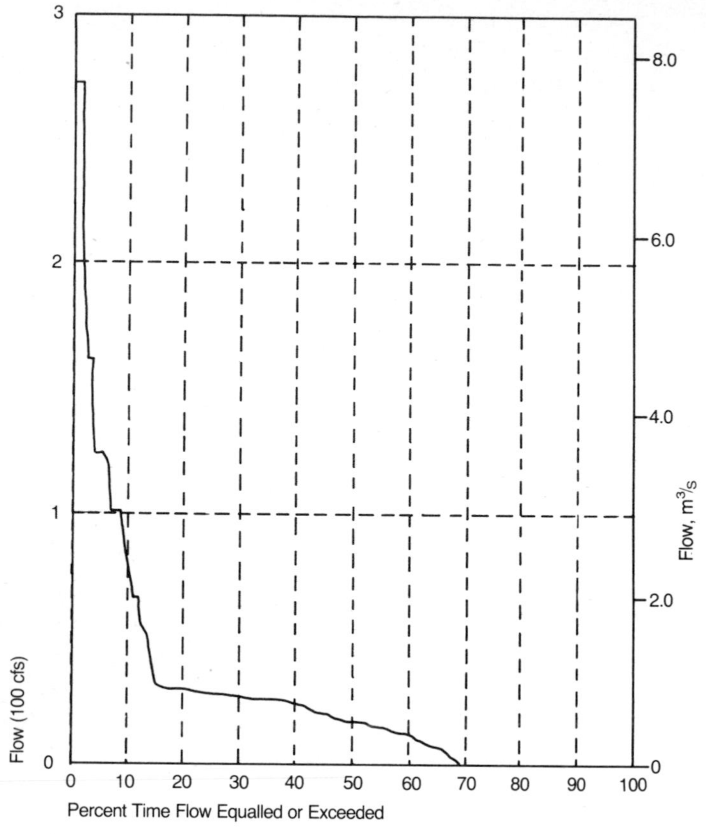

FIG. 12.12 Flow-duration curve for Lower Scotts Flat.

(1.65 m) over the center spillway section. The dam crest is at elevation 2,902 ft (885 m).

Releases from Scotts Flat Reservoir flow about ½ mi (0.8 km) down the Deer Creek natural channel to the upper end of Lower Scotts Flat Reservoir. Backwater analysis using computer program HEC-2 was used to calculate tailwater elevations at the Scotts Flat Power Plant Site for a variety of streamflows. A control elevation of 2,896 ft (883 m) was used for Lower Scotts Flat Reservoir. Deer Creek downstream of Lower Scotts Flat Dam has no control point. Tailwater at this site was taken as the normal depth of the channel.

Table 12.12 presents a summary of the physical fcatures from the feasibility study.

Plant Characteristics

Four different alternative projects at Scotts Flat and one project at Lower Scotts Flat were considered. The alternatives for both sites are presented in Table 12.13.

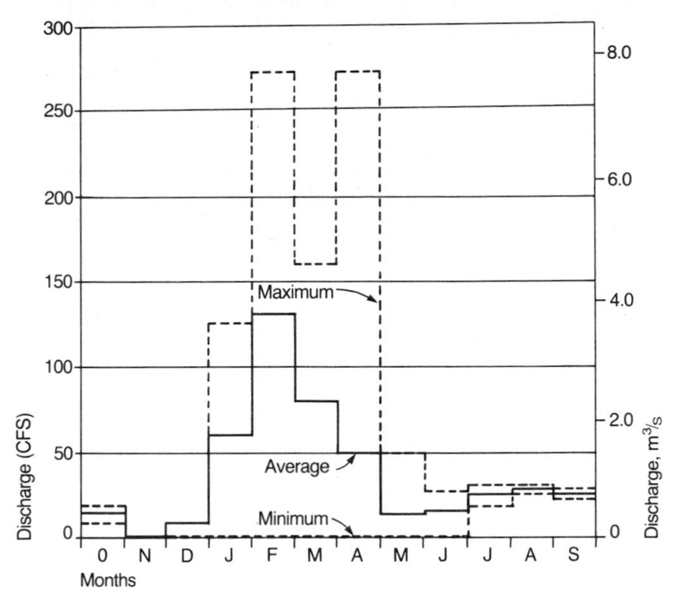

FIG. 12.13 Monthly discharges (1968–1980) from Lower Scotts Flat to Deer Creek.

TABLE 12.12 Summary of Physical Features for Scotts Flat and Lower Scotts Flat

Physical feature	Scotts Flat	Lower Scotts Flat
Turbine type	Horizontal Francis	Horizontal Francis
Installed capacity, kW	825	205
Dependable capacity, kW	234	0
Net head, maximum/minimum:		
feet	163.4/70.5	91/86.2
meters	49.8/21.5	27.8/26.3
Rated head:		
feet	140	91
meters	42.7	27.8
Flow,* maximum/minimum:		
cubic feet per second	83/13	30/5
cubic meters per second	2.35/0.37	0.85/0.14
Design flow:		
cubic feet per second	82	30
cubic meters per second	2.32	0.85
Average annual energy, 10^6 kWh/yr	4.76	0.84

*Flow through turbine.

1. Scotts Flat: The four alternative projects investigated were (1) a powerhouse at the base of the dam with a run-of-river operation, (2) a powerhouse at the base of the dam with a peaking operation, (3) a powerhouse at the base of the dam with a partial-peaking operation, and (4) a powerhouse at the upper portion of Lower Scotts Flat Reservoir with a pumped-storage peaking operation.

For each alternative, a power study and conceptual layout was done. Based on the conceptual layout, construction and operation and maintenance (O&M) costs were estimated. The alternatives were then evaluated by comparing the costs (based on 35-year bond) with the revenues estimated from the power study.

The recommended alternative indicated that a power plant be constructed at the existing outlet to Scotts Flat Dam. The plant will utilize flows of Deer Creek which presently pass through Scotts Flat. The existing 36-in (0.9-m) outlet pipe and bifurcation will be used. From the bifurcation, water would pass through the powerhouse and into the existing discharge channel. The powerhouse foundation would be excavated into the bedrock of the existing streambed channel and will be an extension of the existing valve house. The normal tailwater would be at elevation 2,911.0 ft (887.9 m) and the centerline of the horizontal Francis turbine would be at elevation 2,915.8 ft (889.3 m).

Approximately 800 ft^2 (74.3 m^2) of the new powerhouse would be added to the existing valve house. The powerhouse would contain the turbine, generator, gate controller, turbine shut-off valve, and other related equipment. The turbine would be a horizontal Francis unit with a nameplate capacity of 825 kW. The design head is 140 ft (42.7 m), and the design flow is 82 ft^3/s (2.32 m^3/s). It is expected that an average of 4.76 million kilowatthours per year would be produced based on the hydrology previously described.

Figure 12.14 shows the plan and section view of the Scotts Flat Power Plant. The following design considerations were chosen.

a. Headwater: The headwater for the turbine is the water surface elevation of Scotts Flat Reservoir. Nevada Irrigation District has maintained daily gage height readings at the enlarged Scotts Flat Reservoir since it started filling in 1964. These records were used for power computation and design. The maximum observed water surface elevation was 3,076 ft (938 m) which occurred for several days in 1974. The minimum observed elevation since the enlarged reservoir was filled was 2,956.3 ft (901.7 m) from October 6 to 11, 1977.

b. Tailwater: Lower Scotts Flat Reservoir downstream of Scotts Flat Dam is operated at a normal water surface elevation of about 2,896 ft (883 m). This elevation was used as a control elevation for the HEC-2 backwater analysis of Deer Creek to the power plant site. Under normal nonflood conditions the tailwater elevation would be between 2,910.5 and 2,911.5 ft (887.7 and 888.0 m). The design head is 140 ft (42.7 m).

c. Operating Flows: The mean discharge from Scotts Flat Reservoir was 68 ft^3/s (1.93 m^3/s) during the study period from 1968 through 1980. The outflow was estimated by computer model using storage-elevation records from Scotts Flat and Lower Scotts Flat Reservoir along with records of diversions into and out of the Deer Creek basin above and below Scotts Flat Dam. A design flow of 82 ft^3/s (2.32 m^3/s) was chosen with capabilities of maximum and minimum flows of 85 ft^3/s (2.41 m^3/s) and 13 ft^3/s (0.37 m^3/s) respectively.

d. Manpower: The Scotts Flat power plant will be unstaffed and remotely monitored from the District's dispatch center. As at other remotely monitored sites, visits will be made at regular intervals to ensure normal operation of the plant.

2. Lower Scotts Flat: Because of the configuration of the dam, spillway, and canyon, the only location considered for the powerhouse was at the base of the left abutment with a penstock connecting the existing outlet with the powerhouse. Releases from Lower Scotts Flat Reservoir into Deer Creek were utilized in the operation studies.

The study recommended that at the Lower Scotts Flat Dam site, a 180-ft (54.9-m) length of 2-ft (0.6-m) diameter penstock be constructed from the existing outlet at elevation 2,886 ft (880 m) to the streambed. A 200-kW horizontal Francis turbine would be installed in a 500 ft^2 (46.5 m^2) powerhouse. Approximately 1¼ miles (2013 m) of new transmission lines would have to be constructed to connect with the existing distribution lines at Scotts Flat Reservoir. A new access road would also be necessary prior to construction.

The Lower Scotts Flat Reservoir is normally maintained at the maximum water surface elevation to allow for diversion into the D-S Canal. This elevation, 2,896 ft (883 m), was assumed to be average constant headwater. Normal depth was taken to be the tailwater in Deer Creek downstream of the Lower Scotts Flat Dam. The outflow from Lower Scotts Flat was taken as the releases made to Deer Creek plus any spills. The mean discharge from the reservoir was 40 ft^3/s (1.13 m^3/s) during the study period from 1968 through 1980. A design flow of 30 ft^3/s (0.85 m^3/s) was chosen for maximum efficiency during planned releases. The plant would be capable of generating an average of 840,000 kWh per year. A figure of $0.087 per kilowatthour (the 1985 estimated value of energy) would produce an average annual revenue of $73,000.

3. Project Energy Production: Power production by any hydraulic turbine is a function of installed capacity, turbine and generator operating characteristics, flow passing through the turbine, and net hydraulic head available.

Net hydraulic head was computed by subtracting the tailwater elevation and the head losses from the reservoir water surface elevation. For Scotts Flat, tailwater elevations were determined by backwater calculations from Lower

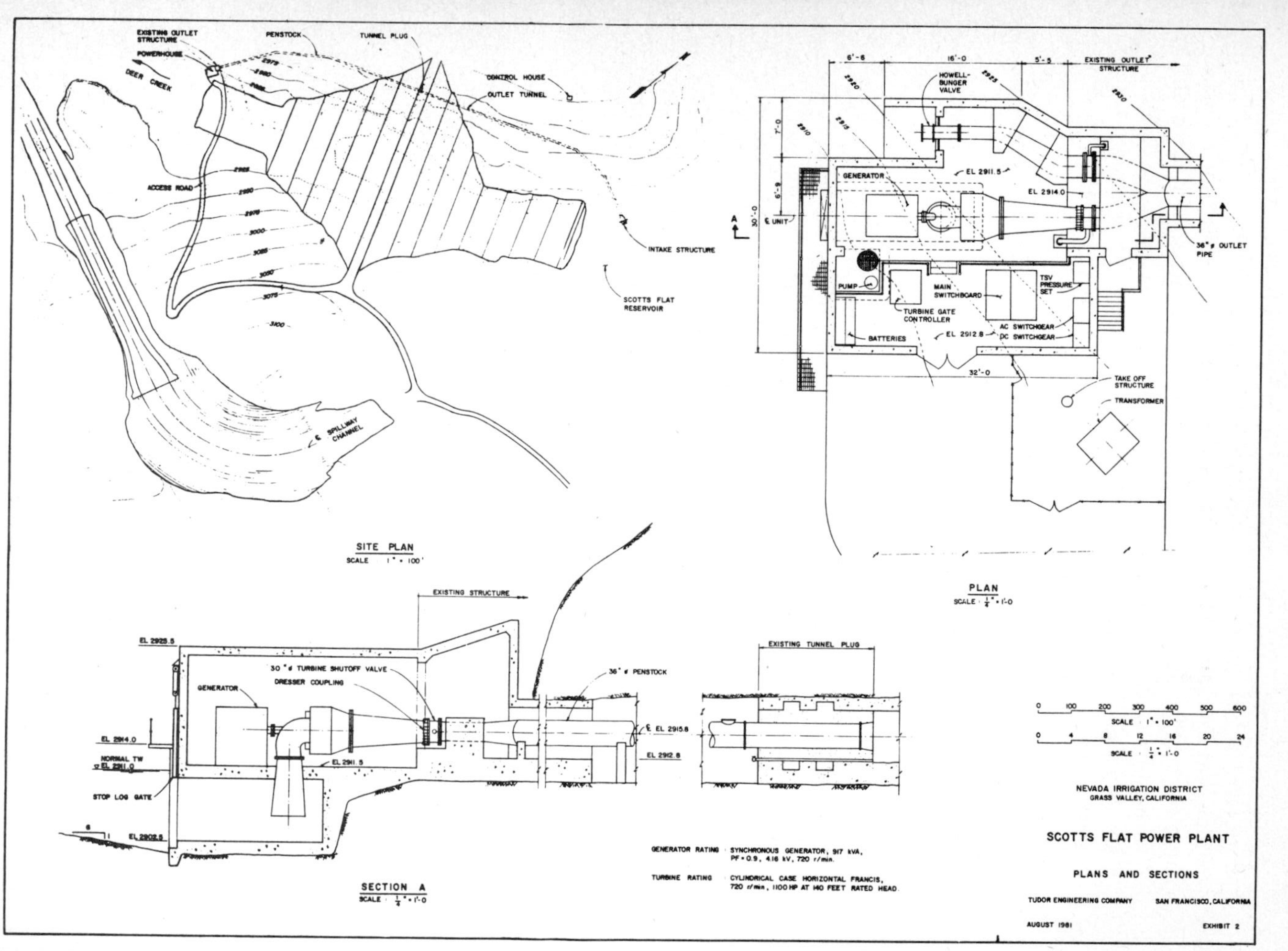

FIG. 12.14 General arrangement for the Scotts Flat Power Plant.

TABLE 12.13 Alternative Developments for Scotts Flat and Lower Scotts Flat

Description	Turbine type	Number of units	Design, H ft (m)	Design, Q ft^3/s (m^3/s)	Design capacity, kW	Benefits 10^6 kWh	Benefits $/yr*	Annual† cost	Difference in 1st-year benefits and costs
Scotts Flat Dam									
Project using existing 36″ outlet; run-of-river operation	Francis	1	140 (42.7)	82 (2.32)	825	4.76	414,000	261,000	153,000
Project using new 48″ outlet; peaking operation	Francis	1	140 (42.7)	325 (9.20)	3,000	5.30	491,000	637,000‡	−146,000
Project using new 48″ outlet; summer peaking operation, winter run-of-river operation	Francis	2	140 (42.7)	145 (4.11)	1,600	5.18	458,000	471,000	−13,000
Project using new 66″ penstock downstream of dam; pumped-storage operation	Francis pump/ turbine	1	140 (42.7)	425 (12.04)	3,600	10.59§ −12.05¶	−25,000	1,400,000‡	−1,425,000
Lower Scotts Flat									
Project below south abutment; penstock running from existing high-level outlet; run-of-river operation	Francis	1	91 (27.8)	30 (0.85)	200	0.84	220,000	73,000	−147,000

*Based on estimated 1985 value of energy = 87 mills/kWh.
†Based on 13 percent interest, 35-year payback.
‡Does not include costs for developing downstream storage.
§Peak generation.
¶Off-peak pumping.

Scotts Flat Reservoir. For Lower Scotts Flat, normal depth at the rated flow was used.

Reservoir water surface elevations were determined from gages on both Scotts Flat and Lower Scott Flat Reservoirs. Monthly water surface elevations were taken as end of the month values.

Head losses for the penstock were computed using the Darcy-Weisbach pipe friction equation with an average friction factor of 0.014. Entrance, valve, and bend losses were added to give an overall head-loss coefficient k of 0.00095 for Scotts Flat and 0.0019 for Lower Scotts Flat. The equation $H = kQ^2$ was used, where H is head loss and Q is flow.

Turbine efficiency depends on the type of unit, the flow through the unit, and the net head at that flow. The relationship between these variables was used to develop a set of dimensionless performance curves for appropriately sized horizontal Francis turbines. The turbine efficiency was obtained from these curves for specific flow and head conditions observed each month of the study period.

Because of the large number of computations required, a computer program for the hydroelectric operation study was used to evaluate the sites. A data file consisting of monthly outflow was evaluated on a monthly basis. The turbine curve was input as a set of empirically determined operating points and was subsequently converted to a single equation using multidimensional linear regression. For each month, an effective head was calculated by taking the average water surface elevation in the reservoir and subtracting the tailwater elevation and the estimated head loss.

During periods when reservoir releases were less than the minimum operating flow through the turbine, no power was assumed to be generated. During high-flow periods, the maximum operating flow through the turbine was used for power generation. It was assumed that all excess flow would bypass the plant.

The effective head, together with the average monthly flow, were compared to the limits of turbine operation. Conditions of insufficient head or excessive flow were evaluated to determine if operational changes would allow generation. If the flow and head were within the operating limits, or could be adjusted to lie within the operating limits, then the power, energy, and efficiency were calculated. A summary of computer study results is given in Table 12.14.

TABLE 12.14 Summary of Energy Output for Scotts Flat and Lower Scotts Flat

Energy output, 1968 to 1980	Scotts Flat	Lower Scotts Flat
Average annual energy, 10^6 kWh	4.76	0.84
Maximum annual energy, 10^6 kWh	6.56	1.07
Year of maximum	1974	1973
Minimum annual energy, 10^6 kWh	1.45	0.59
Year of minimum	1977	1972

Benefits and Costs

Project costs consist of initial capital costs and continuing annual costs. Costs for the Scotts Flat power plant are estimated at the price levels of August 1981. These costs are escalated at an annual rate of 12 percent for 24 months to adjust them to the bid date of August 1983.

1. Capital Costs: The power plant direct construction costs for Scotts Flat are $769,000 and for Lower Scotts Flat, $645,000 (see Table 12.15). This is the August 1981 cost and does not include engineering, escalation, net interest during construction, contingencies, etc. This cost estimate is based on bid prices for similar construction work and on manufacturers' quotes for principal mechanical and electrical equipment.

Contingencies were added to the direct construction cost to include items which were not foreseeable at that time. Tudor's experience in small hydroelectric development was used in selecting contingency percentages. The contingency allowance was obtained by applying a factor of 15 percent to the turbine-generator and other equipment costs and 20 percent to all other costs.

The engineering costs include the final FERC license application, environmental documentation, surveys, foundation investigations, miscellaneous permits, final design, contract plans and specifications, and engineering services during construction. The cost of final design and engineering services during

TABLE 12.15 Total Project Costs Lower Scotts Flat Power Plant

Description	Cost
Direct costs:	
Civil	$180,000
Penstock	38,000
Turbine-generator	179,000
Accessory electrical equipment	189,700
Miscellaneous mechanical	15,000
Transmission line	44,000
Total direct costs	$645,000
Contingencies	
15% for turbine/generator and electrical equipment	49,000
20% for other costs	64,000
Total contingencies	$758,000
Engineering Services, 30%	227,000
Administration, legal, and financial, 10%	76,000
Total construction costs	$1,061,000
Escalation 8/81 to 8/83 (bid date) at 12% per year for 24 months, 25.4%	270,000
	$1,331,000
Net interest during construction, 15%	200,000
Total project capital costs	$1,531,000

construction is based on a level of effort considered to be consistent with the size and cost of the power plant. For power plants less than 1 MW, the engineering costs are estimated to be 30 percent of the direct construction cost plus contingencies.

An allowance is also made for the owner's administrative, legal, and financial expenses. For a power plant less than 1 MW, this amount is estimated to be 10 percent of direct construction costs and contingencies.

Interest must be paid on the construction loan funds during the construction period prior to the time the plant is started. This interest payment will vary depending on the interest rate, the period of construction, and the type of financing adopted. To partially offset this expense, the money obtained at the onset of construction can be invested until it is needed.

As stated previously, construction costs were estimated using August 1981 prices. The costs were escalated to be 12 percent per year from the *Engineering News-Record* Cost Index for various components of hydroelectric development. The estimate was made by graphically extending this historic data to the projected bid date.

2. Annual Costs: The annual costs for the project consist of the costs of operation, maintenance, insurance, interim replacements, and administration, and the financial costs for loan amortization and interest. These costs are estimated based on experience and research. For projects of a similar size, this has been determined to be approximately 1.2 percent of the total capital costs.

At the time of this study, tax-free municipal bonds were being sold at approximately 13 percent. This interest rate exceeds the statutory maximum of 12 percent for tax-free municipal bond sales. It is expected that the California State Legislature will raise the statutory maximum by the time this project's bonds are ready to be sold. The debt service for a loan of $1,812,000, which is the total project cost for Scotts Flat, would be $239,000 using the 13 percent interest rate and a project life of 35 years. For Lower Scotts Flat, total project cost is $1,531,000 and debt service is $202,000. Annual costs which are referred to as operation and maintenance (O&M) are meant to include not only operation and maintenance of the plant, but also equipment replacement allowances, licensing fees, and administrative costs. The annual O&M costs are estimated to be $22,000 for Scotts Flat and $18,000 for Lower Scotts Flat respectively. A summary of the annual costs is provided in Table 12.16.

Economic and Financial Analysis

This section discusses the financial alternatives available to the Nevada Irrigation District (NID) to develop the project, presents the proposed alternative, and shows expected energy costs. It is proposed that the Scotts Flat Power Plant be financed through the sale of tax-free revenue bonds guaranteed by a

TABLE 12.16 Cost Summary at 13 Percent Discount Rates for Scotts Flat and Lower Scotts Flat

Type of cost	Scotts Flat	Lower Scotts Flat
Total capital cost	$1,812,000	$1,531,000
Annual costs:		
Debt service	239,000	202,000
O&M	22,000	18,000
Total annual costs	$ 261,000	$ 220,000

power purchaser or through a bank loan in conjunction with an *avoided cost* power purchase contract.

1. Financial Alternatives: There are advantages to each type of financing. With the bond guarantee there is very little risk to NID as the developer. A disadvantage is that with a power purchaser agreement and bond guarantee by the power purchaser, NID would profit less from future increases in the value of energy, i.e., NID would share these increases with the power purchaser. With a conventional bank loan, NID would carry the risk but all the revenues belong to NID. The time involved in selling bonds would be eliminated, enabling the project to be completed sooner. The specific details follow.

a. **Bond Sale:** The capital cost of the project financed through the sale of revenue bonds is the conventional type of funding for small hydroelectric projects and is commonly used. Federal tax laws in the past had provided that bonds could be exempt from federal income tax only if the power from the project were used by a public utility company. Recent federal legislation contained in the Crude Oil Windfall Profits Tax Act of 1980 (P.L. 96-223, effective April 2, 1980), provides for the sale of tax-exempt bonds to finance hydroelectric projects regardless of the power purchaser.

b. **Conventional Loan:** NID could finance the project through a conventional bank loan. Currently, rates for such a loan are 10 percent and the term is 5 years. The balance of the loan must be refinanced at the end of 5 years.

c. **Power Purchase Agreement:** With a power purchase agreement, the total capital cost would be financed through the sale of revenue bonds as described in the conventional loan option. The power purchaser would support the project by advancing funds for all or portions of the indirect costs and by providing an incentive payment to NID for the energy and power to be generated. The debt-service payment of the revenue bond would be guaranteed by the power purchaser, regardless of the amount of energy generated by the project. There are other alternatives such as investor financing, U.S. Bureau of Reclamation P.L. 984 loans, Farmers Home Administration loans, and Rural Electrification Association (REA) loans. Current legislation would have to be consulted to determine what specific provisions are in force at the time for each of these possibilities.

2. Energy Production Cost: The energy production cost (EPC) is the annual debt service plus the annual operation and maintenance costs divided by the energy produced annually, or simply:
EPC for Scotts Flat

$$\frac{\$239{,}000 + \$22{,}000}{4.76 \times 10^6 \text{ kWh}} = \frac{\$261{,}000}{4.76 \times 10^6 \text{ kWh}} = \$0.0548/\text{kWh}$$

EPC for Lower Scotts Flat

$$\frac{\$202{,}000 + \$18{,}000}{0.84 \times 10^6 \text{ kWh}} = \frac{\$220{,}000}{0.84 \times 10^6 \text{ kWh}} = \$0.2619/\text{kWh}$$

These values represent the cost of production during 1985, the first year of operation, assuming financing by bonds.

3. Benefit/Cost Analysis: The benefit/cost evaluation of the project is a comparison of the present worths of the benefits and costs over the life of the project. A computerized evaluation of costs and benefits for Scotts Flat using bond financing is given in Table 12.17. The following points are the background for the computations. The economic life of the project is taken to be 35 years.

- The debt service is a product of the total project capital cost and the capital recovery factor (at 13 percent for 35 years, the capital recovery factor is 0.1318). The total project capital costs include the direct construction costs; engineering, legal, administrative, and financial fees; interest during construction, and escalation of prices to the bid date.
- The annual operation, maintenance, and replacement costs (O&M) were estimated to be 1.2 percent of the total project capital cost.
- Power revenues are the product of the energy produced and the value of that energy. The average energy produced by Scotts Flat and Lower Scotts Flat is 4.76×10^6 and 0.84×10^6 kWh per year respectively. Using PG&E guidelines, an energy value of \$0.087 per kilowatthour was estimated for 1985.
- The power revenues and operation and maintenance costs were escalated at 10 percent for the first 10 years and at 0 percent thereafter because escalation rates beyond 10 years are uncertain. Overall, this provides a conservative estimate of the costs and benefits to be expected. The present worth of all costs and benefits for each year is computed. They are then added to deter-

mine the project benefit/cost (B/C) ratio. The B/C ratio for Scotts Flat was computed at 2.6 and for Lower Scotts Flat at 0.54.

- A discount rate of 13 percent was selected. That was the current rate of interest NID would have to pay on bonds sold at the time to finance the project. In recent years the interest rate on municipal bonds has varied from 7 percent in July 1979 to more than 13 percent in August 1981. Even though the California statutory maximum interest rate for municipal bonds then stood at 12 percent, it was expected that the Legislature would raise the limit by the time the bonds would be sold.
- The unit capacity cost is the total project capital cost divided by the installed capacity. The unit capacity cost for Scotts Flat Power Plant is $1,812,000 divided by 825 kW or $2,196 per kilowatt. For Lower Scotts Flat the unit capacity cost is $1,531,000 divided by 200 kW or $7,655 per kilowatt.

TABLE 12.17 Benefit/Cost Analysis for Scotts Flat Power Project Bond Financing (8/23/81)

	Cost streams		Benefit stream		Present worth	Present worths	
Year	Debt service	OM&R	Energy	Capacity	factor	Cost	Benefit
1985	238,900	22,000	414,100	15,400	1.00000	260,900	429,500
1986	238,900	24,200	455,500	15,400	0.88496	232,800	416,800
1987	238,900	26,600	501,100	15,400	0.78315	207,900	404,500
1988	238,900	29,300	551,200	15,400	0.69305	185,800	392,700
1989	238,900	32,200	606,300	15,400	0.61332	166,300	381,300
⋮	⋮	⋮	⋮	⋮	⋮	⋮	⋮
2015	238,900	51,900	976,500	0	0.02557	7,400	25,000
2016	238,900	51,900	976,500	0	0.02262	6,600	22,100
2017	238,900	51,900	976,500	0	0.02002	5,800	19,600
2018	238,900	51,900	976,500	0	0.01772	5,200	17,300
2019	238,900	51,900	976,500	0	0.01568	4,600	15,300
						2,369,600	6,193,000

Benefit/cost ratio = 6,193,000/2,369,600 = 2.61

NOTES:
OM&R (operation, maintenance, and replacement) escalation rate = 10% for years 1–10.
OM&R escalation rate = 0% for years 11–35.
Power revenue escalation rate = 10% for years 1–10.
Power revenue escalation rate = 0% for years 11–35.
Discount rate = 13% for years 1–35.

A preliminary cash flow statement for bond financing is presented in Table 12.18. The table shows that there is a positive cash flow during the first complete year of operation. A Department of Energy (DOE) loan for the feasibility study is included.

TABLE 12.18 Preliminary Cash Flow Statement—Scotts Flat: 13 Percent Bond Financing (Thousands of Dollars)

	1981	1982	1983	1984	1985	1986	1987	1988	1989	1990	1991	1992	1993	1994	1995
Expenditures:															
Legal, financial, administrative	10	31	44	19	19										
Engineering, Const. Mgmt.	63	63	129	85	43										
Construction (including contingencies):															
Equipment				364	365										
Civil				198	198										
Interest during construction				157	79										
Debt service on bonds					120	240	240	240	240	240	240	240	240	240	240
Debt service on DOE loans					9	9	9	9	9	9	9	9	9	9	0
O&M escalated at 10% per year					10	24	27	29	32	35	39	43	47	52	52
Total expenditures	73	94	173	823	843	273	276	278	281	284	288	292	296	301	292
Income:															
DOE loan	45														
Bond sales			1,812												
Interest returned on bond money				121	60										
Capacity credit					7	15	15	15	15	15	15	15	15	15	15
Power revenue escalated at 10% per year					207	455	501	551	606	667	733	807	887	976	976
Total income	45	0	1,812	121	274	470	516	566	621	682	748	822	902	991	991
Net	−28	−94	1,639	−702	−569	197	240	288	340	398	460	530	606	690	699

Development Plan

The general plan to implement the Scotts Flat power project is shown as a bar graph schedule in Fig. 12.15. The first activity, feasibility study preparation, is completed with the preparation of the report.

Several additional activities must be completed prior to final design of the project. The first is to prepare the FERC license-exemption application requiring approximately 3 months to finish. FERC must act on the license-exemption application within 4 months of accepting the application for filing. It was estimated that approval of the exemption application would be complete by August 1982. NID should also begin the process of negotiating a power sales agreement with a power purchaser.

As soon as approval is obtained from FERC, equipment procurement specifications may be prepared. As shown in Fig. 12.15, an equipment contract for design for the turbine-generator would be awarded by Dec. 1, 1982. A contract for fabrication of the equipment will be awarded immediately after bonds are sold on Dec. 1, 1983. Delivery of these items requires approximately 14 months after the fabrication contract is awarded. Final project design can begin after the equipment contract is bid and should take place from December 1982 through July 1, 1983. The civil works contract can be bid after sale of the revenue bonds, with a notice to proceed in December 1983. It is expected that project construction will take place from Dec. 1, 1983 through May 1985, with equipment testing in April and May 1985. As shown, the proposed project development plan would bring power on-line in June 1985.

The Nevada Irrigation District (NID) must obtain several licenses and permits, both federal and state, before construction may take place:

1. Federal Energy Regulatory Commission (FERC) license exemption (federal)
2. Water-rights revision (federal)
3. Water Quality Control Board certification (California)
4. Initial environmental study (California)
5. U.S. Army Corps of Engineers permit (federal)
6. Division of Safety of Dams approval (California)

Based on the results of this study, the proposed power plant at Scotts Flat Dam is technically, economically, and environmentally feasible. In review, the proposed Scotts Flat Power Plan consists of an 825-kW horizontal Francis-type turbine-generator unit capable of generating an average annual energy of 4.76 million kWh. The benefit/cost ratio for the recommended development is 2.61, based on a 35-year, 13 percent bond and conservative inflation rates. The benefit/cost ratio for the project using a 10-year, 10 percent bank loan would be 3.34. The unit capacity cost will be $2,196 per installed kilowatt and the

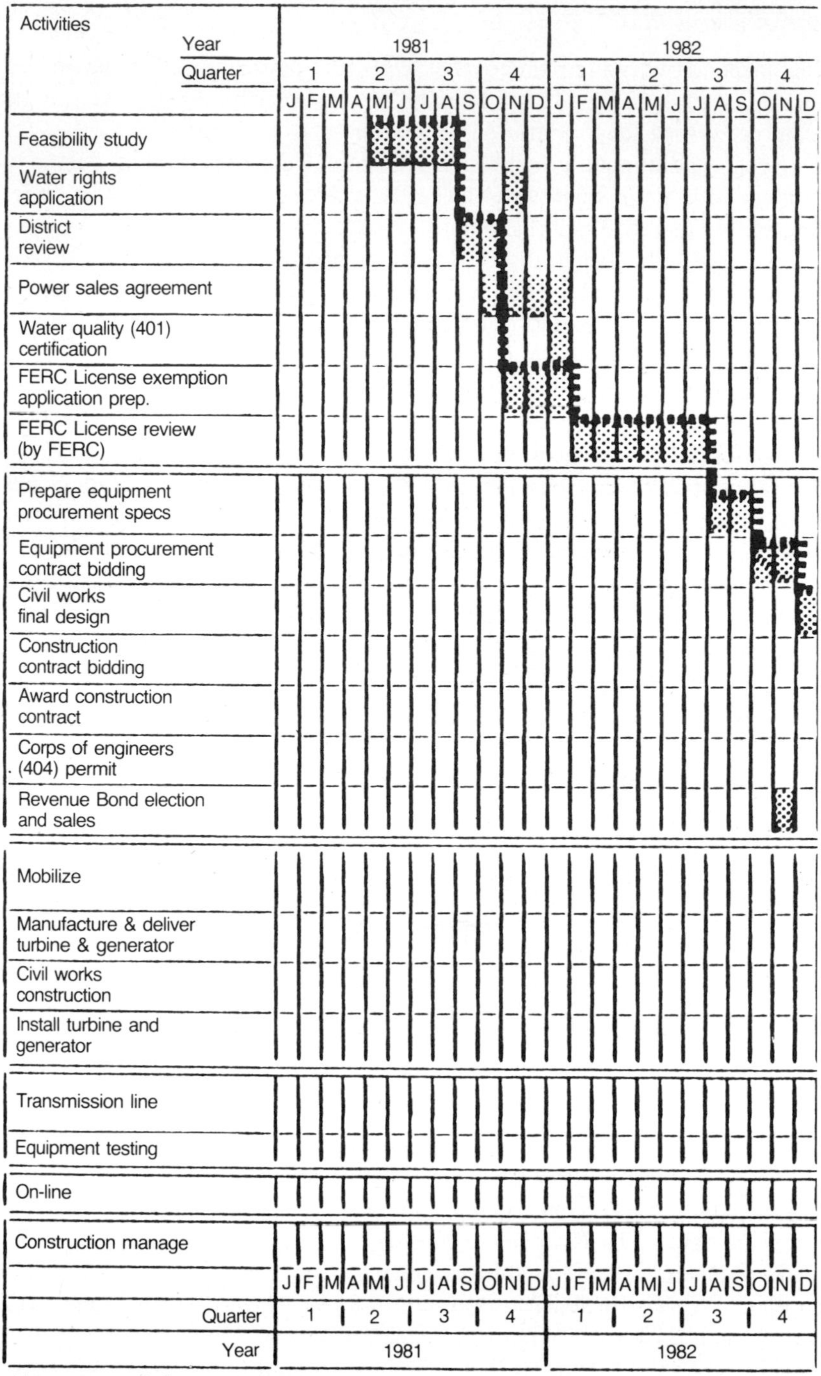

Activities	1981												1982											
Quarter	1			2			3			4			1			2			3			4		
	J	F	M	A	M	J	J	A	S	O	N	D	J	F	M	A	M	J	J	A	S	O	N	D
Feasibility study																								
Water rights application																								
District review																								
Power sales agreement																								
Water quality (401) certification																								
FERC License exemption application prep.																								
FERC License review (by FERC)																								
Prepare equipment procurement specs																								
Equipment procurement contract bidding																								
Civil works final design																								
Construction contract bidding																								
Award construction contract																								
Corps of engineers (404) permit																								
Revenue Bond election and sales																								
Mobilize																								
Manufacture & deliver turbine & generator																								
Civil works construction																								
Install turbine and generator																								
Transmission line																								
Equipment testing																								
On-line																								
Construction manage																								
	J	F	M	A	M	J	J	A	S	O	N	D	J	F	M	A	M	J	J	A	S	O	N	D
Quarter	1			2			3			4			1			2			3			4		
Year	1981												1982											

Note: ▪▪▪▪▪ Indicates critical path

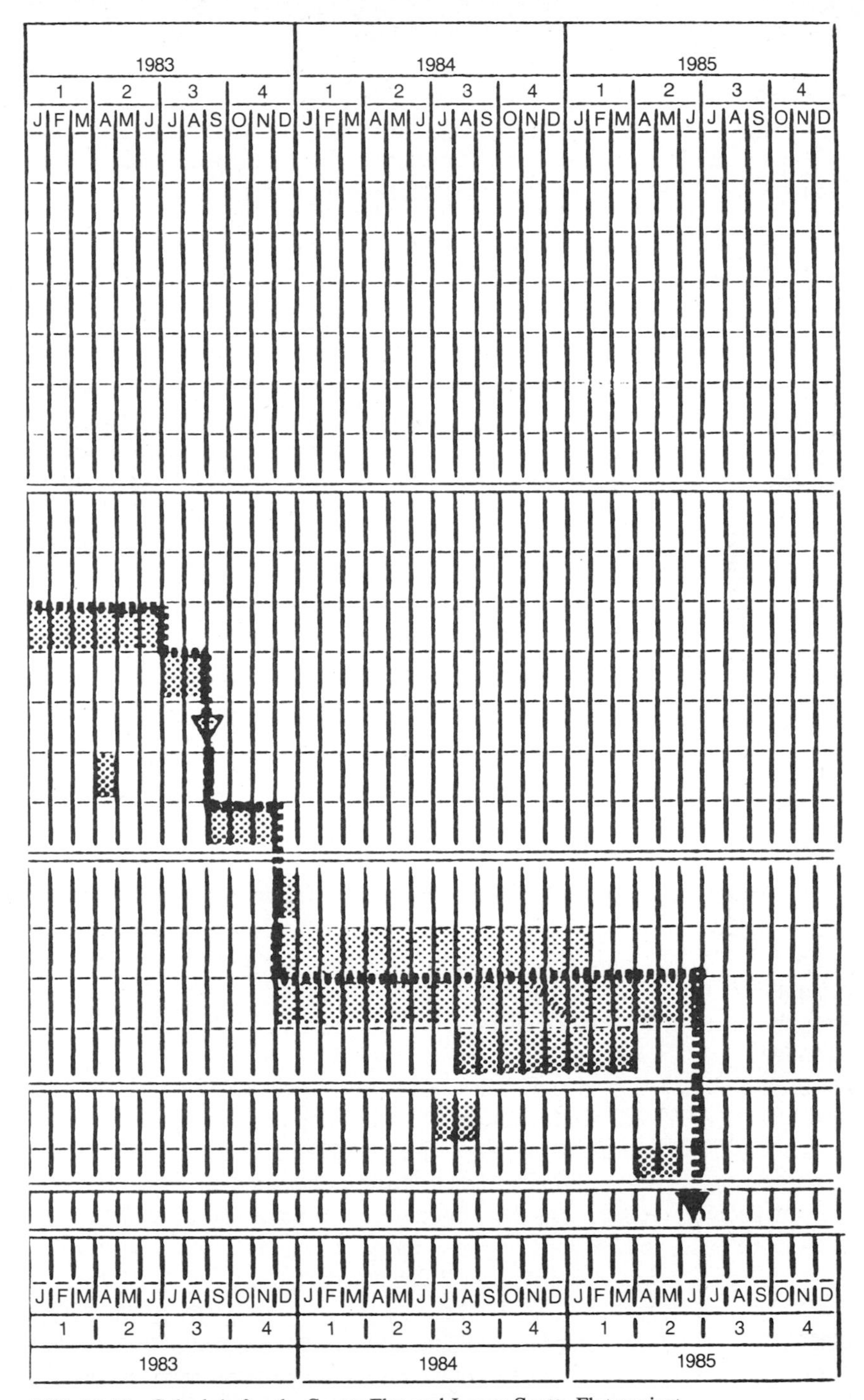

FIG. 12.15 Schedule for the Scotts Flat and Lower Scotts Flat project.

energy production cost will be $0.055 per kilowatthour. The estimated energy value at the time the project goes on-line is $0.087 per kilowatthour.

The Lower Scotts Flat power plant, a 200-kW unit, although technically feasible, is not feasible for NID to finance through conventional methods. However, it may be of interest to private investors who would derive special tax advantages by financing all, or a portion of, such a project.

Discussion

This study sought to determine the feasibility of two hydro plants at consecutive dams in an irrigation system requiring both headwater and tailwater level analysis. However, irrigation and flood-control procedures clearly dictate flow availability and subsequent energy production. Since energy rates had been established, revenues became a function of flow, making the plant at Lower Scotts Flat uneconomical.

The analysis of four equipment alternatives at Scotts Flat and of financing options was complete. Since it was difficult to predict discount rates and federal loan covenants, a clearly favorable financing option did not emerge.

Fula Rapids Project, Juba, Sudan (1981)[9]

Information courtesy of the United Nations Department of Technical Cooperation for Development, New York, New York. Principal Investigators: R. N. Bhargava, G. M. Haggstrom.

Background

The Democratic Republic of Sudan is the largest African nation. It covers an area of nearly 2.5×10^6 km^2 (965,251 mi^2) and has a population estimated at about 18 million. The Public Electricity and Water Corporation (PEWC), a parastatal organization with headquarters in Khartoum, is responsible for all generation, transmission, and distribution of electricity in the public sector of the country.

Southern Sudan is known as Eastern Equatoria Province. The largest center of activity is the town of Juba, with an estimated population of about 85,000. The other two important towns are Torit and Yei. Juba is situated on the banks of the White Nile; Torit and Yei are located on the banks of the Kinyeti River and the Yei River, respectively. The three towns of Juba, Torit, and Yei are located in the most remote part of the country, where the southern province borders Ethiopia, Kenya, Uganda, and Zaire. (See Fig. 12.16)

The landscape in the north of the Yei-Juba-Torit area is almost flat while toward the south, up to the border of Uganda, the landscape gradually becomes mountainous. Maximum altitudes in the Imatong Mountains near the border are more than 3,000 m (9,836 ft) above sea level.

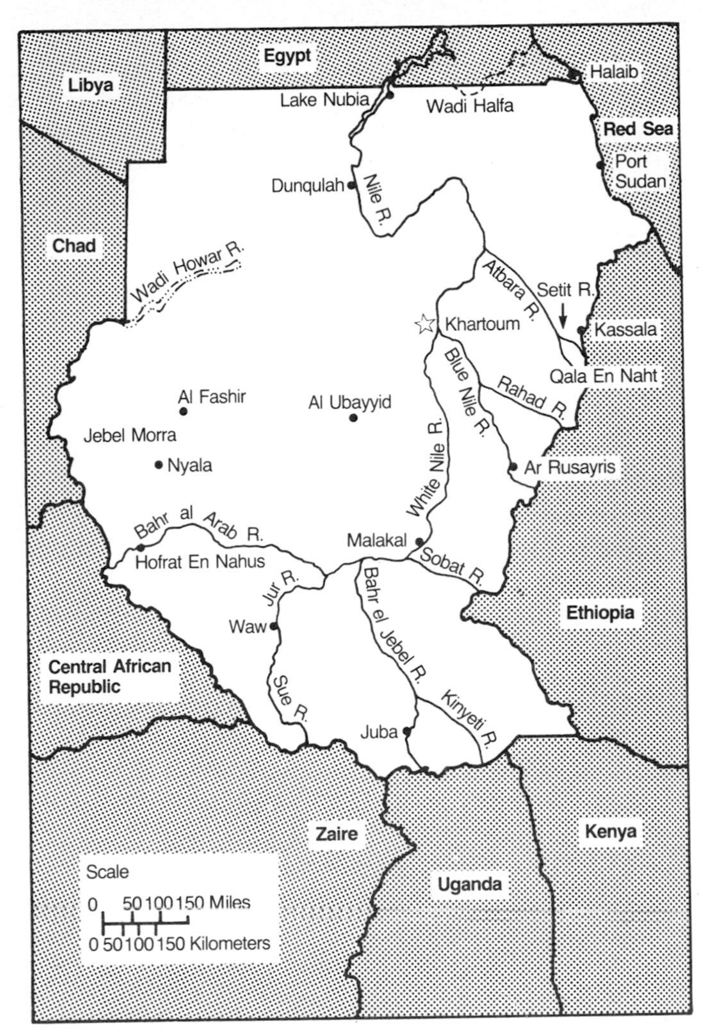

FIG. 12.16 Site of the Fula Rapids project in Sudan.

In Juba, an isolated diesel generating power station with a total installed capacity of about 1,000 kW (derated 750 kW) is owned and managed by PEWC, supplying electricity to the town and its adjoining areas. The demand for power and energy exceeds supply with load shedding as a regular feature for the last several years. The irregular availability and delivery of diesel fuel over long distances and the shortage of spare parts for the generating sets add to the difficulties in maintaining a regular electricity supply. A large number of small privately owned and operated diesel generating sets with a total installed capacity of about 4,300 kW meet a noncoincidental peak of about 1,000 kW. The rated capacity of these sets varies from 10 to about 350 kVA.

Outside Juba, the supply system is isolated and generation is through 8 diesel generating sets. The electricity at diesel stations operated by PEWC is generated at 415 V and stepped up to 3.3 kV for primary distribution. The system supplies areas located within a radius of about 5 km (3.1 mi) from the power station. The supply of electricity to consumers connected with the distribution mains of PEWC is far from satisfactory. Consumers normally receive the energy at about 360 to 375 V, against the rated 415 V at three phase.

A new diesel power station with a capacity of 5 MW is under construction. It was scheduled for commissioning in 1977 but was not completed because of delays in deliveries of material, plant, and equipment. The plant was expected to be commissioned in late 1982.

In the town of Juba, the installed capacity of the isolated power station and the energy generated per year during the period 1976 to 1981 is shown in Table 12.19.

Analysis of the available data indicates practically no growth in demand for power and energy during the last 5 years. This inference, however, is not correct. PEWC was unable to cope with the demand and supply for power and energy, with the result that some consumers had to install their own diesel generating sets. This is indicated by the fact that the diesel generating capacity in the private sector is much higher than the derated generating capacity of the isolated PEWC power station.

The forecast for power and energy shown in Table 12.19 for the period 1986 to 1991 was estimated by the Directorate of Electricity, Planning and Operation Department, PEWC. It anticipates that the demand of 2.7 MW for power capacity and 10.5 GWh in energy in the year 1981–82 will rise to 6.0 MW and 23.3 GWh in the year 1985–86. This estimate appears realistic, particularly since most of the generation presently under the private sector is likely to be connected to the PEWC supply system after commissioning of the new 5-MW diesel power station in 1982.

TABLE 12.19 Power Capacity and Energy Production for Juba (1976–81 and 1986–91)

	Period				
Actual	1976–77	1977–78	1978–79	1979–80	1980–81
Derated installed capacity, MW	1.0	1.0	1.0	1.0	1.0
GWh per year	3.9	4.0	3.68	4.3	—

	Period				
Projected	1986–87	1987–88	1988–89	1989–90	1990–91
Power, MW	6.5	7.0	7.6	8.25	9.0
Energy, GWh	25	27	29.3	32	35
Growth rate per annum, %	8.3	7.7	8.5	8.5	9

Site Description

The main inflow into the White Nile, a river called Bahr-el-Jabel, enters Sudan from Uganda at Nimule. The Fula Rapids commence about 7 km (4.3 mi) downstream of Nimule. A drop of about 10 m (33 ft) at Fula Rapids appears to be the nearest hydropower potential site that could be developed economically for the generation of 10 to 15 MW of electrical power to supply Juba, approximately 135 km (84 mi) away. Nimule, not yet electrified, and about 7 km (4.3 mi) from the proposed power station, where the demand for electricity is rather small, could also be supplied with electricity from the proposed power station.

The profile of the Bahr-el-Jabel River was identified by studies of available topographic maps; 1:250,000 with contour intervals at 50 m (164 ft). The site was recently surveyed by a team of engineers from Hafslund Consulting Engineers, Oslo, Norway, and identified for hydro development.

Two stream gaging stations are located on the river at Nimule and Mongalla and have been in operation since 1915 and 1912, respectively. The maximum discharge along the Fula Rapids recorded at Nimule normally varies between 700 and about 1000 m^3/s (25,000 and 35,310 ft^3/s), with a mean annual discharge of 600 m^3/s (21.186 ft^3/s). The maximum recorded flood discharge at Nimule was 960 m^3/s (24,364 ft $^3/s$) in 1937, whereas the calculated maximum, using data available at Mongalla for the period 1916 to 1973, and a return period of 100 years, was estimated at 2,800 m^3/s (98,870 ft^3/s). With a return period of 1,000 years it was at almost 4,000 m^3/s (141,243 ft^3/s). A low 10-day mean discharge of 320 m^3/s (11,300 ft^3/s) was recorded in 1923.

Plant Characteristics

By utilizing a design discharge of 125 m^3/s (4,414 ft^3/s) and installation of three turbine-generating sets of 3,000 kW each suitable for operation at a net head of 9 m (30 ft), the proposed run-of-river power station at Fula Rapids should be able to harness 9,000 kW of power and generate 47 GWh of energy per year with an average annual capacity factor of 60 percent. This generating capacity could easily be increased when considered necessary.

Allowing for consumption in station auxiliaries and losses in transmission and distribution, it is anticipated that nearly 35.5 GWh of energy per year will be available for sale to prospective consumers at Juba and adjoining rural areas. The town of Nimule would be supplied with electricity by extension of an 11-kV system from the power station.

The scheme is visualized as a run-of-river plant to include a small dam about 10.5 m (34.4 ft) high. The power station will be an integral part of the concrete dam structure at the end of the headrace channel. Intakes will be fitted with trashracks and gates. The power station will be a surface structure with a 66-kV switchgear on one side and will house three bulb-type turbines with three-phase synchronous generators capable of delivering 3,000 kW each at 0.85 power factor at 11 kV with an inbuilt overload capacity of about 10 percent.

The power thus generated is to be transmitted to the load center at Juba over a 135-km (84-m), 66-kV three-phase transmission line for use at 400/230 V ac at 50 Hz.

In view of the provision of a single-circuit 66-kV transmission line, it would be desirable to keep some diesel generating capacity at Juba as standby. Provision for the depreciated cost has not been considered at this stage.

Benefits and Costs

In view of the absence of detailed topographical and geological data and site surveys, it is not possible to prepare detailed cost estimates. Effort has, however, been made to assess the total investment costs at mid-1981 prices on the basis of lump-sum costs.

Estimates for project works on this basis for generation, transmission, and distribution works appear in Table 12.20. The total investment, excluding interest charges during the construction period is estimated to be about US$25 million at mid-1981 prices. The rate of interest at which money for this development is borrowed should be low since it would come from an international donor. A nominal interest rate of 10 percent was used in the study.

It is presumed that construction of the project will take 3 years with the cumulative phased requirement of capital investment 10 percent in the first year, 60 percent in the second year, and 100 percent during the last year of construction. It has also been assumed that the interest at 10 percent per annum on the phased capital investment will be payable at the close of each year for 2 years until commissioning of the plant which will occur at the end of the third year.

The annual overall maintenance charges for the civil, electrical, and mechanical works are based on an average of 1 percent of capital investment for generation and 2 percent for transmission, transformation, and distribution works totaling US$303,000.

The annual operation charges are based on 0.5 percent of investment costs at US$126,000. Annual administration expenses including insurance and taxes have also been taken as 0.5 percent of investment cost at US$126,000.

The total annual fixed costs for generation, transmission, transformation, and distribution comes to US$3,368,000.

The cost of delivered energy can be computed. Total annual energy production from the 9-MW plant at 60 percent capacity factor is 47 GWh. Assuming a consumption of about 1.5 percent in station auxiliaries and 23 percent losses in transmission, transformation, and distribution at 11 kV, the energy available for sale at 11-kV busbars at Juba is 35.5 GWh annually. The energy produc-

TABLE 12.20 Capital Cost Estimates for Fula Rapids Project*

Generation

Item	Rate per kW, US$	Cost, US$, millions
Civil works for 9,000-kW hydropower station	1,600	14.4
Hydro generating sets, consisting of three 3,000-kW turbine-generators, complete with associated switchgear, erection, testing, and commissioning	600	5.4
		19.8

Transmission, transformation, distribution

Item	Number or quantity	Unit cost, US$, thousands	Cost, US$, thousands
11/66-kV transformer substations, 12.5 MVA each, complete including erection, testing, and commissioning	2	500	1,000
11/0.4-kV transformer substation, 500 kVA, complete including erection, testing, and commissioning (power station)	1	30	30
66-kV transmission line on wood poles, complete with terminal equipment	135 km (83.9 mi)	30/km	4,050
11-kV line on wood poles with terminal equipment (additional length only)	10 km (6.2 mi)	15/km	150
			5,230

Summary of costs†

Item	US$, millions
Generation works	19.8
Transmission, transformation, distribution	5.2
	25.0

*All costs at mid-1981 price levels *excluding* escalation.
†Excluding interest during construction.

tion cost was computed at US$0.056 per kilowatthour and the delivered cost at US$0.087 per kilowatthour.

Economic and Financial Analysis

The cash flows for total investments, operation, maintenance, and overhead expenditures, without depreciation and interest charges during construction and return on capital investments have been calculated for a period of 30 years for hydro and for an alternative diesel power scheme in Tables 12.21 and 12.22.

TABLE 12.21 Investment Cash Flow Fula Rapids Hydro Project, US$ (millions)

Year after start of project*	Capital investment† Power generation	Transmission distribution	Administrative‡ overhead, operation, and maintenance	Total cash flow
1	(10%) 1.98	(10%) 0.52		2.50
2	(50%) 9.90	(50%) 2.61		12.51
3§	(40%) 7.92	(40%) 2.10		10.02
4			0.555	0.555
5			0.555	0.555
23			0.555	0.555
24			0.555	0.555
29			0.555	0.555
30			0.555	0.555

*Useful life of generating plant, transmission, transformation, and distribution on 11-kV assumed as 30 years.

†Excluding interest during construction period.

‡All expenditure is taken to be incurred at the end of the year.

§It is presumed that construction will extend over a period of 3 years and power station shall be commissioned at the end of third year.

TABLE 12.22 Investment Cash Flow Fula Rapids Alternative Diesel Power Scheme, US$ (millions)

Year after start of hydroproject	Capital investment Generation works[a]	Transmission and distribution[b]	Administrative[c] overhead, operation and maintenance	Fuel cost[d] at US$0.19 per kWh	Total cash flow
1[e]					
2	(20%) 1.26	(50%) 0.045			1.3
3[f]	(40%) 2.52	(50%) 0.045			2.56
4	(5%) 0.31		0.142	1.7	2.15
5	(35%) 2.21		0.142	3.4	5.75
6			0.221	5.1	5.32
7[g]			0.221	6.8	7.02
15			0.221	6.8	7.02
16[h]			0.221	6.8	7.02
17	2.52		0.221	6.8	9.54
18	(Plant only)		0.221	6.8	7.02
19	2.21		0.221	6.8	9.01
20			0.221	6.8	7.02
30			0.221	6.8	7.02

[a]Installed capacity 2 × 4,500 kW with 2 years' gap between first and second set. Investment cost assumed as $700.00 per kW.

[b]As per hydro project except 66-kV system and 4 km (2.5 mi) of 11-kV line.

[c]Maintenance costs at 2.5 percent of capital investment for unit one and rest at 1 percent of total capital for both units and 2 percent for transmission and distribution system until year 6. Thereafter, 2.5 percent maintenance costs for both units besides overheads and operation at 1 percent for total investment for generation and 2 percent for transmission and distribution.

[d]Based on price of diesel fuel at US$0.57 per liter delivered at site.

[e]Interest charges on capital investment during construction have not been taken into account at this stage.

[f]Commercial service at the end of year for unit one and at end of year 5 for unit two.

[g]kWh generated at an average annual load factor of about 45 percent and utilization assumed as 25 percent in the fourth year, 50 percent in the fifth year, 75 percent in the sixth year, and 100 percent in the seventh year.

[h]Useful life of diesel plant assumed as 15 years.

TABLE 12.23 Computation of Benefit Fula Rapids Hydro vs. Diesel, US$ (thousands)

Year after start of hydro project	Cash flows: Hydro	Cash flows: Diesel	Present worth factor at 10 percent per annum	Present worth (PW): Hydro	Present worth (PW): Diesel
1	2,500	—	0.909	2,272	—
2	12,510	1,300	0.826	10,333	1,073
3	10,020	2,560	0.751	7,525	1,922
4	555	2,150	0.683	379	1,468
5	555	5,750	0.621	345	3,570
6	555	5,320	0.565	313	3,005
7	555	7,020	0.513	284	3,601
8	555	7,020	0.467	259	3,278
⋮	⋮	⋮	⋮	⋮	⋮
17	555	9,540	0.198	110	1,889
18	555	7,020	0.180	100	1,264
19	555	9,010	0.164	91	1,478
20	555	7,020	0.149	83	1,046
⋮	⋮	⋮	⋮	⋮	⋮
28	555	7,020	0.069	38	484
29	555	7,020	0.063	35	442
30	555	7,020	0.057	32	400
				Total 23,982	47,482

Present worth of investment cash flows for hydro and diesel schemes have been ascertained for a period of 30 years at a discounted rate of 10 percent per annum and appear in Table 12.23. The benefit/cost (B/C) ratio was computed as follows

$$\text{B/C ratio} = 1 + \frac{(\text{PW diesel} - \text{PW hydro})}{\text{PW Hydro}} = 1.98$$

Energy costs of hydro and diesel generation can also be compared as follows

$$\frac{\text{PW of hydro costs}}{\text{PW of energy produced}} = \text{US\$0.0716/kWh}$$

$$\frac{\text{PW of diesel costs}}{\text{PW of energy produced}} = \text{US\$0.1418/kWh}$$

Development Plan

No information was given.

Discussion

This case study is an example of a reconnaissance evaluation at a site for which little data exists. Being typical of a developing country study, the only known information may be mean, maximum, and minimum flows, head, and general topography. The investigation must therefore be carried out on the basis of roughly forecast energy demand and capacity factor, an expected donor interest rate, and general cost guidelines of the type published by the U.S. Army Corps of Engineers.[3]

The project included transmission and distribution elements since a nearby grid did not exist. Also, the yardstick for economic comparison was a diesel-generator, the most likely alternative. Clearly before an investment decision can be made, further fieldwork and analysis must be done.

West Creek Hydroelectric Project, Skagway, Alaska (1982)[10]

Information courtesy of R. W. Beck and Associates, Seattle, Washington. Project Director: Donald R. Melnick, Partner. Project Manager: Wilson V. Binger, Jr., Principal Engineer.

Background

In 1981, the Alaska Power Authority (APA) authorized a detailed study to determine the technical, economic, and environmental feasibility of a hydroelectric project on West Creek as a means of meeting the future electrical requirements of Haines and Skagway. APA's decision was based on two previous studies carried out in the region. The West Creek project site is about 6 mi (9.7 km) northwest of Skagway. West Creek flows into the Taiya River about 3 mi (4.8 km) above its mouth.

The city of Haines is located in southeast Alaska on the Chilkoot Peninsula near the mouth of Taiya Inlet. Approximately 16 mi (25.8 km) north of Haines, the city of Skagway is located at the head of Taiya Inlet at the mouth of the Skagway River. The two cities are about 80 mi (128.8 km) north of Juneau. Roads from both cities connect with the Alaskan Highway in Canada. The White Pass Railway and Yukon Railroad run north from Skagway to Whitehorse, Canada.

The city of Skagway corporate limits encompass 431 mi^2 (1,116 km^2) and include the West Creek Project site. Skagway had a 1980 census population of 768. Most of the employment in Skagway is associated with the White Pass Railway or summer tourism. Haines had a 1980 census population of 993. Haines is located about 20 mi (32.2 km) south of the West Creek Project site and serves as a shipping point for the interior via the Haines Highway. Haines has an economy which is typical of southeast Alaska with employment primarily in lumber production, fisheries, transportation, trade, and services.

A 4,000-kW wood-waste generation plant is currently being installed at the Schnabel Mill in Haines and will be supplying electricity to Haines under a 5-

year contract beginning early in 1983. Based on the terms of the fuel contract, wood-waste generation is expected to provide power at a lower cost than diesel generation. However, wood-waste generation will not supply the needs of Skagway; nor are the costs assured after the 5 years covered in the contract.

Previous studies of the Haines-Skagway Region concluded that the most economical hydroelectric project in the region would be the West Creek project with a storage reservoir and a transmission intertie between Haines and Skagway.

Site Characteristics

The Haines-Skagway Region is characterized by steep, rugged ridges divided by rivers and streams running from high glaciers and snowfields down to ocean inlets. These inlets are generally steep sided with depths to greater than 1,000 ft (305 m). Mountain peaks are at the 5,000 to 6,000-ft (1,525 to 1830-m) elevation. Level ground is limited to the flood plains of the rivers.

The climate in the Haines-Skagway Region is influenced by interior and maritime weather patterns and by the surrounding topographical features. Southeast Alaska is located in the maritime climatic zone experiencing considerable precipitation. Climatological records in Haines are typical of the region, with a mean annual precipitation of 61 in (1550 mm) and a mean annual snowfall of 133 in (3380 mm).

West Creek, a stream 9.0 mi (14.5 km) long, originates in an arm of the Chilkoot Glacier at about the 1,600-ft (488-m) elevation. The stream skirts another arm of the glacier and enters a broad valley approximately 0.5 mi (0.8 km) wide with a moderate gradient through which it flows for about 5 mi (8.0 km). The proposed dam site is located at the lower end of the valley at approximately river mile (RM) 2.8 (4.5 km) measured from the confluence of West Creek with the Taiya River.

The project area is underlain by granitic crystalline rocks, primarily granodiorite. The rocks are part of a large multistage batholith known as the Coast Range Plutonic Complex. Bedrock is exposed intermittently throughout the project area. In the dam site area, rock is at or near the ground surface. A thin mantle of overburden ranging up to approximately 10 ft (3.0 m) in thickness covers the abutment areas. Recent alluvium underlies the active stream channel and is estimated to be about 20 ft (6.1 m) thick.

Below the dam site the stream gradient increases sharply as the creek passes through an upper gorge approximately 0.3 mi (0.48 km) long. For the next mile (1.6 km) the gradient moderates before the stream enters a steep lower gorge that extends 0.6 mi (0.96 km). The creek then enters the Taiya River Valley and flows through a low-gradient braided channel to its confluence with the Taiya River. The proposed powerhouse site lies at the west edge of the Taiya River floodplain about 1,200 ft (365.85 m) south of West Creek.

At the powerhouse site, the bedrock surface slopes downward beneath the

Taiya River floodplain at approximately the same slope as the valley wall. The floodplain consists of sand, gravel, and interbedded sand and silt. Immediately to the south of West Creek is an area of overburden consisting of cobbles and boulders with a sand and silt matrix. The deposit extends up to about elevation 300 ft (91.5 m) and south about 600 ft (183 m).

The project area is generally coniferous forest with riparian shrub and marsh and an area of deciduous woodland near the powerhouse. The West Creek Valley is populated by black bear, mountain goat, and various furbearers. West Creek has few fisheries resources. The steep gradient and high velocities in the lower and upper gorges make upstream fish passage extremely difficult if not impossible. The Creek is not used for fishing by residents in Dyea or Skagway and fish-sampling attempts yielded only a low density.

Most of the land at or near the project site is government owned. In the West Creek Valley, where the reservoir is to be located, all of the land is state-selected land, which is in the process of transfer from federal control. In the Taiya River Valley, almost all the valley floor lies within the boundary of Klondike Gold Rush National Historical Park. Land within the boundary is owned by the National Park Service, the state of Alaska, and private individuals. The entire Taiya Valley and West Creek Valley lie within the corporate limits of the City of Skagway.

Hydrologic investigations of the West Creek project are based on the 15 years of streamflows recorded at the West Creek gage near the mouth of the creek. Analysis of these together with climatological records at Haines and Skagway resulted in the development of basin-yield estimates, flood frequency, and the probable maximum flood (PMF).

The U.S. Geological Survey has streamflow records for the West Creek near Skagway gaging station from May 1962 to September 1977. The flow data indicate an average annual flow for West Creek at the gaging station of 334.9 ft^3/s (9.48 m^3/s). This is an average runoff of 7.75 $ft^3/(s \cdot mi^2)$ [0.085 $m^3/(s \cdot km^2)$]. Table 12.24 lists West Creek average monthly discharges from 1966 to 1977 at the gaging station.

To derive long-term streamflow data that could be used to conduct power studies and to estimate flood flows for spillway and diversion facility sizing, an effort was made to extend the available flow data by statistical means. This involved correlating recorded West Creek flow data with other recorded flow data from nearby drainage basins as well as with long-term precipitation data in the region.

The resulting correlations for monthly and annual streamflow data were found to be poor. Thus, since 15 years is a reasonable length for the sizing of a project, it was decided to use the West Creek record as the best representation of the long-term average flows to be expected. The flow data indicate an average annual flow for West Creek at the dam site of 288 ft^3/s (81.6 m^3/s). The average monthly inflow into the West Creek storage dam reservoir has been taken as 86 percent of the gaged flows at the former gage based on a ratio

TABLE 12.24 West Creek Monthly Average Discharges, 1966 to 1977

Month	Gaging station		Dam site	
	ft^3/s	(m^3/s)	ft^3/s	(m^3/s)
October	254.2	7.20	219	6.20
November	120.0	3.40	103	2.92
December	46.0	1.30	40	1.13
January	75.9	2.15	22	0.62
February	27.4	0.78	24	0.68
March	29.6	0.84	25	0.71
April	53.0	1.50	46	1.30
May	212.1	6.00	182	5.15
June	629.5	17.83	541	15.32
July	983.9	27.86	846	23.96
August	953.1	26.99	818	23.17
September	661.9	18.76	569	16.11

of the corresponding drainage areas. The drainage area at the dam site is 37.2 mi^2 (96.3 km^2).

Flood-frequency studies on West Creek have been conducted using the Gumbel Extreme-Value Type 1 distribution as well as the Log Pearson Type III method applied to the recorded momentary maximum discharges. Peak flows for each year of record at the West Creek gage are shown in Table 12.25. For the purpose of sizing construction diversion facilities, a peak discharge with a 10-year recurrence interval was considered to be reasonable for the 2½-year construction period and for the type of construction contemplated. This resulted in a peak discharge of 5,500 ft^3/s (155.8 m^3/s) to be used to size construction diversion facilities.

A unit hydrograph was developed for West Creek using U.S. Bureau of Reclamation methodology. The synthetic hydrograph was found to have a significantly shorter time to peak than the unit hydrograph derived from U.S. Geological Survey data. The more conservative synthetic hydrograph was used to derive the probable maximum flood.

The reservoir, with normal maximum elevation 705 ft (215 m), will have a surface area of about 1 mi^2 (2.6 km^2) which is 2.6 percent of the total drainage basin area. To properly evaluate inflow from storm runoff, rain falling directly on the reservoir area was considered to have a zero time of concentration. The unit hydrograph developed (Fig. 12.17) has a precipitation duration of 30 minutes and a peak of 13,835 ft^3/s (391.8 m^3/s).

Available precipitation data for Haines and Skagway was reviewed. It appeared reasonable that the spillway design storm would occur in the late summer or early fall. This would produce a flood with a higher peak inflow than a flood which might occur during the spring snowmelt season, although a spring flood would have a much longer duration and volume. On this basis, the

TABLE 12.25 Peak Flows at West Creek Gage

		Peak flow	
Year	Date	ft^3/s	(m^3/s)
1962	September 25	3,760	106.5
	August 23	2,490	70.5
1963	August 8	2,850	80.7
	July 18	2,500	70.8
1964	June 9	1,680	47.6
1965	August 11	1,630	46.2
1966	October 2	2,760	78.2
	September 26	1,900	53.8
1967	September 15	9,800	277.5
	August 9	3,200	90.6
1968	September 28	2,350	66.6
	September 5	2,090	59.2
1969	August 8	3,190	90.3
	November 1	2,280	64.6
1970	September 27	1,660	47.0
1971	August 2	2,130	60.3
	July 16	1,840	52.1
1972	August 6	2,910	82.4
	August 26	2,290	64.9
1973	August 12	2,670	75.6
1974	September 13	2,480	70.2
	September 25	2,430	68.8
1975	September 13	3,310	93.7
	July 10	2,440	69.1
1976	September 27	3,000	85.0
	October 1	2,390	67.7
1977	September 12	2,760	78.2
	August 20	2,400	68.0

probable maximum precipitation (PMP) design storm was assumed to occur in the fall (October).

Considering the range of temperatures that can be expected for the month of October, it was considered possible to have a snowpack condition antecedent to the PMP, at least in the higher elevations of the basin which would contribute to runoff. The combination of an October precipitation together with some snowmelt runoff was considered appropriate and reasonable in formulating the spillway design flood.

The probable maximum flood (PMF) is the flood resulting from the occurrence of the PMP in combination with snowmelt as described. The PMP, based on 22.5 in (571 mm) of rain and 2.0 in (61 (51 mm) of snowmelt, has a peak inflow of 59,700 ft^3/s (1690.7 m^3/s) and a volume of 118,400 acre-feet (146.0 $\times 10^6$ m^3). The PMP and its effect on the PMF is shown in Fig. 12.18.

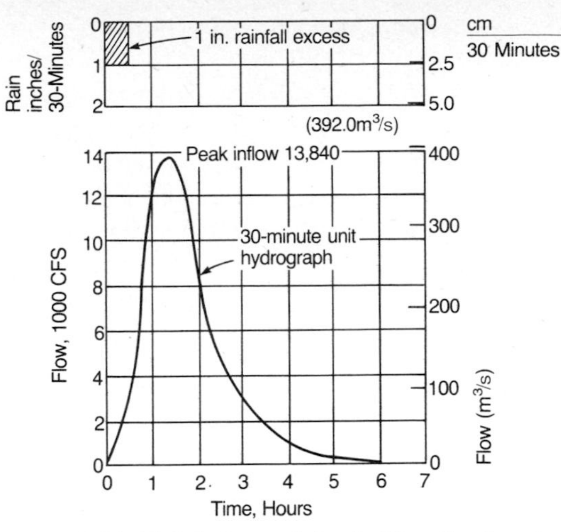

FIG. 12.17 Unit hydrograph for the West Creek project based on a 30-minute rainfall.

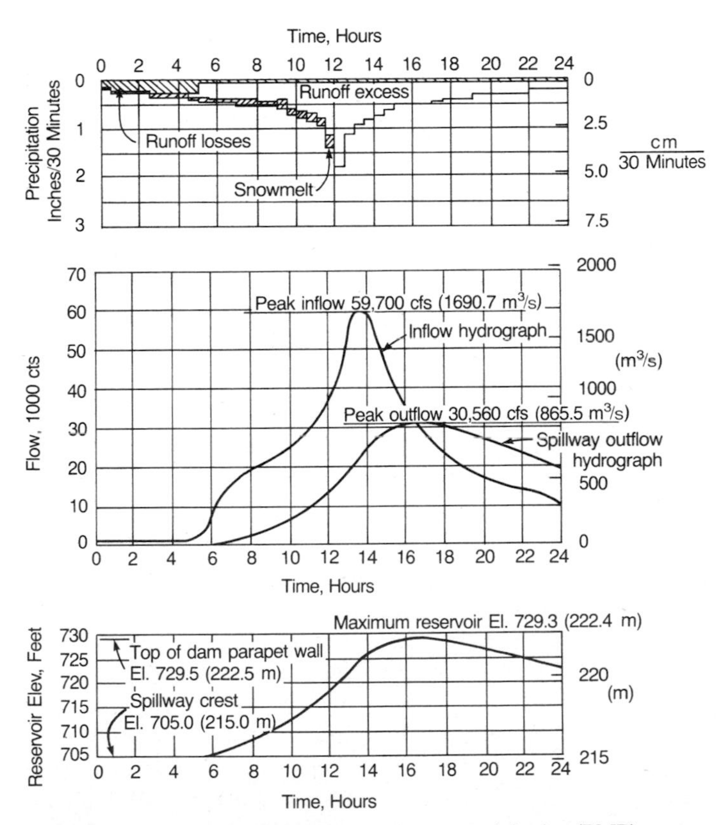

FIG. 12.18 The effect of probable maximum precipitation (PMP) on reservoir elevation and West Creek Flow.

Sediment yield from West Creek was based upon suspended sediment measurements made by the U.S. Geological Survey between the years 1963 and 1977. Using these measurements, a suspended-sediment rating curve was developed and used with the flow-duration curve to estimate average annual suspended-sediment quantities. For an assumed sediment density of 70 lb/ft^3 (2.95 kg/m^3) and a bed-load factor of 1.25, the annual sediment yield was estimated to be 10 acre-feet (12,335 m^3). On this basis, the sediment yield to the reservoir for a 100-year period would be 1,000 acre-feet (12.335×10^5 m^3). At the dam, the 100-year sediment level was calculated to be elevation 640 ft (195.2 m) which corresponds to a sediment depth of 7.5 ft (2.3 m).

Plant Characteristics

For most hydroelectric projects, the scale of the various features are optimized to produce a project which makes the best use of the hydrologic, topographic, and geologic conditions of the site. This presupposes that there is a market for any and all power produced by the project. However, in the case of the West Creek Project, the output can be used only in Haines and Skagway. Thus, the size of the project is limited by the forecast loads and energy requirements rather than site conditions.

For planning purposes, the West Creek Project was sized to meet the 1996 peak load and energy requirements forecast minus the portion of the load which would be met by the existing hydroelectric units in Skagway. The resulting 1996 annual requirements are 23,630 MWh with a peak of 5,400 kW for a system capacity factor of 50 percent. The installed capacity was set at 6,000 kW which allows some margin should the system capacity factor change. In addition, the reservoir was sized to meet the 1996 energy requirements during the driest year of record. Figure 12.19 shows the forecast demand for the region using three growth scenarios.

The operation studies showed that a reservoir with a usable storage of 18,130 acre-feet (22.4×10^6 m^3) would be necessary. This requires a normal maximum level of elevation 705 ft (215.0 m) and a minimum operating level of elevation 662 ft (201.9 m). Average gross power head, based on an average reservoir elevation of 697.6 ft (212.8 m) for the period of study and tailwater at elevation 38 ft (11.6 m) is 659.6 ft (201.2 m). Maximum gross head is 667 ft (203.4 m) and minimum gross head is 624 ft (190.3 m).

These results indicate that secondary energy could be generated during the dry months when the inflows are higher than those occurring in the driest year of record. This secondary energy would only have an economic value after 1996 when the load exists. To provide a conservative analysis, no credit has been taken for secondary energy during the dry months.

The operation studies also show that during months of high flow, June through October, there is sufficient excess water to generate up to the installed capacity of the plant. This would be a total energy of 22,000 MWh as com-

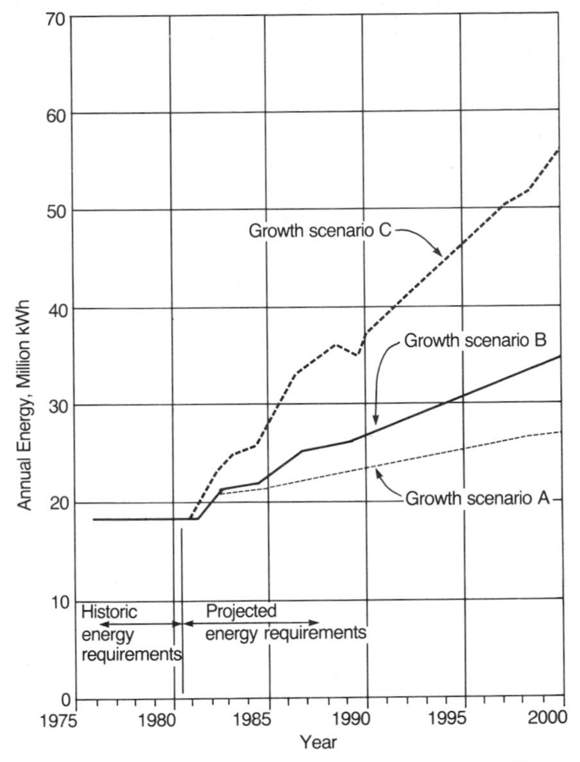

FIG. 12.19 Forecast energy demand for the Haines-Skagway region.

pared with the 9,381 MWh required during those months in the 1996 forecast. The only limitation on use of this energy is the availability of a load which can make use of the energy. Thus, as the load grows after 1996 the energy from the project will increase during the high-flow months.

The major physical features of the project are the reservoir, dam, power intake, power conduit and surge shaft, powerhouse, site access facilities, and transmission system. Figure 12.20 shows a general arrangement of the powerhouse.

The reservoir formed by the dam will provide 18,130 acre-feet (22.4×10^6 m^3) of active storage above the minimum pool at elevation 662 ft (201.9 m). At normal maximum reservoir elevation 705 ft (215.0 m), the surface area will be 635 acres (257.2 hectares). Some of the areas on both sides of West Creek are densely forested so that it will be necessary to clear about 330 acres (133.7 hectares) within the area of the reservoir. The remaining area is covered with low brush with a good root structure which should hold the brush in place after filling.

The selected dam type is a concrete-faced rockfill embankment. The down-

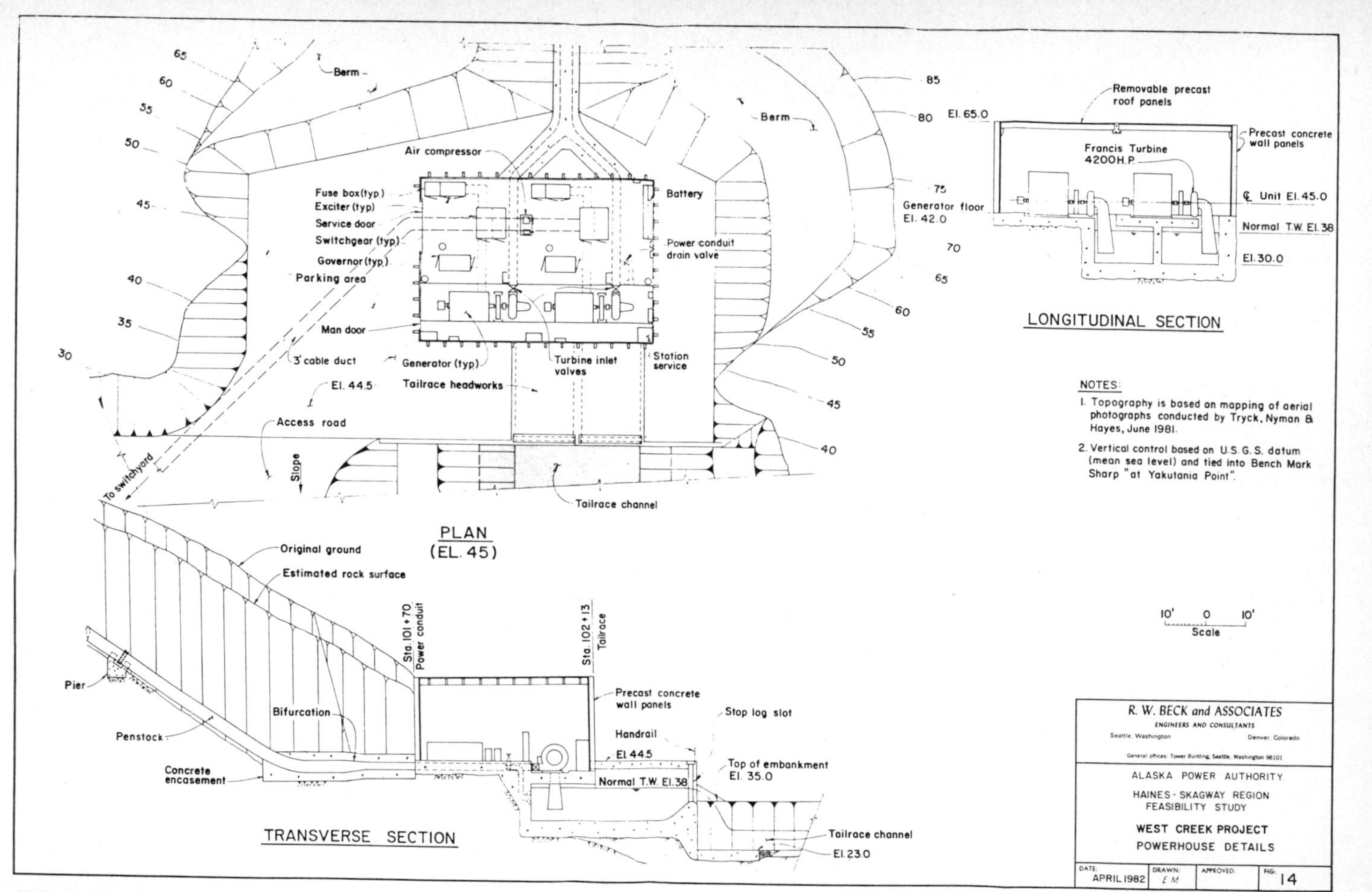

FIG. 12.20 General arrangement for the West Creek Power Plant.

stream face will have a slope of 1.5:1. The upstream face will have a slope of 1.6:1. The embankment will be constructed entirely of compacted rock excavated from the spillway and intake except for a 10-ft (3.0-m) thick layer of processed gravel on the upstream face which will serve as a bedding for the reinforced-concrete face.

The dam will have a crest at elevation 729.5 ft (222.5 m), including a 3.5-ft (1.07-m) high parapet wall, and an overall height of 117 ft (35.7 m). During the probable maximum flood (PMF), water level would be elevation 729.5 ft (222.5 m). The dam will have a volume of 254,000 yd^3 (194,310 m^3) of rock and 30,500 yd^3 (23,333 m^3) of processed gravel. An unlined open channel spillway 75 ft (22.9 m) wide will be excavated in the right abutment. It will discharge into West Creek at a point 450 ft (137.3 m) downstream of the toe of the dam. The power intake will be located on the right abutment, adjacent to the dam. The power tunnel will be excavated using a tunnel-boring machine 9.5 ft (2.9 m) in diameter. The tunnel will be left unlined except for a 50-ft (15.3-m) length at the upstream end and a 180-ft (54.9-m) length at the downstream end.

From the downstream portal, a steel penstock 7.5 ft (2.3 m) in diameter will extend out of the tunnel about 50 ft (15.3 m) to a transition section 15 ft (4.6 m) long, reducing the diameter to 3 ft (0.9 m). Before entering the powerhouse, the penstock bifurcates into two branches 24-in (610 mm) in diameter.

The powerhouse will be an indoor-type structure founded on rock located about 1,200 ft (366 m) south of West Creek. It will contain two horizontal-shaft 1,200-r/min Francis turbine-generator units. Each turbine will deliver 4,200 hp (3,133 kW) at a rated net head of 635 ft (193.7 m) at best gate. The corresponding discharge through each turbine will be 65 ft^3/s (1.84 m^3/s). Each turbine will drive a synchronous, three-phase ac generator rated at 3,450 kVA, at 0.9 power factor, 12.47 kV, and 1,200 r/min with a 60°C temperature rise. The plant installed capacity will be 6,000 kW. Each generator will be enclosed and will include a solid-state exciter, voltage regulator, surface coolers, a CO_2 fire-protection system, and all necessary auxiliary features. The plant control system will be designed for fully automatic operation from a remote location to be determined later. The plant will also be controllable locally from control panels in the powerhouse.

The generator step-up transformer will be located in the switchyard just south of the powerhouse. The transformer will be rated 7,200 kVA, 13.8 to 34.5 kV, three-phase, forced-air cooled, 60 Hz. The switchyard will also include a 34.5-kV, three-phase circuit breaker; two 34.5-kV, three-phase, gang-operated disconnect switches; and buswork.

Benefits and Costs

Project capital and annual costs were estimated. Direct construction cost estimates for the project were prepared for the selected arrangement based on the preliminary design layouts and details. The total investment cost was determined by summing the direct construction costs of each major project component and adding indirect costs and interest during construction. The effect of escalation was considered, and the total investment cost was adjusted to reflect a project on-line date in January 1982.

Annual operating costs such as operation and maintenance, owner administration, insurance, and interim replacements were also estimated. How these costs were determined follows:

1. **Project Capital Costs**
   a. **Direct Construction Cost:** This cost includes the total of all costs directly chargeable to the actual construction of the project, which can be considered equivalent to a contractor's bid based on a January 1982 bid-price level, and reflects estimated inflation of costs which would occur during the 2½-year construction period. This corresponds to the project entering into service in July 1983.
   b. **Contingencies:** To allow for unforeseen conditions during construction and miscellaneous items not included in the estimate, an allowance of 20 percent for contingencies was applied to the direct construction cost of major civil items, and 15 percent was applied to the direct construction cost of the powerhouse electrical and mechanical equipment and the transmission line.
   c. **Engineering and Owner Administration:** Engineering and owner administration costs are based on actual experience with costs for similar work. This item includes all preliminary engineering work; project feasibility and environmental studies; field investigations; applications for, and processing of, required permits and licenses; final design and preparation of construction; and owner administration. An allowance of 15 percent of the sum of the direct construction cost plus contingencies is considered a reasonable estimate for this item.
   d. **Total Construction Cost:** The total construction cost includes the direct construction cost plus contingencies and engineering and owner administration. The total construction cost for the project with a bid date of January 1982 is estimated to be $61,324,000.
   e. **Interest During Construction:** Interest during construction is dependent on the interest rate at which money is available for the project and the cash flow during construction. Using the inflation-free interest rate of 3 percent and typical cash flow patterns for projects similar to the West Creek project, it was calculated to be 3.6 percent of the total construction cost.
   f. **Total Investment Cost:** The total investment cost is the sum of the total construction cost and interest during construction and is therefore $63,532,000. This total investment cost is for a project bid in January 1982 and coming on-line in June 1984.

g. **Escalation Adjustment:** The cost estimates discussed include estimates of escalation over the 2½-year construction period. Thus, they represent a project coming on-line in June 1984. To compare project feasibility in the manner prescribed by the APA, it was necessary to take escalation out of the estimate. This would give an inflation-free estimate for a project which would theoretically be built at January 1982 costs of material and labor. Escalation was estimated to be 12 percent of the total investment cost. Thus, the total investment cost for an inflation-free project would be $55,908,000. Construction cost summary is given in Table 12.26.)

2. **Annual Costs**

The principal annual cost of any hydro project is the debt service on the capital cost. Since the financial terms for the project are still uncertain, these costs were not computed except as part of the economic analysis.

Annual costs for operation and maintenance, administration, insurance, and interim replacement of project components were estimated based on experience on similar projects as well as on FERC guidelines. The variable annual costs are estimated to be $634,000 at the January 1982 cost level.

Economic and Financial Analysis

To economically compare various generation options, three alternative plans were developed, each of which could meet the region's scenario for forecast

TABLE 12.26 West Creek Project Construction Cost Estimate Summary

Item	Estimated cost
Physical works:	
Preparatory Work	$ 2,786,000
Dam and Reservoir	11,571,000
Power Conduit	12,348,000
Powerhouse	5,382,000
Switchyard and transmission line	13,273,000
Direct construction cost (Bid 1/82, on-line 6/84)	$45,360,000
Contingencies	8,139,000
Subtotal	$53,499,000
Engineering and owner administration	7,825,000
Total construction costs (Bid 1/82, on-line 6/84)	$61,324,000
Interest during construction	2,208,000
Total investment cost (Bid 1/82, on-line 6/84)	$63,532,000
Escalation during construction	7,624,000
Total investment cost (Inflation-free on 1/82)	$55,908,000

loads. The first, or base-case plan, is a continuation of the current status quo with the existing diesel and hydro units meeting loads up to their capacity and new diesel generation being added as required. The second plan assumes that the West Creek Project would be built and used up to its capacity. Existing hydro units in Skagway would continue in operation and the existing diesel generators would initially serve as reserves and later meet needs beyond the capacity of the West Creek Project. The third plan assumes that the wood-waste generation being installed at the Schnabel Mill would be used to meet part of the Haines load. Under this plan, diesel generators would still be needed to meet part of both the Haines and Skagway loads.

The method of analysis used for the study computed the total present worth cost over a period from the present through the 50-year economic life of the West Creek Project (1982–2036). Specific criteria used for conducting project economic analyses as established by the APA, were as follows:

1. Constant dollars assumed (zero inflation).
2. Inflation-free present worth discount rate of 3 percent.
3. Petroleum fuel cost escalated at 2.6 percent per year for 20 years and then held constant.
4. Electrical energy demand forecast according to load growth scenarios described and then held constant.
5. Interest during construction calculated using a 3 percent interest rate.
6. No financing expenses included.
7. No debt service on existing diesel or hydroelectric generation included.
8. Period of economic analysis was 55 years (1982 through 2036).
9. Economic life of hydroelectric plant was 50 years.
10. Economic life of diesel generation facility was 20 years.
11. Economic life of wood-waste generation facility was 20 years.

The comparison of total present worth costs of the alternative plans showed that the least expensive plan is the West Creek Project. Table 12.27 shows the economic analysis in tabular form. This plan has a total cost of $88,217,000 and a 1.32 base case to alternate plan ratio. The wood waste generation plan is also less expensive than the base case. When the annual generation from wood waste decreases to 12,000 MWh, the alternative becomes more expensive than the base case. The final costs were

1. Base-case plan: $116,562,000
2. West Creek plan: $88,217,000
3. Wood-waste plan: $101,337,000

Thus, West Creek was determined to be the most economical alternative, and the report recommends that development of this project proceed immediately.

TABLE 12.27 Haines-Skagway Region Feasibility Study: Economic Analysis West Creek Project

Year	Total annual generation required, MWh[a,b]	Diesel generation required, MWh[c]	Altern. capital cost, $000[d]	Altern. O&M cost, $000[e]	Diesel O&M cost, $000[f]	Diesel fuel cost, $000[g]	Total annual cost, $000	Present worth of annual cost, $000[h]
1982	11,758	11,758	0	0	901	1,137	2,038	2,008
1983	12,431	12,431	0	0	901	1,233	2,134	2,041
1984	12,794	12,794	0	0	901	1,302	2,203	2,046
1985	13,288	12,288	0	0	901	1,387	2,288	2,063
1986	14,736	14,736	0	0	901	1,579	2,480	2,171
⋮	⋮	⋮	⋮	⋮	⋮	⋮	⋮	⋮
1997	22,982	0	2,173	727	459	0	3,359	2,124
1998	23,778	0	2,173	727	459	0	3,359	2,062
1999	24,636	46	2,173	727	459	7	3,366	2,007
2000	25,491	557	2,173	727	459	85	3,444	1,993
2001	26,381	1,091	2,173	727	459	172	3,531	1,984
2002–2036	(Assuming no additional growth or escalation)				35 Years Cumulative			42,633
					Cumulative Present Worth of Project Annual Costs			88,217

[a]Scenario B, loads less 2,900 MWh annually of existing hydro generation. Includes 10% of Schnabel load. Includes White Pass and Yukon RR.
[b]Based on existing capacity of 7510 kW for both systems.
[c]Assumes 23,630 MWh annually of West Creek generation to begin in 1987 plus 40% of load requirements above 23,630 MWh.
[d]Assumed level debt service at 3% over 50 years. Total investment cost assumed to be $55,908,000 including IDC at 3%.
[e]Based on 1.3% of total investment cost.
[f]Assumes $120 per kW annual expense. Approximately 50% of existing capacity is mothballed at 2% base cost in 1987.
[g]Assumes diesel fuel cost of $1.16/gallon in 1982 escalated at 2.6% per year. Fuel usage assumed to be 12 gallons/kWh.
[h]Discounted to January 1982 at 3% per year.

Development Plan

A design and construction schedule was developed for the project. This schedule contemplates commercial operation of both units by September 1986, which is considered to be the earliest reasonable date considering the time required for additional final design investigations, preparation and processing of the FERC license application, design, and construction.

Construction could start with mobilization and construction of the access road to the dam in the summer of 1984. Clearing and grubbing of the dam abutments and the intake area would also start then. During the first winter, the streambed would be stripped and prepared and the diversion conduits placed. Once those were in place, the upstream and downstream cofferdams would be built.

During the second summer, 1985, the spillway would be excavated and the dam constructed. During the same period, the contractor would excavate the intake channel, prepare the tunnel portal, and start excavating the tunnel from the upstream end. Rock excavated from the intake and spillway would be used in the dam. Rock excavated from the tunnel would be spoiled in the reservoir area. Upon completion of the embankment, the concrete facing would be slip formed. Also during the second summer, the penstock would be constructed, the powerhouse civil works and tailrace channel would be completed, and the transmission line would be started.

Closure of the diversion conduits would be done during the 1985–1986 winter and the pipes plugged with concrete. Also, the electrical and mechanical equipment would be received and installed. Filling of the reservoir would take an estimated 40 days during the spring runoff. Start-up and testing would take place as soon as the reservoir had enough water to allow safe operation.

Discussion

This project was under review at the time this handbook was in preparation. Therefore cost/benefit and financial analyses were not included. Capital investment cost is approximately $15,000 per kilowatt installed. This appears somewhat high; however, construction costs in Alaska are usually on the order of 50 percent higher than those in the "lower 48." Therefore, comparison with costs at installations within the contiguous United States is not valid.

The cost of energy production could not be computed since the financial terms have not been determined. Nevertheless, when compared to the two other alternatives, the base-case and the wood-waste plant, the small hydro alternative remains the most economical.

12.6 CLOSURE

In this chapter we introduced methodologies for both reconnaissance and feasibility studies as well as demonstrated how these studies are organized through five examples. Each case study was unique in that the specific circumstances such as physical site conditions, institutional structures, and economic considerations dictated how the study would be approached. In some cases, the choice of equipment or reservoir level or establishment of a mean streamflow was of primary importance. In other cases, the type of financing, federal loan or local banks, would eventually dictate the economic viability of the project.

From these analyses, it became clear that a successful and economically feasible project is not only a matter of competent engineering analysis but also of finance and organizational structures. Most engineers will have some difficulty dealing with the latter, but the typical consulting engineer's client expects a broad range of services that include finance and operations. It therefore becomes important for the engineer to develop skills in these areas. Since small hydropower is a relatively mature technology, its economic application to meet the electrical energy needs of the United States as well as of many developing countries remains a challenge for the future.

REFERENCES

1. U.S. Army Corps of Engineers, *Feasibility Studies for Small Scale Hydropower Additions,* The Hydrologic Engineering Center, Davis, Calif., July 1979.
2. Tudor Engineering Co., *Reconnaissance Evaluation of Small, Low-Head Hydroelectric Installations,* San Francisco, July 1980.
3. U.S. Army Corps of Engineers, *Hydropower Cost Estimating Manual,* Institute of Water Resources, Portland, Ore., May 1979.
4. H. Egoavil, "Planning Small Electric Systems and Mini-Hydropower Plants," *Small Hydroelectric Power Plants,* NRECA, Washington, D.C., 1980.
5. G. Calderon, T. C. Hough, and D. R. Limaye, "Computer Model for Evaluation of Small Scale Hydroelectric Projects in Latin America," *Proceedings Waterpower 81,* U.S. Army Corps of Engineers, Washington, D.C., 1981.
6. Douglas W. Barr Consulting Hydraulic Engineers, *Feasibility of New Power Units at Port Arthur Dam, Ladysmith, Wisconsin,* Minneapolis, Minn., November 1964.
7. Tudor Engineering Co., *Final Report on Potential Hydroelectrical Power, Mora Canal Drop,* San Francisco, December 1978.
8. Tudor Engineering Co., *Feasibility Report on the Potential Hydroelectric Development, Scotts Flat and Lower Scotts Flat Power Plants,* San Francisco, October 1981.
9. R. N. Bhargava and T. M. Haggstrom, *Evaluation of Small Hydro-Power Sites in Sudan,* U.N. Department of Technical Co-Operation for Development, TCD/INT-80-R47/12, New York, 1981.
10. R. W. Beck & Associates, Inc., *Haines-Skagway Region Feasibility Study,* Seattle, Wash., June 1982.

Appendix

TABLE A.1 Conversion Factors

Length:	Pressure:
1 mm = 0.0394 in	1 kg/cm^3 = 14.22 lb/in^2
1 m = 3.28 ft	= 0.968 atm
1 km = 0.621 mi	= 98.07 kPa
Area:	Flow:
1 m^2 = 10.764 ft^2	1 m^3/s = 35.31 ft^3/s
1 km^2 = 0.386 m^2	
1 ha = 2.471 acres	Power:
	1 kW = 1.3405 hp
Volume:	= 0.293 $\times$ 10^{-3} Btu/h
1 m^3 = 35.31 ft^3	= 737.6 ft·lb/s
= 1.308 yd^3	
= 0.81 $\times$ 10^{-3} acre-feet	Energy:
1 L = 0.264 gal	1 kWh = 2.655 $\times$ 10^6 ft·lb
	= 1.341 hp·h
Weight:	= 3,415 Btu
1 kg = 2.205 lb	= 0.278 MJ
Velocity:	
1 m/s = 3.281 ft/s	
1 km/h = 0.621 mi/h	

TABLE A.2 Present Worth Factors

Years,	Discount rates					
n	5	10	15	20	25	30
1	.952	.909	.870	.833	.800	.769
2	.907	.826	.756	.694	.640	.592
3	.864	.751	.658	.579	.512	.455
4	.823	.683	.572	.482	.410	.350
5	.784	.621	.497	.402	.328	.269
6	.746	.565	.432	.335	.262	.207
7	.711	.513	.376	.279	.210	.159
8	.677	.467	.327	.233	.168	.123
9	.645	.424	.284	.194	.134	.094
10	.614	.386	.247	.162	.107	.072
15	.481	.239	.123	.065	.035	.020
20	.377	.149	.061	.026	.012	.005
25	.295	.092	.030	.011	.004	.001
30	.231	.057	.015	.004	.001	
35	.181	.036	.008	.002		
40	.142	.022	.004	.001		
45	.111	.014	.001			
50	.087	.009	.001			

Index